T0176007

A COURSE ON
GROUP THEORY

JOHN S. ROSE

DOVER PUBLICATIONS, INC.
Mineola, New York

Bibliographical Note

This Dover edition, first published in 1994, and reissued in 2012, is an unabridged and unaltered republication of the work first published by the Cambridge University Press, Cambridge, England, in 1978.

Library of Congress Cataloging-in-Publication Data

Rose, John S.
  A course on group theory / John S. Rose. — 1st Dover ed.
      p. cm.
  Originally published: Cambridge : Cambridge University Press, 1978.
  Includes bibliographical references and index.
  ISBN-13: 978-0-486-68194-8
  ISBN-10: 0-486-68194-7
    1. Group theory I. Title.

QA174.2.R66 1994
512.2—dc20

94-20435
CIP

Manufactured in the United States by LSC Communications
68194709   2023
www.doverpublications.com

CONTENTS

PREFACE

In his presidential address to the London Mathematical Society in 1908 (published in the article [a12]), William Burnside remarked that 'It is undoubtedly the fact that the theory of groups of finite order has failed, so far, to arouse the interest of any but a very small number of English mathematicians . . .' And he ended with the words 'I wish, in conclusion, to appeal to those who have the teaching of our younger pure mathematicians to do something to stimulate the study of group-theory in this country. If, when advice is given for the course of study to be pursued, the importance of some knowledge of group-theory for a pure mathematician (which is generally recognized elsewhere) were insisted on, there is little doubt but that a demand for the serious teaching of the subject would soon arise.'

Seventy years on, such a plea would be scarcely necessary: the central importance of group theory is now fully recognized and reflected in the teaching of mathematics in universities and colleges. No doubt this is due in no small measure to the profound influence of Burnside's own masterly book on the subject of groups ([b3]). Nowadays it is customary in British universities to provide introductory courses of lectures on groups and other algebraic systems for undergraduates in their first year of study. The present work offers material for a further course of study on group theory. It is based on courses of lectures given by the author at the University of Newcastle upon Tyne to third year honours undergraduates and to candidates for the Master's degree.

The reader is supposed to be familiar with the contents of the kind of introductory course mentioned above. Specifically, knowledge is presupposed of the notions of isomorphism classes of groups, cyclic and abelian groups, subgroups and cosets, Lagrange's theorem, orders of elements, symmetric groups and the decomposition of a permutation as a product of disjoint cycles; and of the most elementary properties of vector spaces, linear maps and matrices, fields and rings. A rather terse summary of the facts about groups which are presupposed is contained in the preliminary chapter 0, which also serves to establish notation used throughout the book.

Chapter 0 is followed by another short chapter, chapter 1. This is intended as a curtain-raiser to the whole book and attempts to trace various important themes in terms comprehensible to the reader on the basis of the presupposed knowledge. The aim is to provide a motivation for some of the technical definitions and procedures to be treated in detail later on. The emphasis in chapter 1 is entirely on *finite* groups, which are in fact the primary objects of study of the whole book. Nevertheless, an attempt has been made to avoid finiteness restrictions wherever their imposition does not materially simplify the discussion; and certain important results on infinite groups which arise naturally in context have been included, especially in chapters 7 and 8.

The systematic treatment begins in chapter 2, where many basic examples which recur throughout the book are introduced. Subsequent chapters deal in a fairly leisurely way with a selection of the most important lines of development in the subject, and an attempt has been made to give a unified rather than a piecemeal treatment. An indication of the selection made is given by the chapter titles. In brief, one may say that chapters 3, 7, 8 and 9 deal with the *normal* structure of groups and chapters 4, 5 and 6 with the *arithmetical* structure, while chapters 10 and 11 treat aspects of the interplay between normal and arithmetical structures.

A particular emphasis is placed on the idea of *group actions*. This is conceptually important as typifying the way in which groups occur in mathematics, as well as providing a powerful method within group theory itself. The basic facts about group actions are given in chapters 4 and 9, with applications in chapters 5 and 10.

Chapter 10 calls for special comment. It is devoted to an exposition of the beautiful treatment of the classical notions of *transfer* and *splitting* by means of group action arguments which was given by Professor Helmut Wielandt in a lecture at the Mathematisches Forschungsinstitut, Oberwolfach, in May 1972. This provides a very impressive illustration of the power of group action techniques within group theory. This material has not previously appeared in print and I am very greatly indebted to Professor Wielandt for allowing me to include it here.

Any book which attempts to give a general account of group theory must inevitably be selective. To the practitioner of the subject who inspects the book, the omissions are probably more striking than the topics chosen for inclusion. In offering the present work for scrutiny, I am especially conscious of two major omissions: the representation theory of finite groups (and the associated theory of group characters) and the theory of defining relations of groups. Of these two important topics, the first may seem a surprising omission from a book which stresses group actions, since representation theory may be viewed as the theory of group actions on vector spaces (as is explained at the beginning of chapter 9). I feel, however, that in both cases an adequate treatment would lengthen the book unacceptably. Moreover, a proper discussion of defining relations would involve the theory of free groups which, although fundamental, is rather different in spirit from the topics treated here. Good accounts of both representation theory and defining relations are to be found in several of the general texts on group theory listed among the references at the end of this book. A number of specific references to representation theory are given at the beginning of chapter 9. On the subject of defining relations, I wish also to mention explicitly the important established work of reference by H. S. M. Coxeter and W. O. J. Moser [b6], and the recent book by D. L. Johnson [b22].

The present work is arranged in short sections which are numbered consecutively through each chapter. In many instances, a section is devoted to the statement and proof of a single result, to which reference may be made in other parts of the book by citing the appropriate section number. The more important results are designated 'Lemma' or 'Theorem'.

The exercises form an essential constituent of the book. They are numbered consecutively from 1 to 679, their numbers appearing in bold type. Exercises 1 to 12 appear at the end of chapter 0 and are meant to be accessible to the reader with the presupposed knowledge. There are no exercises in chapter 1. From chapter 2 onward, the exercises are set at roughly equal intervals throughout the text. The aim is to give exercises which illustrate and extend the material of the formal course as soon as the relevant facts have been established in the text. There are many cross-references to exercises, particularly in the later chapters of the book: these references are given merely by citing exercise numbers; it is hoped that the regular distribution of the exercises and the bold type of their numbers will make these easy to locate without the additional citation of page numbers. The statements of many of the exercises omit the conventional imperatives 'prove that' or 'show that'; they are nevertheless assertions to be proved. I imagine that few readers of the book are likely to attempt all the exercises. However, a number of definitions and results given in exercises are needed in the main text, and these are indicated by asterisks against their numbers (e.g. *1); these exercises are mostly straightforward. Many of the more difficult exercises are accompanied by suggested hints for their solution.

Dates have been attached to some results in order to indicate a historical perspective to the development of the subject. For the same reason, various references to early articles and books have been included in the list at the end of the book. However, no pretence of historical scholarship is made: this is a book on group theory, not on the history of the subject. For a scholarly account of the early development of group theory, the reader may consult the book by H. Wussing [b40].

The references to works of other authors are divided into articles, with numbers prefixed by the letter 'a', and books, with numbers prefixed by the letter 'b'. The works listed are mainly those to which reference is made in the text and in no way constitute a comprehensive bibliography of the subject. Many of the authors quoted have written other important works on group theory, and there are of course also many important works by authors who are not quoted. An impression of the scope and bulk of publication on group theory in the years 1940 to 1970 may be obtained from the recently published volumes of reviews taken from the periodical *Mathematical Reviews: Reviews on Finite Groups* (ed. D. Gorenstein, Amer. Math. Soc. 1974) and *Reviews on Infinite Groups* (2 vols., ed. G. Baumslag, Amer. Math. Soc. 1974).

I was fortunate in the mathematicians who provided my first comprehensive impressions of group theory, which were formed by listening to courses of lectures given by Professor W. L. Edge in Edinburgh, Professor P. Hall and Dr D. R. Taunt in Cambridge, Professor B. Huppert in Mainz and Professor H. Wielandt in Tübingen. Although they might not care to recognise the views or the modes of treatment of the subject which I have chosen to adopt here as deriving from their own accounts, I wish to place on record my sense of indebtedness to them for providing the original stimulus which led me to pursue the subject. Of my particular indebtedness to Professor Wielandt regarding the content of chapter 10, I have already spoken.

During the writing of the book, I have been helped by many friends and colleagues. I wish especially to offer thanks to Dr R. H. Dye, whose misfortune it has been to occupy an office adjacent to my own and who has shown exemplary patience in listening to my trial expositions of many arguments which have been incorporated (and others which have not) and frequently brought clarity to my confusion; to Professor A. Mann for several valuable suggestions which I have adopted; and to Dr S. E. Stonehewer, who most generously undertook to read the whole manuscript, performed this task with characteristic conscientiousness, and helped me with many pertinent comments. I wish also to express warm thanks to Miss Joyce Edger who cheerfully and skilfully transformed the manuscript to typescript; and to Dr and Mrs R. H. Dye for their help with proof-reading.

Newcastle upon Tyne
26 May 1976

0

SOME CONVENTIONS AND SOME BASIC FACTS

In this book, the capital letters G, H, J, K, L (sometimes with subscripts and superscripts) will always denote groups. The reader is supposed to be familiar with the elementary basic facts about groups which are summarized in this chapter. Further details are to be found for instance in Green [b14], or chapters 1, 2 of Ledermann [b29], or chapters 1, 2, 3 of Macdonald [b30].

A *subgroup* of G is a subset of G which itself forms a group with respect to the operation defining G. Then, if X is a non-empty subset of G, X is a subgroup of G if and only if $x_1 x_2^{-1} \in X$ whenever x_1 and x_2 are elements of X.

We shall usually denote the identity element of a group by 1. Then if G is the group in question, the subset $\{1\}$ consisting only of the identity element of G forms a subgroup which we call the *trivial* subgroup of G. Strictly we ought to preserve the notational distinction between the element 1 and the subgroup $\{1\}$ of G, but in practice the same symbol 1 without brackets is used for both element and subgroup. A subgroup H of G is said to be *non-trivial* if $H \neq 1$. We say that G itself is trivial if G has just one element (in which case $G = 1$); and similarly that G is non-trivial if G has more than one element. Sometimes we refer to an element of G distinct from 1 as a *non-trivial element*.

Let $g \in G$. If the elements $g, g^2, g^3, \ldots$ of G are all distinct, we say that g is an element *of infinite order* in G and we write $o(g) = \infty$. If on the other hand there are distinct positive integers r, s such that $g^r = g^s$, we say that g is *of finite order* in G: then there is a positive integer n such that $g^n = 1$ and we call the least such n the *order* of g, denoted in this book by $o(g)$. If g is of finite order and m is an integer, then $g^m = 1$ if and only if $o(g)$ divides m.

Elements g_1 and g_2 of G are said to *commute* if $g_1 g_2 = g_2 g_1$. A non-empty subset X of G is said to be a *commuting set of elements* if $x_1 x_2 = x_2 x_1$ whenever x_1 and x_2 are elements of X. If G is itself a commuting set of elements then G is called an *abelian* group (in honour of Niels Henrik Abel, 1802–29).

If there is an element g of G such that every element of G is expressible as a power g^m of g (where m is an integer), we say that G is a *cyclic* group and that g *generates* G: then we write $G = \langle g \rangle$. Any cyclic group is abelian.

Let X be any set. If X is infinite we write $|X| = \infty$. If X is finite, we denote by $|X|$ the number of elements in X. Sometimes we write $|X| < \infty$ to signify that X is finite. For any group G, we call $|G|$ the *order* of G. In particular, if G is a finite cyclic group with, say, $G = \langle g \rangle$, then $|G| = o(g)$: explicitly, if $o(g) = n$ then $G = \{1, g, g^2, \ldots, g^{n-1}\}$. For an arbitrary group, the trivial subgroup is the only subgroup of order 1.

In a few passing remarks throughout the book we refer to infinite cardinal numbers. When X is an infinite set, $|X|$ may be interpreted as the cardinality of X. With this interpretation, various statements to be made can be refined to give corresponding results which distinguish between different types of infinite sets. However, the reader who is unfamiliar with infinite cardinal numbers may ignore all such remarks without impairing his understanding of the rest of the text.

For sets X, Y we use the notation

$Y \subseteq X$ to mean Y is a subset of X,
$Y \subset X$ to mean Y is a subset of X and $Y \neq X$.

In the latter case we say that Y is a *proper* subset of X. When $Y \subseteq X$ we denote by $X \setminus Y$ the set of elements of X which do not belong to Y. The empty set is denoted by $\emptyset$. In group theory we write

$H \leqslant G$ to mean H is a subgroup of G,
$H < G$ to mean H is a proper subgroup of G.

(Warning. Ledermann and Macdonald use 'proper' to mean 'proper and non-trivial', but we do not follow this usage.)

According to Lagrange's theorem, if G is a finite group and $H \leqslant G$ then $|H|$ divides $|G|$. This fact is of crucial importance for finite group theory. We recall that the theorem is proved by partitioning G as the union of a number of disjoint subsets of G each containing $|H|$ elements. For these subsets we may select the *right cosets of H in G*, that is the subsets $Hg = \{hg : h \in H\}$ with $g \in G$ (where each g determines a subset Hg). Alternatively, we may select for the subsets the *left cosets of H in G*, that is the subsets $gH = \{gh : h \in H\}$ with $g \in G$. If g_1 and g_2 are elements of G, we have $Hg_1 = Hg_2$ if and only if $g_1 g_2^{-1} \in H$, and $g_1 H = g_2 H$ if and only if $g_1^{-1} g_2 \in H$.

For an arbitrary (not necessarily finite) group G and $H \leqslant G$, the same argument still applies. We may partition G as the disjoint union of right cosets of H in G or of left cosets of H in G. If there are only finitely many distinct right cosets of H in G then there are only finitely many distinct left cosets of H in G, and conversely, and the numbers of them are the same. This number is called the *index of H in G* and denoted by $|G : H|$. Note that $|G : H| = 1$ if and only if $H = G$. If G is a finite group then $|G : H| = |G|/|H|$. But it can also happen that an infinite group has proper subgroups with finite indices. If on the other hand there are infinitely many distinct right (or, equivalently, left) cosets of H in G, we write $|G:H| = \infty$. Later on (in chapter 4) we shall place this idea of partitioning a group as a union of disjoint subsets in a much more general context, and derive other important results by arguments of the same kind. It is assumed that the reader is familiar with the idea of an *equivalence relation* on a set, and the connexion between this and the partitioning of a set as a union of disjoint subsets.

If $g \in H \leqslant G$, then the order of g in H is the same as the order of g in G. The element g generates a cyclic subgroup $\langle g \rangle$ of G, and this subgroup is finite if and only if $o(g) < \infty$, in which case $|\langle g \rangle| = o(g)$. Now if G is a finite group with, say, $|G| = n$, and $g \in G$, then $\langle g \rangle$ is finite and so $o(g)$ is finite; and, by Lagrange's theorem, $o(g)$ divides n. Hence $g^n = 1$ for all $g \in G$.

We use the notation $\varphi : X \to Y$ to mean that φ is a map (the words mapping and function are synonyms for map) of a set X into a set Y. Frequently we write $x\varphi$ or x^φ for the image under φ of an element $x \in X$. This convention has the advantage that if $\varphi : X \to Y$ and $\psi : Y \to Z$ then the *composite* map of X into Z, defined by applying to each $x \in X$ first φ and then ψ, is denoted by $\varphi\psi$: thus, by definition, $x(\varphi\psi) = (x\varphi)\psi$. (This corresponds to the European convention of reading from left to right.) With the functional notation customary in analysis, by which the image of x under φ is denoted by $\varphi(x)$, the composite map is denoted by $\psi\varphi : (\psi\varphi)(x) = \psi(\varphi(x))$. However, it is inconvenient to maintain a consistent convention for notation of maps, and either convention will be adopted according to circumstances.

If $\varphi : X \to Y$ and $x \in X$, we often use the notation $\varphi : x \mapsto x\varphi$ to show the effect of φ on x. The barred arrow $\mapsto$ is used only between *elements* of sets.

A map $\varphi : X \to Y$ is said to be *injective* if $x_1\varphi \neq x_2\varphi$ whenever $x_1, x_2 \in X$ and $x_1 \neq x_2$; and to be *surjective* if every element $y \in Y$ is expressible as $y = x\varphi$ for some $x \in X$. If φ is both injective and surjective, it is said to be *bijective*. Let 1_X denote the *identity map on* X, i.e. the map $1_X : X \to X$ defined by $1_X : x \mapsto x$ for all $x \in X$. Then $\varphi : X \to Y$ is injective if and only if there is a map $\psi : Y \to X$ such that $\varphi\psi = 1_X$. Also, if 1_Y denotes the identity map on Y then $\varphi : X \to Y$ is surjective if and only if there is a map $\psi : Y \to X$ such that $\psi\varphi = 1_Y$. Finally, $\varphi : X \to Y$ is bijective if and only if there is a map $\psi : Y \to X$ such that $\varphi\psi = 1_X$ and $\psi\varphi = 1_Y$, that is, if and only if φ is *invertible*. A bijective map $\varphi : X \to X$ is often called a *permutation* of X. If X is a finite set then an injective map $\varphi : X \to X$ is necessarily a permutation of X, as is a surjective map $\psi : X \to X$; here the condition of finiteness of X is essential. The reader is supposed to be familiar with the elementary properties of permutations. These will be summarized when the symmetric groups are introduced in chapter 2.

Elements of a set are sometimes called *points*. A map $\varphi : X \to X$ is said to *fix* a point $x \in X$ if $x\varphi = x$.

Two groups are said to be *isomorphic* if there is a bijective, structure-preserving map from one to the other (see 2.6). Then the relation of isomorphism is an equivalence relation on any set of groups. This relation is fundamental to group theory in the sense that group theory is concerned with classifying groups 'to within isomorphism'. We cannot expect to distinguish *group-theoretically* between groups which are isomorphic but have no elements in common. For the sake of brevity we shall call an isomorphism class of groups a *type*. We write $G_1 \cong G_2$ to denote that groups G_1 and G_2 are isomorphic.

If $\varphi : X \to Y$ and S is a subset of X, then the *restriction of φ to S* is the map $\varphi_1 : S \to Y$ defined by $\varphi_1 : s \mapsto s\varphi$ for all $s \in S$. This is sometimes denoted by $\varphi|_S$. It may happen that there is a subset T of Y such that $s\varphi \in T$ for all $s \in S$. Then we may want to refer to the map $\psi : S \to T$ defined by $\psi : s \mapsto s\varphi$ for all $s \in S$. Logically this is a different map from φ_1 if $T \subset Y$. We shall say that ψ is obtained from φ *by restriction*.

If $\varnothing \subset S \subseteq X$ and 1_X denotes the identity map on X then $1_X|_S : S \to X$ is the *inclusion map of S in X*.

Throughout this book,

p always denotes a prime number,

ϖ a set of prime numbers,

C the field of complex numbers,

R the field of real numbers,

Q the field of rational numbers,

Z the ring of integers.

For any positive integer n,

$\mathbf{Z}_n$ denotes the ring of integers modulo n.

In particular, $\mathbf{Z}_p$ is a field, sometimes denoted by GF(p) (a so-called *Galois field*).

If a and b are integers and n is a positive integer,

$$a \equiv b \bmod n \text{ means that } n \text{ divides } a - b.$$

We sometimes write (a, b) for the greatest common divisor of a and b: this is defined provided that a and b are not both 0. We say that a and b are *co-prime* if $(a, b) = 1$. If a and b are co-prime integers then there exist integers a' and b' such that $aa' + bb' = 1$.

*1 Any group of prime order is cyclic.

*2 Any two cyclic groups of the same finite order are isomorphic.

*3 If $g^2 = 1$ for every $g \in G$ then G is an abelian group.

4 Let $g_1, g_2 \in G$. Then $o(g_1 g_2) = o(g_2 g_1)$.

*5 (i) Let $g \in G$ with $o(g) = n < \infty$. Then, for every integer m, $o(g^m) = n/(m, n)$.

(ii) If G is a cyclic group of finite order n then the number of distinct elements which generate G is $\varphi(n)$, where φ is Euler's function: that is, $\varphi(n)$ is the number of positive integers not exceeding n which are co-prime to n.

*6 Let $g_1, g_2 \in G$ with $o(g_1) = n_1 < \infty$, $o(g_2) = n_2 < \infty$. If n_1 and n_2 are co-prime and g_1 and g_2 commute then $o(g_1 g_2) = n_1 n_2$.

7 Let $g \in G$ with $o(g) = n_1 n_2$, where n_1 and n_2 are co-prime positive integers. Then there are elements $g_1, g_2 \in G$ such that $g = g_1 g_2 = g_2 g_1$ and $o(g_1) = n_1$, $o(g_2) = n_2$. Moreover, g_1 and g_2 are uniquely determined by these conditions.

8 By considering orders of elements in the multiplicative group of all non-zero elements of the field Z_p, prove Fermat's theorem: for every integer a not divisible by p, $a^{p-1} \equiv 1 \bmod p$.

9 Let G be an abelian group of finite order n. Show that the product of the n distinct elements of G is equal to the product of all the elements of G of order 2 (where the latter product is interpreted as 1 if G has no element of order 2). By applying this result to the multiplicative group of all non-zero elements of the field Z_p, prove Wilson's theorem for prime numbers: $(p - 1)! \equiv -1 \bmod p$.

10 Let $H \leqslant G$ and let $g_1, g_2 \in G$. Then $Hg_1 = Hg_2$ if and only if $g_1^{-1}H = g_2^{-1}H$.

*11 (i) Let $J \leqslant H \leqslant G$. Then $|G : J|$ is finite if and only if $|G : H|$ and $|H : J|$ are both finite, and if so, $|G : J| = |G : H||H : J|$.
(ii) Let $J \leqslant H \leqslant G$ with $|G : J| = p$, prime. Then either $H = J$ or $H = G$.

12 Let $H \leqslant G$ and $g \in G$. If $o(g) = n$ and $g^m \in H$, where n and m are co-prime integers, then $g \in H$.

1

INTRODUCTION TO FINITE GROUP THEORY

The ideal aim of finite group theory is to 'find' all finite groups: that is, to show how to construct finite groups of every possible type, and to establish effective procedures which will determine whether two given finite groups are of the same type. The attainment of this ideal is of course quite beyond the reach of present techniques (though the corresponding aim for finite abelian groups was achieved a hundred years ago: see 8.24, 8.41). But what kind of programme might be devised towards the fulfilment of such an aim?

To each finite group G there is associated the positive integer $|G|$. We note two elementary facts.

1.1. *For each positive integer n, there is at least one type of group of order n.*

For instance, the set of complex nth roots of unity forms a (cyclic) group of order n under multiplication: see 2.14.

1.2. *For each positive integer n, there are only finitely many different types of groups of order n.*

To see this we observe that for any group G of order n and any set X of n elements, X can be given the structure of a group isomorphic to G. All that is needed is to choose some bijective map $\varphi : G \to X$ and then to define multiplication in X by the rule $(g_1\varphi)(g_2\varphi) = (g_1 g_2)\varphi$ for all $g_1, g_2 \in G$. It is straightforward to check that this multiplication on X satisfies the group axioms; then also, by definition, φ becomes an isomorphism. This means that groups of order n of all possible types appear among all possible assignments of a binary operation to any particular set of n elements. But the number of different such assignments is n^{n^2}, and so this is also an upper bound for the number of types of groups of order n.

(For another proof of 1.2, see 4.24.) For each positive integer n, let $v(n)$ denote the number of types of groups of order n. Very little is known about $v(n)$ in general (see **301** for a sharper upper bound on $v(n)$); but one simple remark can be made immediately. It follows from Lagrange's theorem that a group of prime order must be cyclic (**1**). Since any two cyclic groups of the same order are isomorphic (**2**), we have

1.3. *For each prime number p, $v(p) = 1$.*

There are numbers n other than primes for which $v(n) = 1$. We mention a result which characterizes these numbers – though the result is not of importance in group theory, but merely a curiosity (see 575).

1.4. *Let* $n = p_1^{m_1} p_2^{m_2} \ldots p_s^{m_s}$, *where* $s, m_1, \ldots, m_s$ *are positive integers and* $p_1, \ldots, p_s$ *distinct primes. Then* $v(n) = 1$ *if and only if* $m_1 = m_2 = \ldots = m_s = 1$ *and for all* $i, j = 1, \ldots, s$, $p_i - 1$ *is not divisible by* p_j.

(Thus for example $v(15) = 1$; see **215**.)

Now, for each positive integer n, let $v_a(n)$ denote the number of types of abelian groups of order n: then $v_a(n) \leqslant v(n)$. From theorems on the structure of finite abelian groups (see 8.43), we have

1.5. *Let* $n = p_1^{m_1} p_2^{m_2} \ldots p_s^{m_s}$, *where* $s, m_1, \ldots, m_s$ *are positive integers and* $p_1, \ldots, p_s$ *distinct primes. Then*

$$v_a(n) = v_a(p_1^{m_1}) v_a(p_2^{m_2}) \ldots v_a(p_s^{m_s}),$$

and, for each $j = 1, \ldots, s$,

$v_a(p_j^{m_j})$ *is the number of partitions of* m_j; *that is, the number of ways of expressing* m_j *as a sum of positive integers (the order of components being disregarded). In particular,* $v_a(p_j^{m_j}) \geqslant m_j$.

This shows that there is no upper bound for $v_a(n)$ which is independent of n; and hence also no upper bound for $v(n)$ independent of n.

A natural approach to the problem of constructing finite groups is to seek an inductive method in terms of group orders. Thus we should try to describe each finite group in terms of groups of smaller orders: then in principle we might hope to start with certain basic groups and to build up a description of all types of finite groups step by step.

Therefore we naturally think about subgroups. What subgroups does a group G of order n have? By Lagrange's theorem, the order of any subgroup of G is a divisor of n. However, it is not necessarily true that G has a subgroup of order m for each divisor m of n (see **185**). The best general result about existence of subgroups of prescribed orders is the following consequence of Sylow's theorem (see 5.32).

1.6. *Let* G *be a group of finite order* n. *For each prime* p *and power* p^m *of* p *which divides* n, G *possesses a subgroup of order* p^m.

This result directs attention to groups of prime power orders. Such groups have helpful special properties and play an important part in the analysis of general finite groups.

Among the subgroups of a group G there are some which are especially useful in deriving information about G: the so-called *normal* subgroups. We use the notation '$K \trianglelefteq G$' to mean 'K is a normal subgroup of G'. The explicit definition of the term is given in 3.2, but for the present discussion we merely state the following key facts (see 3.20–3.22).

1.7. *Suppose that* $K \trianglelefteq G$. *Then we can define a corresponding group* G/K

(*not a subgroup of G*) *which is called the* quotient group *of G by K. In some sense, G is built up from the two groups K and G/K. In particular, if G is finite then so are K and G/K, and* $|G| = |K|.|G/K|$.

One always finds among the normal subgroups of G the group G itself and 1, the trivial subgroup; and $G/G \cong 1$ and $G/1 \cong G$. But the interesting normal subgroups are the ones different from these, if they exist.

If, in 1.7, G is finite and $K \neq G$ and $K \neq 1$ then we get a description of G in terms of two groups K and G/K of smaller orders than G. This description cannot be regarded as complete, because knowledge of the types of the groups K and G/K is not in general enough to determine uniquely the type of G (see **116**). This then raises the *extension problem*: given groups K and Q, determine the types of all groups G such that $K \trianglelefteq G$ and $G/K \cong Q$. Note that if K and Q are finite groups then the number of such types is finite, because all such groups G have the same finite order $|K|.|Q|$ and therefore the number of types is at most $\nu(|K|.|Q|)$. Although the extension problem is hard, it is much more amenable to attack by currently known methods than some of the problems mentioned earlier.

Assuming that we can deal with this extension problem, we are encouraged to try the following programme. The notation '$K \lhd G$' is used to mean 'K is a normal subgroup of G and $K \neq G$'. For any finite group G we consider chains of subgroups

$$1 = K_0 \lhd K_1 \lhd K_2 \lhd \ldots \lhd K_{s-1} \lhd K_s = G \tag{i}$$

which cannot be *refined*: that is, such that we cannot insert a subgroup H with $K_{j-1} \lhd H \lhd K_j$ for any $j = 1, \ldots, s$. Then we try to describe G in terms of the quotient groups

$$K_1/K_0, K_2/K_1, \ldots, K_s/K_{s-1}$$

which we have made as small as possible. Note that

$$|K_1/K_0|.|K_2/K_1|\ldots|K_s/K_{s-1}| = |G|.$$

Such a chain (i) which cannot be refined is called a *composition series* of G.

1.8. *Any finite group G possesses at least one composition series.*

To prove this, we argue by induction on $|G|$. If $|G| = 1$ then $1 = K_0 = G$ is a composition series of G. Suppose that $|G| > 1$. Choose $K \lhd G$ with $|K|$ as large as possible (it may be that $K = 1$). Then $|K| < |G|$, so that by the induction hypothesis K has a composition series

$$1 = K_0 \lhd K_1 \lhd \ldots \lhd K_{s-1} = K,$$

where s is a positive integer. Then

$$1 = K_0 \lhd K_1 \lhd \ldots \lhd K_{s-1} \lhd K_s = G$$

is a composition series of G. The induction argument goes through.

The most important fact about composition series is contained in the Jordan–Hölder theorem (see 7.9):

1.9. *Let G be a finite group and let*

$$1 = K_0 \lhd K_1 \lhd \ldots \lhd K_s = G$$

and

$$1 = H_0 \lhd H_1 \lhd \ldots \lhd H_r = G$$

be composition series of G. Then $r = s$, and the two sequences of s quotient groups $K_1/K_0, K_2/K_1, \ldots, K_s/K_{s-1}$ and $H_1/H_0, H_2/H_1, \ldots, H_s/H_{s-1}$ contain groups of exactly the same types with the same multiplicities (possibly in different orderings). We call s the composition length *of G and the groups $K_1/K_0, K_2/K_1, \ldots, K_s/K_{s-1}$ the* composition factors *of G.*

Some groups will have composition length 1: such groups are called simple. Explicitly, a group G (not necessarily finite) is *simple* if $G \neq 1$ and if the only normal subgroups of G are 1 and G. One can show (see 7.2)

1.10. *Every composition factor of a non-trivial finite group is a simple group.*

Now we may think of the Jordan–Hölder theorem 1.9 as an analogue for finite groups of the fundamental theorem of arithmetic for positive integers. Finite simple groups play a corresponding rôle to prime numbers. The Jordan–Hölder theorem says that every non-trivial finite group G is a kind of 'product' of simple groups, and that these simple factors are uniquely determined by G (apart from ordering). Of course formation of a 'product' in this context is not a uniquely determined process as it is for numbers. Nevertheless, the results quoted effect a division of the original classification problem into two parts: (i) find finite simple groups of all possible types (the 'building blocks' of finite group theory), (ii) solve the extension problem (that is, find how the building blocks fit together).

A great deal of effort has been devoted during the last ten years to problem (i) and, although the obstacles ahead look formidable and the goal is not yet in sight, significant advances have been made. At this stage we merely mention (see 3.6, 3.60, 3.61, 5.24, 5.28)

1.11. *The only abelian simple groups are the groups of prime orders. There are also infinitely many types of non-abelian finite simple groups.*

How might we attempt to investigate the structure of a non-abelian simple group? We have no non-trivial proper normal subgroup, by means of which we might hope to express the structure of the group in terms of the structures of two smaller groups; but we should still like to work from smaller groups up to larger groups. Therefore we think about subgroups which are not normal. A starting point is provided by the following result, which is one of the most striking achievements of the modern period.

1.12 (W. Feit and J. G. Thompson [a23]). *Every non-abelian finite simple group has even order.*

The result was a conjecture of Burnside in 1911: see [b3] p. 503. In fact, in the first edition of his book in 1897, Burnside had already recommended an investigation of the existence or non-existence of a non-abelian simple group of odd order, without predicting the outcome. The question was finally settled in 1963 with the publication of article [a23]. A proof of this theorem is unfortunately beyond the scope of any textbook at present, though D. Gorenstein's book [b13] gives an account of many of the techniques involved. The importance of the Feit–Thompson theorem is that it ensures the existence of elements of order 2 in non-abelian finite simple groups.

1.13. *Let G be any group of even order. Then G possesses at least one element of order 2. (Any such element is called an* involution.*)*

The result is an immediate consequence of 1.6. Alternatively, there is a more elementary proof as follows. Let $T = \{x \in G : x^2 = 1\}$ and $U = \{x \in G : x^2 \neq 1\}$. Then T and U are subsets of G such that $G = T \cup U$ and $T \cap U = \emptyset$. Now we count the elements of U. Possibly $U = \emptyset$, in which case $|U| = 0$. If not, choose $x_1 \in U$. Then $x_1 \neq x_1^{-1} \in U$. Possibly $U = \{x_1, x_1^{-1}\}$, in which case $|U| = 2$. If not, choose another element $x_2 \in U$. Then $x_2 \neq x_2^{-1} \in U$, and also $x_2^{-1} \neq x_1$ and $x_2^{-1} \neq x_1^{-1}$ (since $x_2 \neq x_1^{-1}$ and $x_2 \neq x_1$). Possibly $U = \{x_1, x_1^{-1}, x_2, x_2^{-1}\}$, in which case $|U| = 4$. If not, ... We continue until all elements of U are exhausted. We see that in any case $|U|$ is even. Since also $|G|$ is even and $|G| = |T| + |U|$, it follows that $|T|$ is even. But $T \neq \emptyset$, since $1 \in T$. Hence $|T| \geq 2$. Therefore there is an element $t \in T$ with $t \neq 1$. Such an element t is an involution.

Now let G be any group. For each $x \in G$ we define

$$C_G(x) = \{g \in G : gx = xg\}.$$

It is easy to check that $C_G(x)$ is a subgroup of G; it is called the *centralizer* of x in G (see 4.25). Also

$$\bigcap_{x \in G} C_G(x) = \{g \in G : gx = xg \text{ for all } x \in G\}$$

is a subgroup of G, called the *centre* of G and denoted by $Z(G)$ (see **117**). Note that, immediately from these definitions, if $x \in G$ and $H = C_G(x)$ then $x \in Z(H)$.

A good deal of the recent discussion of finite simple groups has been concerned with centralizers of involutions. An important reason for this lies in the following result (see 6.9).

1.14 (R. Brauer and K. A. Fowler [a8], 1955). *Let G be a non-abelian* *finite simple group (so that, by 1.12, $|G|$ is even) and let t be an involution in G. Then $C_G(t) \neq G$, and if $|C_G(t)| = m$ then $|G| \leqslant (\frac{1}{2}m(m+1))!$*

This raises the possibility of characterizing a simple group G in terms of the structure of the centralizer of an involution, a group of smaller order than G. Specifically, from 1.14 we deduce

1.15. *Let H be a group of even order with an involution $u \in Z(H)$. Then there are at most finitely many types of finite simple group G possessing an involution t with $C_G(t) \cong H$.*

To prove this, we note first that if such a simple group G is abelian then, by 1.11, $|G| = 2$. On the other hand, any non-abelian such group G has, by 1.14, order at most $(\frac{1}{2}|H|(|H|+1))!$, a number dependent only on the given group H. By 1.2, there are only finitely many different types of groups of any given order; hence also there are only finitely many different types of groups of orders not exceeding any given number.

The following scheme has been used repeatedly. Start with a known non-abelian finite simple group E and an involution $u \in E$, and let $H = C_E(u)$. Then consider finite simple groups G having an involution t with $C_G(t) \cong H$. By 1.15, there are only finitely many types of such groups G. Try to prove that there is actually only *one* type, in other words, that $G \cong E$ necessarily. If this succeeds, a *characterization theorem* for E has been established: a characterization of E in terms of the structure of the centralizer of one of its involutions (a group of smaller order than E). Many such characterization theorems are known. But if the attempt fails because there are groups not isomorphic to E among the groups G, there may be previously unknown simple groups among the groups G. This procedure has been a source of discovery of several new finite simple groups during the past few years.

2

EXAMPLES OF GROUPS AND HOMOMORPHISMS

It will be convenient to start with the notion of a semigroup since, as we shall see, many important examples of groups arise in a natural way from semigroups. However, we shall not in this book develop the extensive algebraic theory of semigroups.

2.1 Definition. A *semigroup* is a non-empty set S, together with an associative binary operation on S. The operation is often called *multiplication* and, if x, $y \in S$, the product of x and y (in that ordering) is written as xy.

The associativity in S is notationally very helpful. It permits us to write unambiguously $x_1 x_2 x_3$ for $(x_1 x_2)x_3 = x_1(x_2 x_3)$, where x_1, x_2, x_3 are any elements of S. Furthermore it follows that we may refer unambiguously to the product of any finite number of elements, taken in a definite ordering: brackets may be arbitrarily inserted in or removed from a product $x_1 x_2 \ldots x_n$ without altering the result. (For a formal proof of this, see Ledermann [b29] pp. 3–4 or Macdonald [b30] pp. 18–19.) There follow the standard *power laws*: if $x \in S$ and m, n are any positive integers then $x^m x^n = x^{m+n}$ and $(x^m)^n = x^{mn}$. Of course it is not in general permissible to alter the ordering of elements in a product, for we may have $x_1 x_2 \neq x_2 x_1$.

2.2 Definitions. Let S be a semigroup.

(i) An element $e \in S$ is called an *identity element* of S if $ex = x = xe$ for all $x \in S$. If S has an identity element then it is unique: for if e and f are identity elements of S then $f = ef = e$.

(ii) Suppose that S has an identity element e, and let $x \in S$. An element $y \in S$ is called an *inverse* of x if $yx = e = xy$. If x has an inverse then it is unique: for if y and z are inverses of x then $y = ye = yxz = ez = z$.

Now a group is just a semigroup with an identity element such that every element has an inverse.

(iii) A *subgroup* of S is a subset of S which forms a group with respect to the operation defining S.

13 Give an example of a semigroup without an identity element.

14 Give an example of an infinite semigroup with an identity element e such that no element except e has an inverse.

2.3. *Let S be a semigroup with an identity element e. An element of S which has an inverse is called a* unit *in S. Then the set of all units in S forms a subgroup of S, called the* group of units *of S.*

Proof. Let U be the set of all units in S. Then $U \neq \emptyset$, since $e \in U$. Since the operation defining S is associative, so is the operation on U. Let $x_1, x_2 \in U$. Then there are elements $y_1, y_2 \in S$ such that

$$x_1 y_1 = y_1 x_1 = e = x_2 y_2 = y_2 x_2.$$

Then

$$(x_1 x_2)(y_2 y_1) = x_1(x_2 y_2)y_1 = x_1 e y_1 = x_1 y_1 = e = (y_2 y_1)(x_1 x_2)$$

similarly. Hence $x_1 x_2 \in U$. Clearly e is an identity element for U and $y_1 \in U$. Thus y_1 is an inverse of x_1 in U. Hence U is a subgroup of S.

15 Let S be a semigroup and let $x \in S$. Show that $\{x\}$ forms a subgroup of S (of order 1) if and only if $x^2 = x$. Such an element x is called an *idempotent* in S. (Warning. A semigroup may have several different subgroups of order 1: see **16**. Why does a group have only one subgroup of order 1?)

16 Let X be any non-empty set. Let S be the set of all subsets of X. Show that S is a semigroup with respect to the operation $\cap$. Does S have an identity element, and if so, what are the units in S? Show that every element of S is an idempotent (**15**). Deduce that for all $Y \in S$, $\{Y\}$ is a subgroup of S, and that every subgroup of S has order 1.

What happens if $\cap$ is replaced by $\cup$?

2.4. *If X is any non-empty set, the set M_X of all maps of X into itself forms a semigroup with respect to the operation of composition. There is in M_X an identity element 1, defined by $1 : x \mapsto x$ for all $x \in X$. The units in M_X are just the permutations of X. The group of units of M_X is denoted by Σ_X and called the* (unrestricted) symmetric group *on X.*

Proof. Certainly $M_X \neq \emptyset$; and M_X is closed with respect to composition, because the composite of two maps $\varphi_1 : X \to X$ and $\varphi_2 : X \to X$ is defined to be a map $\varphi_1 \varphi_2 : X \to X$. Moreover, composition is associative: for if $\varphi_1, \varphi_2, \varphi_3 \in M_X$, the equation $(\varphi_1 \varphi_2)\varphi_3 = \varphi_1(\varphi_2 \varphi_3)$ merely expresses the fact that for every $x \in X$,

$$x((\varphi_1 \varphi_2)\varphi_3) = (x(\varphi_1 \varphi_2))\varphi_3 = ((x\varphi_1)\varphi_2)\varphi_3 = (x\varphi_1)(\varphi_2 \varphi_3) = x(\varphi_1(\varphi_2 \varphi_3)).$$

The map 1 defined above belongs to M_X and is obviously an identity element for M_X. Finally, if $\varphi, \psi \in M_X$, then ψ is an inverse of φ in M_X if and only if $\psi \varphi = 1 = \varphi \psi$. For a given φ, there is such a ψ if and only if φ is a bijective map, that is, if and only if φ is a permutation of X. Thus the group of units of M_X consists of all permutations of X.

Remark. The *restricted* symmetric group on a non-empty set X will be defined in **110**.

2.5. *If X is a finite set with, say, $|X| = n > 0$, then $|M_X| = n^n$ and $|\Sigma_X| = n!$*

The reader will know the standard notation for permutations of finite sets: see for instance Ledermann [b29], pp. 20–6. Thus, if n is a positive integer,

$$\begin{pmatrix} 1 & 2 & \dots n \\ a_1 & a_2 & \dots a_n \end{pmatrix}$$

denotes the permutation σ of the set $\{1, 2, \dots, n\} = \{a_1, a_2, \dots, a_n\}$ for which

$$\sigma : j \mapsto a_j \quad \text{for each } j = 1, \dots, n.$$

Any permutation of a finite set X can be expressed as a product of *cycles* on disjoint subsets of X, and the expression is unique apart from the ordering of the cycles. Moreover, any two disjoint cycles commute. Thus, for example,

$$\begin{pmatrix} 1 & 2 & 3 & 4 & 5 & 6 & 7 & 8 \\ 2 & 6 & 1 & 7 & 5 & 3 & 8 & 4 \end{pmatrix} = (1263)(478) = (478)(1263),$$

where, for instance, (1263) denotes the permutation $\begin{pmatrix} 1 & 2 & 6 & 3 \\ 2 & 6 & 3 & 1 \end{pmatrix}$, and we adopt the usual convention of suppressing on the right points which are fixed by the permutation – in this case the point 5. The number of distinct points which occur in a cycle is called the *length* of the cycle. So, for example, the cycle (1263) has length 4.

The notation in terms of cycles is a very convenient one for making explicit calculations in finite symmetric groups. For instance, if $X = \{1, 2, 3\}$ then, in Σ_X,

$$(12)(13) = (123), \qquad \text{while } (13)(12) = (132).$$

This shows that Σ_X is a non-abelian group.

Here, and always in this book, when we are discussing permutations of a set (rather than maps of other kinds), we shall place the symbols which denote the permutations on the right of the points to which they apply. Thus permutations are multiplied together from left to right.

17 Let X be a set with $|X| = 2$. Do two elements of M_X necessarily commute? Do two elements of Σ_X necessarily commute?

18 Express the permutation $\sigma = \begin{pmatrix} 1 & 2 & 3 & 4 & 5 & 6 & 7 \\ 1 & 5 & 7 & 4 & 6 & 2 & 3 \end{pmatrix}$ as a product of disjoint cycles.

Find expressions for σ^2 and σ^{-1} as products of disjoint cycles.

We shall often be concerned with particular instances of the following situation: there are sets X, Y (possibly with $X = Y$) and a set T of maps of X into Y. Some or all of the sets X, Y, T will usually be groups. We have noted one such situation in 2.4, and now consider another.

Suppose that we have two *groups* G and H (possibly the same). We should expect that among all the maps $\varphi : G \to H$, those which are *structure-preserving* would be of particular significance.

2.6 Definitions. A map $\varphi : G \to H$ is said to be a *homomorphism* if

$$(g_1 g_2)\varphi = (g_1\varphi)(g_2\varphi) \quad \text{for all } g_1, g_2 \in G.$$

If in addition φ is bijective, it is said to be an *isomorphism*.

Suppose that $\varphi : G \to H$ is an isomorphism and let $\psi : H \to G$ be the inverse of the bijective map φ. Then ψ is also an isomorphism: for ψ is bijective, and if $h_1, h_2 \in H$ then there are elements $g_1, g_2 \in G$ such that $h_1 = g_1\varphi, h_2 = g_2\varphi$; hence

$$\begin{aligned}(h_1 h_2)\psi &= ((g_1\varphi)(g_2\varphi))\psi = ((g_1 g_2)\varphi)\psi = (g_1 g_2)(\varphi\psi)\\ &= g_1 g_2 = (g_1\varphi\psi)(g_2\varphi\psi)\\ &= ((g_1\varphi)\psi)((g_2\varphi)\psi) = (h_1\psi)(h_2\psi).\end{aligned}$$

If there is an isomorphism of G onto H we say that G and H are *isomorphic groups*, or that G and H are *of the same type*, and write $G \cong H$. The relation $\cong$ is an equivalence relation on groups (**20**). If G and H are not isomorphic we write $G \not\cong H$.

*19 If $\varphi : G \to H$ and $\psi : H \to J$ are homomorphisms then the composite map $\varphi\psi : G \to J$ is also a homomorphism.

*20 Verify that if $\mathscr{G}$ is any set of groups then $\cong$ is an equivalence relation on $\mathscr{G}$.

It is often convenient to regard isomorphic groups whose elements form disjoint sets as the *same* group. However, we must be careful in dealing with subgroups of a group G. It may happen that G has subgroups H and K which are distinct sets with $H \cong K$. Then it is of course not permissible to say that $H = K$. In this situation we are concerned with H and K not merely as abstract groups but also with their relation to the containing group G.

2.7. *If X and Y are non-empty sets such that there is a bijective map $\varphi : X \to Y$ then $\Sigma_X \cong \Sigma_Y$. In particular, in studying the finite symmetric groups it is enough to consider the permutations of a single set of n elements, for each positive integer n. We often choose for this the set $\{1, 2, \ldots, n\}$, and we write Σ_n (in many books, S_n) for the symmetric group on n points. Then Σ_n is called* the symmetric group of degree n.

Proof. Since $\varphi : X \to Y$ is bijective, there is an inverse map $\psi : Y \to X$; thus

$$\varphi\psi = 1_X, \text{ the identity map on } X,$$

and $\qquad \psi\varphi = 1_Y, \text{ the identity map on } Y.$

For each $\sigma \in \Sigma_X, \psi\sigma\varphi : Y \to Y$. In fact, $\psi\sigma\varphi \in \Sigma_Y$; the map $\psi\sigma^{-1}\varphi : Y \to Y$ is such that

$$(\psi\sigma\varphi)(\psi\sigma^{-1}\varphi) = 1_Y = (\psi\sigma^{-1}\varphi)(\psi\sigma\varphi).$$

Now it is easy to check that the map $\theta : \Sigma_X \to \Sigma_Y$, defined by

$$\theta : \sigma \mapsto \psi\sigma\varphi \quad \text{(for all } \sigma \in \Sigma_X),$$

is an isomorphism.

2.8 Definition. If there is an injective homomorphism (sometimes called a *monomorphism*) of G into H, we say that G *can be embedded in* H.

For any groups G and H there is at least one homomorphism $\varphi : G \to H$, namely the *trivial* homomorphism $\varphi : g \mapsto 1$ for all $g \in G$. But of course in general G cannot be embedded in H.

*21 (i) Calculate the products $(12)(13)(14)$ and $(12)(13)(14)(15)$ in the group Σ_n, where n is an integer with $n \geqslant 5$.

(ii) A permutation such as (12), which interchanges two points and fixes all others, is called a *transposition*. For any integer $n > 1$, show that the permutation $(123 \dots n)$ can be expressed as a product of $n - 1$ transpositions.

(iii) For any integer $n > 1$, prove that every non-trivial element of Σ_n can be expressed as a product of at most $n - 1$ transpositions.

22 Let n be a positive integer and let $\sigma \in \Sigma_n$. Suppose that σ can be expressed as a product of s disjoint cycles of lengths $n_1, n_2, \dots, n_s$ respectively, where $s, n_1, \dots, n_s$ are positive integers such that $n_1 + \dots + n_s = n$. Then $o(\sigma)$ is the least common multiple of $n_1, n_2, \dots, n_s$.

23 Let n and m be positive integers such that m divides n. Let σ be a cycle of length n in Σ_n. Then σ^m is the product of m disjoint cycles of length n/m.

24 Any group which can be embedded in an abelian group is abelian.

25 Prove that if X is a non-empty set and Y a non-empty subset of X then Σ_Y can be embedded in Σ_X. (In particular, whenever m and n are positive integers such that $m \leqslant n$ then Σ_m can be embedded in Σ_n.) Hence or otherwise show that if $|X| > 2$ then Σ_X is non-abelian.

The next two results give certain basic properties of homomorphisms.

2.9. *Let* $\varphi : G \to H$ *be a homomorphism. Then* $1\varphi = 1$, *and for every* $g \in G$, $g^{-1}\varphi = (g\varphi)^{-1}$. *The set* $\{g\varphi : g \in G\}$ *is a subgroup of* H *which we call the image of* φ *and denote by* Im φ *or* $G\varphi$. *Moreover, for every subgroup* K *of* G, *the set* $K\varphi = \{k\varphi : k \in K\}$ *is a subgroup of* Im φ.
Proof. Since in G $1.1 = 1$, it follows that in H

$$(1\varphi)(1\varphi) = 1\varphi,$$

hence that $$1\varphi = (1\varphi)(1\varphi)^{-1} = 1.$$

(In the last equation, on the left 1 denotes the identity element of G while on the right 1 denotes the identity element of H.) Also in G

$$g^{-1}g = 1,$$

and so in H $$(g^{-1}\varphi)(g\varphi) = 1\varphi = 1,$$

hence $$g^{-1}\varphi = (g\varphi)^{-1}.$$

Let $J = \{g\varphi : g \in G\}$. Then J is a non-empty subset of H. Moreover, if $g_1, g_2 \in G$ then

$$(g_1\varphi)(g_2\varphi)^{-1} = (g_1\varphi)(g_2^{-1}\varphi) = (g_1 g_2^{-1})\varphi \in J.$$

Hence $J \leqslant H$.

For any $K \leqslant G$, the map

$$\varphi|_K : K \to H$$

is a homomorphism, and $\operatorname{Im} \varphi|_K = K\varphi$. Hence $K\varphi \leqslant H$, by what we have proved; and then, since $K\varphi \subseteq \operatorname{Im} \varphi$, $K\varphi \leqslant \operatorname{Im} \varphi$.

We note the following immediate consequence of 2.9.

2.10. (i) *If $\varphi : G \to H$ is an injective homomorphism then $G \cong G\varphi$, and for every subgroup K of G, $K \cong K\varphi$,*

(ii) *G can be embedded in H if and only if G is isomorphic to a subgroup of H.*

26 If $\varphi : G \to H$ is a homomorphism and G is abelian then $\operatorname{Im} \varphi$ is abelian.

*27 Suppose that G_1 and G_2 are isomorphic groups, and let $\varphi : G_1 \to G_2$ be an isomorphism. If $K_1 \leqslant G_1$ and $K_2 = K_1\varphi$ then $|G_1 : K_1| = |G_2 : K_2|$.

28 If G and H are finite then a necessary condition that G can be embedded in H is that $|G|$ divide $|H|$. Show by an example that this is not a sufficient condition.

It will be convenient at this stage to gather together some important examples of groups and homomorphisms.

2.11. Any ring R forms an abelian group R^+ under addition: the *additive group* of R. The identity element of R^+ is the zero element 0 of R, and the inverse in R^+ of $a \in R$ is the element $-a$. Note in particular the groups

$$\mathbf{Z}^+ < \mathbf{Q}^+ < \mathbf{R}^+ < \mathbf{C}^+.$$

For any $a \in R$ (an arbitrary ring) the maps

$$\lambda_a : R^+ \to R^+ \quad \text{and} \quad \rho_a : R^+ \to R^+,$$

defined by $\qquad \lambda_a : x \mapsto ax \quad \text{and} \quad \rho_a : x \mapsto xa$

for all $x \in R^+$, are homomorphisms: this follows from the distributive laws for the ring R. If multiplication in R is commutative then $\lambda_a = \rho_a$ for every $a \in R$; but if the multiplication is non-commutative then $\lambda_a \neq \rho_a$ for some $a \in R$.

If $R = F$, a field, and $0 \neq a \in F$, then $\lambda_a : F^+ \to F^+$ is an isomorphism: for the map λ_a is invertible, with inverse $\lambda_{a^{-1}}$.

If $R = \mathbf{Z}$ and $0 \neq n \in \mathbf{Z}$, then the map $\lambda_n : \mathbf{Z}^+ \to \mathbf{Z}^+$ is an injective homomorphism. However, $\operatorname{Im} \lambda_n = \{nx : x \in \mathbf{Z}^+\}$, which is a proper subgroup of $\mathbf{Z}^+$ unless $n = \pm 1$. (For instance, $\operatorname{Im} \lambda_2$ is the proper subgroup of $\mathbf{Z}^+$ consisting of all even integers.) So when $|n| \geqslant 2$, $\lambda_n : \mathbf{Z}^+ \to \mathbf{Z}^+$ is an injective homomorphism which is not an isomorphism; and, by 2.10, Im λ_n is a proper subgroup of $\mathbf{Z}^+$ which is isomorphic to $\mathbf{Z}^+$ itself (cf. **30**). Note that $\mathbf{Z}^+$ is a cyclic group with $\mathbf{Z}^+ = \langle 1 \rangle$. None of the groups $\mathbf{Q}^+, \mathbf{R}^+, \mathbf{C}^+$ is cyclic.

*29 (i) Any infinite cyclic group is isomorphic to $\mathbf{Z}^+$.

(ii) If G is a non-trivial group of which the only subgroups are 1 and G then G is cyclic of prime order.

30 (i) If G is finite then no proper subgroup of G is isomorphic to G.

(ii) If G is a group such that, whenever $J < H \leqslant G, J \not\cong H$ then every element of G has finite order. (An example of an infinite group with this property will be given in **144**.)

31 Show that $\mathbf{Z}^+$ has infinitely many distinct subgroups. Deduce that every infinite group has infinitely many distinct subgroups.

*32 No two of the groups $\mathbf{Z}^+, \mathbf{Q}^+, \mathbf{R}^+$ are isomorphic. (Remark. It is in fact true that $\mathbf{R}^+ \cong \mathbf{C}^+$. This can be proved by regarding $\mathbf{R}$ and $\mathbf{C}$ as vector spaces over $\mathbf{Q}$, and using vector space theory and facts about infinite cardinals.)

33 Let $\mathrm{Hom}(G, A)$ denote the set of all homomorphisms of a group G into an abelian group A. Define a binary operation $+$ on $\mathrm{Hom}(G, A)$ as follows: for $\varphi, \psi \in \mathrm{Hom}(G, A)$,

$$\varphi + \psi : G \to A$$

is defined by

$$\varphi + \psi : g \mapsto g^\varphi g^\psi$$

for all $g \in G$. Verify that $\varphi + \psi \in \mathrm{Hom}(G, A)$ and that, with respect to $+$, $\mathrm{Hom}(G, A)$ acquires the structure of an abelian group.

Show that for any abelian group A, $\mathrm{Hom}(\mathbf{Z}^+, A) \cong A$.

2.12. Any ring R forms a semigroup under multiplication. If R has a multiplicative identity element 1 then, by 2.3, the elements of R which have multiplicative inverses in R form a group under multiplication; this is called the *group of units* of R. We denote this group by $R^\times$. Note in particular the groups $\mathbf{Z}^\times < \mathbf{Q}^\times < \mathbf{R}^\times < \mathbf{C}^\times$.

For any field F, $F^\times$ is the *multiplicative group* of all non-zero elements of F, and is abelian. We mention without proof the fact that if F is any finite field then $|F| = p^m$ for some prime p and positive integer m, and $F^\times$ is cyclic of order $p^m - 1$ (see Herstein [b19] chapter 7, §1, or Lang [b28] chapter VII, §5, or Zassenhaus [b41] pp. 104–5; see also 9.15(ii)). Note that $\mathbf{Z}^\times = \{1, -1\}$, while $\mathbf{Q}^\times, \mathbf{R}^\times, \mathbf{C}^\times$ are infinite.

34 Show that if $\varphi : G \to H$ is a homomorphism, $g \in G$, and n is a positive integer such that $g^n = 1$, then $(g\varphi)^n = 1$. Hence, by considering the elements z of $\mathbf{C}^\times$ satisfying $z^3 = 1$, or otherwise, show that $\mathbf{C}^\times$ is not isomorphic to either $\mathbf{Q}^\times$ or $\mathbf{R}^\times$.

35 Prove that $\mathbf{Q}^\times$ is not isomorphic to $\mathbf{R}^\times$. (Hint. Note that for any positive real number a and any positive integer n, there is a unique positive real number b such that $a = b^n$.)

36 Prove that there is no field F such that $F^+ \cong F^\times$. (Hint. Assume that there is a field F with an isomorphism $\varphi : F^\times \to F^+$ and consider $(-1)\varphi$.)

37 Let F be a field. Define a binary operation $*$ on F by

$$a * b = a + b - ab \quad \text{for all } a, b \in F.$$

Prove that the set of all elements of F distinct from 1 forms a group F^* with respect to the operation $*$, and that $F^* \cong F^\times$.

2.13. The set of all positive real numbers forms a subgroup $\mathbf{R}_{pos}^\times$ of $\mathbf{R}^\times$. The map

$$\log : \mathbf{R}_{pos}^\times \to \mathbf{R}^+$$

defined by

$$x \mapsto \log x \quad \text{for all } x \in \mathbf{R}_{pos}^\times$$

(where $\log x$ denotes the natural logarithm of x) is an isomorphism. (What is the inverse isomorphism?) Hence

$$\mathbf{R}_{pos}^\times \cong \mathbf{R}^+.$$

The map

$$| \ | : \mathbf{C}^\times \to \mathbf{R}_{pos}^\times$$

defined by

$$| \ | : z \mapsto |z| \quad \text{for all } z \in \mathbf{C}^\times$$

(where $|z|$ denotes the modulus of z) is a surjective homomorphism (sometimes called an *epimorphism*). It is not an isomorphism since, for instance, $|-1| = |1|$. The restriction of this homomorphism to $\mathbf{R}^\times$ is a surjective homomorphism

$$| \ | : \mathbf{R}^\times \to \mathbf{R}_{pos}^\times$$

which is also not an isomorphism.

2.14. For any positive integer n, the set of all complex nth roots of 1 forms a subgroup C_n of $\mathbf{C}^\times$. It is a finite group of order n, and it is cyclic, generated by $e^{2\pi i/n}$:

$$C_n = \{1, e^{2\pi i/n}, e^{4\pi i/n}, \dots, e^{2(n-1)\pi i/n}\}.$$

In particular, $C_2 = \{1, -1\} = \mathbf{Z}^\times$. There is a surjective homomorphism

$$v_n : \mathbf{Z}^+ \to C_n$$

defined by

$$v_n : x \mapsto e^{2\pi x i/n} \quad \text{for all } x \in \mathbf{Z}^+.$$

We call C_n the *cyclic group of order* n, for any cyclic group of order n is isomorphic to C_n (**2**).

For any $z \in \mathbf{C}^\times \setminus \bigcup_{n=1}^{\infty} C_n$, the cyclic subgroup $\langle z \rangle$ of $\mathbf{C}^\times$ is infinite. We sometimes denote an infinite cyclic group by C_∞, for any two such groups are isomorphic (**29**).

38 The set of all positive rational numbers forms a subgroup $\mathbf{Q}_{pos}^{\times}$ of $\mathbf{Q}^{\times}$, and there is a surjective homomorphism

$$|\ \ | : \mathbf{Q}^{\times} \to \mathbf{Q}_{pos}^{\times}$$

which is not an isomorphism.

Is $\qquad\qquad \mathbf{Q}_{pos}^{\times} \cong \mathbf{Q}^{+}$ $\qquad$ (cf. 2.13)?

39 For any positive integer n, the only homomorphism $\varphi : C_n \to \mathbf{C}^{+}$ is the trivial homomorphism.

***40** Let n be a positive integer. Then
   (i) $\mathbf{Z}_n^{+} \cong C_n$.
   (ii) $\mathbf{Z}_n^{\times}$ is an abelian group of order $\varphi(n)$, where φ is Euler's function (see **5**).
   (iii) By considering orders of elements in $\mathbf{Z}_n^{\times}$, prove the Euler–Fermat theorem:

$$m^{\varphi(n)} \equiv 1 \bmod n,$$

whenever m is an integer co-prime to n. (This generalizes **8**.)

41 (i) Let G be a group and n a positive integer such that $g^n = 1$ for all $g \in G$. Show that if $\varphi : G \to \mathbf{C}^{\times}$ is a homomorphism then $\operatorname{Im} \varphi \leqslant C_n$. Hence show that $\operatorname{Hom}(G, \mathbf{C}^{\times}) \cong \operatorname{Hom}(G, C_n)$ (cf. **33**).
   (ii) Deduce that if G is a finite group then so is $\operatorname{Hom}(G, \mathbf{C}^{\times})$. (Remark. If G is a finite abelian group then in fact $\operatorname{Hom}(G, \mathbf{C}^{\times}) \cong G$, but we do not yet have the means of proving this. See **454**.)

2.15. Any vector space V over a field F forms an abelian group V^{+} under addition: the *additive group* of V. For any $a \in F$, the map

$$\lambda_a : V^{+} \to V^{+},$$

defined by $\qquad\qquad \lambda_a : v \mapsto av$

for all $v \in V^{+}$, is a homomorphism. It is an isomorphism if $a \neq 0$. If V has dimension 1 over F then $V^{+} \cong F^{+}$.

2.16. Let V be any vector space $\neq 0$ over a field F. Then the set of all linear maps of V into itself forms a ring $\mathscr{L}(V)$ with respect to the usual operations of addition and multiplication of linear maps. This ring has a multiplicative identity element, namely the linear map

$$1 : v \mapsto v \quad \text{for all } v \in V.$$

The group of units of $\mathscr{L}(V)$ (as defined in 2.12) consists of all invertible (that is, non-singular) linear maps of V to itself. It is called the *general linear group* of V and denoted by $\operatorname{GL}(V)$.

Suppose that F is a finite field with, say, $|F| = p^m = q$, and that V has finite dimension n over F. Then $|V| = q^n$. Let vectors $v_1, \ldots, v_n$ form a base of V. Then a linear map $\theta : V \to V$ is determined by its effect on $v_1, \ldots, v_n$, and θ is invertible if and only if $v_1\theta, \ldots, v_n\theta$ form a base of V. Moreover, for any base $w_1, \ldots, w_n$ of V there is a unique linear map $\theta : V \to V$ such that $v_i\theta = w_i$ for $i = 1, \ldots, n$. Hence

$$|GL(V)| = \text{the number of ordered bases of } V.$$

In forming a base $w_1, \ldots, w_n$ of V we may first choose w_1 to be any non-zero vector of V, then w_2 to be any vector other than a scalar multiple of w_1, then w_3 to be any vector other than a linear combination of w_1 and w_2, and so on. Hence

$$|GL(V)| = (q^n - 1)(q^n - q)(q^n - q^2) \ldots (q^n - q^{n-1}).$$

2.17. Let F be any field and n any positive integer. Then the set of all invertible (that is, non-singular) $n \times n$ matrices with entries in F forms a group with respect to matrix multiplication. This is called the *general linear group of degree n over F* and denoted by $GL_n(F)$.

If V is a vector space of dimension n over F then $GL(V) \cong GL_n(F)$. To see this, it is enough to choose a base of V and then to map each invertible linear map of V into itself to the matrix representing it with respect to the chosen base.

There is a surjective homomorphism

$$\det : GL_n(F) \to F^\times$$

defined by $\det : x \mapsto \det x$ for all $x \in GL_n(F)$ (where $\det x$ denotes the determinant of x). Note that this homomorphism is an isomorphism if and only if $n = 1$; in particular, $GL_1(F) \cong F^\times$.

42 Let G be a finite group such that $g^2 = 1$ for all $g \in G$. Show that $G \cong V^+$ for some finite dimensional vector space V over the field $\mathbf{Z}_2$. Deduce that $|G| = 2^m$ for some integer $m \geqslant 0$. (Hint. By **3**, G is abelian. Let V consist of the same elements as G, with the sum of 2 elements of V equal to their product in G, and scalar multiplication defined in the obvious way.)

43 Let F be a field and let m, n be positive integers with $m \leqslant n$. Then
  (i) $GL_m(F)$ can be embedded in $GL_n(F)$,
  (ii) $GL_n(F)$ is non-abelian for all $n \geqslant 2$.

44 $GL_2(\mathbf{Z}_2) \cong \Sigma_3$.

45 Let G be the set of all matrices of the form $\begin{pmatrix} a & b \\ 0 & c \end{pmatrix}$, where a, b, c are real numbers such that $ac \neq 0$. Prove that G forms a subgroup of $GL_2(\mathbf{R})$, and that the set H of all elements of G in which $a = c = 1$ forms a subgroup of G isomorphic to $\mathbf{R}^+$.

Find all elements in G of order 2. Hence show that the product of two elements of order 2 in G can be an element of infinite order.

Our next example associates to every group G another group Aut G which has an important rôle to play in the following chapters.

2.18. The set of all homomorphisms of a group G into itself (sometimes called the *endomorphisms* of G) forms a semigroup with respect to composition of maps (by **19**): it is a subsemigroup of M_G (see 2.4). This semigroup has an identity element, namely the homomorphism

$$1 : g \mapsto g \quad \text{for all } g \in G.$$

Now the units of the semigroup are just the isomorphisms of G onto G. These are called the *automorphisms* of G. By 2.3, the set of all automorphisms of G forms a group with respect to composition of maps: it will be denoted by Aut G. Note that Aut $G \leqslant \Sigma_G$.

2.19. To each $g \in G$ there is associated an automorphism τ_g of G, defined as follows. For all $x \in G$,

$$\tau_g : x \mapsto g^{-1}xg ;$$

the element $x\tau_g = g^{-1}xg$ is called the *conjugate of x by g*. Certainly τ_g is a well-defined map of G into itself; and if $g, g_1, g_2, x, x_1, x_2 \in G$,

$$(x_1 x_2)\tau_g = g^{-1}x_1 x_2 g = g^{-1}x_1 gg^{-1}x_2 g = (x_1 \tau_g)(x_2 \tau_g),$$

so that τ_g is a homomorphism. Furthermore,

$$x\tau_{g_1 g_2} = (g_1 g_2)^{-1}xg_1 g_2 = g_2^{-1}g_1^{-1}xg_1 g_2 = x\tau_{g_1}\tau_{g_2},$$

and so

$$\tau_{g_1 g_2} = \tau_{g_1}\tau_{g_2}. \tag{i}$$

In particular, since $\tau_1 = 1$,

$$\tau_g \tau_{g^{-1}} = 1 = \tau_{g^{-1}}\tau_g.$$

Therefore τ_g is invertible: thus $\tau_g \in$ Aut G. The automorphism τ_g of G is called the *inner automorphism of G induced by g* (or *conjugation of G by g*). An automorphism of G which is not inner is called an *outer* automorphism of G.

By 2.9 and 2.10, τ_g maps each subgroup K of G to a subgroup $K\tau_g$ of Im $\tau_g = G$ with

$$K \cong K\tau_g = \{g^{-1}kg : k \in K\}.$$

We call $K\tau_g$ the *conjugate of K by g*, and denote it by $g^{-1}Kg$. Thus we have

2.20. *For every $K \leqslant G$ and every $g \in G$, $g^{-1}Kg$ is a subgroup of G isomorphic to K.*

2.21. The map $\tau : G \rightarrow \Sigma_G,$

defined by $\tau : g \mapsto \tau_g$

for all $g \in G$, is a homomorphism. This is shown by equation (i) of 2.19. Moreover, Im $\tau = \{\tau_g : g \in G\} \leqslant$ Aut G. We denote Im τ by Inn G: it is the *group of all inner automorphisms of G*. Thus

$$\text{Inn } G \leqslant \text{Aut } G \leqslant \Sigma_G.$$

When is τ the trivial homomorphism; that is, when is Inn $G = 1$? We have

$$\tau_g = 1 \quad \text{if and only if} \quad g^{-1}xg = x$$

for all $x \in G$, that is, if and only if $xg = gx$ for all $x \in G$. Hence $\tau_g = 1$ for all $g \in G$ if and only if all pairs of elements x, g of G commute. Hence we have

2.22. Inn $G = 1$ *if and only if G is abelian.*

*46 (i) If R is a ring with a multiplicative identity 1 then $R^\times$ can be embedded in Aut R^+. (Hint. See 2.11.)

(ii) Aut $\mathbf{Z}^+ \cong \mathbf{Z}^\times$ and Aut $\mathbf{Z}_n^+ \cong \mathbf{Z}_n^\times$ for every positive integer n.

*47 Let V be a vector space $\neq 0$ over a field F. Then

(i) GL$(V) \leqslant$ Aut V^+. (See 2.15, 2.16.)

(ii) If $F = \mathbf{Z}_p$ and V has finite dimension n over $\mathbf{Z}_p$ then Aut $V^+ \cong$ GL$_n(\mathbf{Z}_p)$.

*48 If $G_1 \cong G_2$ then Aut $G_1 \cong$ Aut G_2 and Inn $G_1 \cong$ Inn G_2.

*49 Define a relation $\sim$ on G by

$$x \sim y \quad \text{if and only if} \quad g^{-1}xg = y \quad \text{for some } g \in G.$$

Show that this relation $\sim$ of *conjugacy* is an equivalence relation on G. (This fact will be placed in a general context in chapter 4.)

50 If $\alpha \in$ Aut G and $x \in G$ then

$$o(x^\alpha) = o(x).$$

In particular, conjugate elements of a group have the same order.

51 Find a group G with a subgroup K and an element g such that $g^{-1}Kg \neq K$.

52 (i) The map defined by $g \mapsto g^{-1}$ for all $g \in G$ is an automorphism of G if and only if G is abelian.

(ii) If G is any group for which $g^2 \neq 1$ for some $g \in G$ then Aut $G \neq 1$. (Remark. In fact Aut $G \neq 1$ if and only if $|G| > 2$. The proof is completed by observing that if $g^2 = 1$ for all $g \in G$, so that by **3** G is abelian, then G has the structure of a vector space V over $\mathbf{Z}_2$, and any invertible linear map $V \to V$ is an automorphism of G. For the finite case, see **42**.)

53 (i) Let $\alpha \in$ Aut G and let

$$H = \{g \in G : g^\alpha = g\}.$$

Prove that H is a subgroup of G: it is called the *fixed point subgroup of G under α.*

(ii) Let n be a positive integer and F a field. For any $n \times n$ matrix y with entries in F, let y' denote the transpose of y. Show that the map

$$* : \text{GL}_n(F) \to \text{GL}_n(F),$$

defined by
$$* : \quad x \mapsto (x^{-1})'$$

for all $x \in \text{GL}_n(F)$, is an automorphism of $\text{GL}_n(F)$, and that the corresponding fixed point subgroup consists of all orthogonal $n \times n$ matrices with entries in F (that is, matrices y such that $y'y = 1$).

Prove (by assuming the contrary and considering determinants) that ∗ is an outer automorphism of $GL_n(F)$ if $F \neq \mathbf{Z}_2$ and $F \neq \mathbf{Z}_3$. (Remark. In fact, ∗ is an outer automorphism of $GL_n(F)$ unless either $F = \mathbf{Z}_2$ and $n \leqslant 2$ or $F = \mathbf{Z}_3$ and $n = 1$.)

54 Let $\alpha \in$ Aut G. Then α is said to be *fixed-point-free* if the fixed point subgroup of G under α is trivial (see **53**); that is, if $g^\alpha \neq g$ whenever $1 \neq g \in G$.

(Remark. The term 'fixed-point-free' is standard. It is perhaps a slight abuse of language, since of course any automorphism of a group fixes the identity element. To say that an automorphism is fixed-point-free means that it fixes no element of the group other than the identity element.)

(i) Suppose that α is a fixed-point-free automorphism of the finite group G. Show that

$$\{g^\alpha g^{-1} : g \in G\} = G.$$

Deduce that if $o(\alpha) = 2$ then, for all $x \in G$,

$$x^\alpha = x^{-1},$$

and that G is abelian of odd order greater than 1.

(ii) Let G be a finite abelian group of odd order greater than 1. Then the map

$$\alpha : x \mapsto x^{-1},$$

defined for all $x \in G$, is a fixed-point-free automorphism of G of order 2.

(Hints. For (i), show that the map of G into itself defined, for all $g \in G$, by $g \mapsto g^\alpha g^{-1}$, is injective. Then use **52**(i) and **1.13**.)

We mention next some examples of groups which arise in geometric contexts.

2.23. Let X be a metric space, with distance function $d : X \times X \to \mathbf{R}$. Then a bijective map $\varphi : X \to X$ is structure-preserving if

$$d(x\varphi, y\varphi) = d(x, y) \quad \text{for all } x, y \in X.$$

Such a map φ is called an *isometry* of X. It is easy to verify that the set of all isometries of X forms a subgroup Isom X of Σ_X.

Now if $\varnothing \subset Y \subseteq X$ then Y is a subspace of X: that is, Y is a metric space with respect to the restriction of d to $Y \times Y$. For each $\varphi \in$ Isom X let

$$Y\varphi = \{y\varphi : y \in Y\} \subseteq X.$$

Then let $\qquad S_X(Y) = \{\varphi \in \text{Isom } X : Y\varphi = Y\}.$

A straightforward verification shows that

$$S_X(Y) \leqslant \text{Isom } X.$$

We call $S_X(Y)$ the *symmetry group* of Y (with respect to the metric space X).

For instance, suppose that $X = E^2$, the euclidean plane, with the usual distance function. We describe without proof some facts about Isom E^2: for details see Coxeter [b5] chapter 3. (The corresponding facts about the group of isometries of the euclidean line E^1 can readily be worked out from the definition, and are given in **56**.) We begin by noting some special isometries of E^2.

(i) If $\varphi : E^2 \rightarrow E^2$ is a map which moves every point of E^2 a fixed distance in a fixed direction then φ is called a *translation* of E^2. Any translation is an isometry. To describe translations in coordinate terms, choose an arbitrary point of E^2 as origin and denote it by O. To each point $s \in E^2$, we associate the directed line segment **Os** from O to s. The set of all such directed line segments forms a vector space V of dimension 2 over **R** in a familiar way: addition of vectors is defined by means of the parallelogram law.

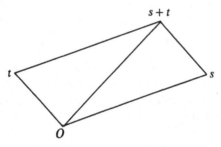

Now for every $s, t \in E^2$, write $s + t$ for the unique point of E^2 corresponding to the vector **Os** + **Ot**. Then a translation of E^2 is just a map

$$\tau_a : s \mapsto s + a \quad \text{for all } s \in E^2,$$

where a is an element of E^2. The set

$$\text{Tr } E^2 = \{\tau_a : a \in E^2\}$$

of all translations is an abelian subgroup of Isom E^2, and in fact, by means of the map defined by

$$\textbf{Os} \mapsto \tau_s \quad \text{. for all } s \in E^2,$$

it is clear that

$$V^+ \cong \text{Tr } E^2.$$

If we fix a cartesian coordinate system for E^2 with origin O and identify each point s of E^2 with the ordered pair (x, y) of its coordinates in that system (where $x, y \in \textbf{R}$) then a translation is just a map of the form

$$(x, y) \mapsto (x + b, y + c),$$

where b, c are real numbers, the same for all x, y.

(ii) A map $\varphi : E^2 \rightarrow E^2$ is called a *reflexion* of E^2 if there is a line l in E^2 such that φ moves each point of E^2 to its mirror image with respect to l. If l is chosen as the x-axis for a cartesian coordinate system then reflexion about l is the map

$$\varepsilon : (x, y) \mapsto (x, -y)$$

for all $x, y \in \mathbf{R}$. Then ε is an isometry of E^2 and $\varepsilon^2 = 1$.

(iii) Let s be any point of E^2 and view E^2 as the Argand diagram with s as zero: that is, represent the points of E^2 by complex numbers in the usual way, with s represented by 0. Then each point is represented in polar form by an expression $re^{i\theta}$, where $r, \theta \in \mathbf{R}$ with $r \geqslant 0$. A *rotation* of E^2 about s is a map

$$\rho_\alpha : re^{i\theta} \mapsto re^{i(\theta + \alpha)}$$

for some $\alpha \in \mathbf{R}$. Then ρ_α is an isometry of E^2. Since $e^{i\alpha} = 1$ if and only if α is an integral multiple of 2π, each rotation about s is uniquely expressible as ρ_α with $0 \leqslant \alpha < 2\pi$; α is the *angle* of the rotation. The set

$$\mathrm{Rot}(E^2 ; s) = \{\rho_\alpha : 0 \leqslant \alpha < 2\pi\}$$

of all rotations about s is an abelian subgroup of Isom E^2. Note that there is one such subgroup of Isom E^2 for each point $s \in E^2$.

Now the structure of Isom E^2 can be described as follows. Let $G = $ Isom E^2 and let H be the set of all translations and rotations of E^2: that is,

$$H = \mathrm{Tr}\ E^2 \cup \bigcup_{s \in E^2} \mathrm{Rot}(E^2 ; s).$$

Then H is a subgroup of index 2 in G, and if ε is any reflexion of E^2,

$$G = H \cup H\varepsilon = H \cup \varepsilon H.$$

It is straightforward to calculate products of elements of G from these definitions.

2.24. As particular symmetry groups, we note the following important examples. Let n be any integer with $n \geqslant 3$, and let P_n denote a regular polygon in E^2 with n edges. We consider the symmetry group

$$S_{E^2}(P_n) = L, \text{ say.}$$

The elements of L are just the n rotations about the centre of the polygon through angles $0, 2\pi/n, 4\pi/n, \ldots, 2(n-1)\pi/n$, together with the n reflexions about the lines joining opposite vertices of P_n and the lines joining mid-points of opposite edges of P_n (if n is even) or about the lines joining vertices of P_n to mid-points of opposite edges (if n is odd). Thus

$$|L| = 2n.$$

Let ρ denote the rotation about the centre of P_n through angle $2\pi/n$, and ε any one of the n reflexions in L. Then one verifies that

$$\rho^n = 1 = \varepsilon^2 \quad \text{and} \quad \varepsilon\rho = \rho^{-1}\varepsilon,$$

and the $2n$ distinct elements of L are

$$1, \rho, \rho^2, \ldots, \rho^{n-1}, \varepsilon, \rho\varepsilon, \rho^2\varepsilon, \ldots, \rho^{n-1}\varepsilon.$$

The group L is called *the dihedral group of order* $2n$ and denoted by D_{2n}. (The definite article is used because the type of D_{2n} is independent of the size of the polygon and of its position in the plane: it depends only on the number n of edges. Some authors use D_n or other notations where we always use D_{2n}.) Note that $\rho^{-1} \neq \rho$, and so D_{2n} is a non-abelian group.

55 Let X be any non-empty set. Let c be any positive real number and define, for all $x, y \in X$,

$$d(x, y) = \begin{cases} 0 & \text{if } x = y \\ c & \text{if } x \neq y. \end{cases}$$

Show that d is a distance function for X, and that for this metric space

$$\text{Isom } X = \Sigma_X.$$

56 View $\mathbf{R}$ as the euclidean line E^1 with the usual distance function (that is, for $x, y \in \mathbf{R}, d(x, y) = |x - y|$).

(i) For each $a \in \mathbf{R}$, the map

$$\tau_a : x \mapsto x + a \quad \text{for all } x \in \mathbf{R}$$

is an isometry of $\mathbf{R}$, called a *translation*, and

$$\mathbf{R}^+ \cong T = \{\tau_a : a \in \mathbf{R}\} \leqslant \text{Isom } \mathbf{R}.$$

(ii) For each $a \in \mathbf{R}$, the map

$$\varepsilon_a : x \mapsto a - x \quad \text{for all } x \in \mathbf{R}$$

is an isometry of $\mathbf{R}$, called a *reflexion*, and

$$\varepsilon_a^2 = 1 (= \tau_0) \quad \text{and} \quad \varepsilon_a \tau_b = \tau_{-b} \varepsilon_a \quad \text{for all } a, b \in \mathbf{R}.$$

(iii) Every isometry of $\mathbf{R}$ is either a translation or a reflexion.

(iv) Isom $\mathbf{R}$ is a non-abelian group and T is an abelian subgroup of index 2.

57 Let the notation be as in **56**. Show that for every $n \in \mathbf{Z}$,

$$\tau_1^n = \tau_n,$$

and that the elements of the symmetry group $S_\mathbf{R}(\mathbf{Z})$ are just the isometries

$$\tau_1^n \quad \text{for all } n \in \mathbf{Z} \quad \text{and} \quad \tau_1^n \varepsilon_0 \quad \text{for all } n \in \mathbf{Z}.$$

Note that $\varepsilon_0^2 = 1$ and $\varepsilon_0 \tau_1 = \tau_1^{-1}\varepsilon_0$. The group $S_\mathbf{R}(\mathbf{Z})$ is called *the infinite dihedral group* and denoted by D_∞.

58 Let n be an integer, $n \geqslant 3$. Then D_{2n} can be embedded in Σ_n. Moreover, $D_6 \cong \Sigma_3$ but, whenever $n > 3, D_{2n} \ncong \Sigma_n$. (It may be assumed that an isometry of E^2 is uniquely determined by its effect on any 3 non-collinear points.)

59 Let n be an integer, $n \geqslant 3$, and let $L = D_{2n}$. Let $J = \langle \rho \rangle$, where ρ is defined as in 2.24. Show that every element of $L \backslash J$ has order 2. Deduce that J is the only cyclic subgroup of L of order n.

60 Every group of order 6 is isomorphic to either C_6 or D_6. Hence, in the notation of chapter 1, $v(6) = 2$.

(Hints. Let G be a non-cyclic group of order 6. By **42**, G has an element x of order 3. Let y be an element of G other than $1, x, x^2$. Then $G = \{1, x, x^2, y, xy, x^2y\}, y^2 = 1$ and $yx = x^2y$.)

61 (i) For any 2 points s, t of E^2 there is a unique translation τ_a of E^2 which maps s to t.

(ii) For any 2 points s, t of E^2, if τ_a is the unique translation of E^2 which maps s to t then

$$\tau_a^{-1} \operatorname{Rot}(E^2; s)\tau_a = \operatorname{Rot}(E^2; t).$$

Thus any 2 groups of rotations are conjugate subgroups of Isom E^2.

62 For any non-empty set X, any map $\varphi : X \to X$ and any $Y \subseteq X$, let $Y\varphi = \{y\varphi : y \in Y\} \subseteq X$. Verify that the following analogues of the symmetry groups introduced in 2.23 are subgroups of the appropriate groups, as stated.

(i) Let X be any non-empty set and Y any non-empty subset of X. Then

$$S_X(Y) = \{\varphi \in \Sigma_X : Y\varphi = Y\} \leqslant \Sigma_X.$$

(ii) Let V be any vector space over a field F and Y any non-empty subset of V. Then

$$S_V(Y) = \{\varphi \in \operatorname{GL}(V) : Y\varphi = Y\} \leqslant \operatorname{GL}(V).$$

(iii) Let G be any group and Y any non-empty subset of G. Then

$$S_G(Y) = \{\varphi \in \operatorname{Aut} G : Y\varphi = Y\} \leqslant \operatorname{Aut} G.$$

2.25. E. Artin [b2] said 'In modern mathematics the investigation of the symmetries of a given mathematical structure has always yielded the most powerful results. Symmetries are maps which preserve certain properties'. And this is why groups play a fundamental rôle in mathematics today. The symmetries of a structure are the permutations of the underlying set X which are structure-preserving; and the set of all such symmetries forms a subgroup of Σ_X. We may view the group associated to the structure in this way as a kind of measure of the 'regularity' of the structure.

We have had several instances of this idea in the preceding examples. For a non-empty set X with no further distinguished structure, the symmetries of X are just the permutations of X and the corresponding group is Σ_X itself. For a vector space V, the symmetries of V are the invertible linear maps $V \to V$ and the corresponding group is $\operatorname{GL}(V)$. For a group G, the symmetries of G are the automorphisms of G and the corresponding group is Aut G. For a metric space X, the symmetries of X are the isometries of X and the corresponding group is Isom X.

Again, if X is a set with a particular structure and Y is a subset of X, there is a corresponding relative symmetry group $S_X(Y)$ of Y with respect to X: as for instance in 2.23 and **62**.

This idea of a group as a measure of the symmetry of a mathematical structure is of fundamental significance. To mention another instance, one in which we may trace the origins of group theory and indeed of abstract algebra, the same idea lies at the heart of the Galois theory of equations: for information about this, we refer to Artin [b1], Herstein [b19] chapter 5, §6, Kaplansky [b25] Part I, Lang [b28] chapter VII, or Rotman [b34] pp. 96–103.

63 Let $X = E^2$, the euclidean plane.
  (i) Let $s \in X$, $J = S_X(s)$ and $K = \text{Rot}(X \,; s)$.
  Then $K \leqslant J$, $|J : K| = 2$ and $J \backslash K$ consists of all reflexions of X about lines through s.
  (ii) Let $s, t \in X$ with $s \neq t$. Then $|S_X(\{s, t\})| = 4$.
  (iii) Let l be any line in X, $L = S_X(l)$, and let M consist of all the translations of X which belong to L. Then $M \leqslant L$, $|L : M| = 4$ and $M \cong \mathbf{R}^+$.

64 Let n be a positive integer, F a field and V a vector space of dimension n over F. If B is a base of V then $S_V(B) \cong \Sigma_n$. Hence Σ_n can be embedded in $\text{GL}_n(F)$.

65 Let X be a set with $|X| \geqslant 2$, and let $x \in X$ and $Y = X \backslash \{x\}$. Then $S_X(x) \cong \Sigma_Y$.

Associated to a group G there are in general many other groups of different types: for instance, G determines a set of groups as its subgroups (though it is often impracticable to find these explicitly). The following observation is immediate from the definition of a subgroup.

2.26. *If $H \leqslant G$ and $K \leqslant G$ then $H \cap K \leqslant G$. More generally, if $\{H_i : i \in I\}$ is any set of subgroups of G (indexed by a set I) then $\bigcap_{i \in I} H_i \leqslant G$.*

*66 If $H \leqslant G$ and $K \leqslant G$, with $|G : H| < \infty$ and $|G : K| < \infty$, then $|G : H \cap K| < \infty$. (This is often called Poincaré's theorem. Hint. Apply **11**.)

67 Let G be a finite group.
  (i) Any two distinct subgroups of G of order p intersect in 1.
  (ii) The total number of elements of order p in G is a multiple of $p - 1$.

Note that intersections of subgroups are always non-empty, for 1 belongs to every subgroup. Of course, an intersection of non-trivial subgroups may turn out to be trivial.

Now if X is any subset of G, there is certainly at least one subgroup of G which contains X, namely G itself. Therefore the intersection of all subgroups of G containing X is a well-defined subgroup of G.

2.27 Definition. Let $X \subseteq G$. Then the intersection of all subgroups of G which contain X is called *the subgroup of G generated by X* and denoted by $\langle X \rangle$ (in some books by $\text{Gp}\{X\}$). It is the unique smallest subgroup of G containing X in the sense that, whenever $X \subseteq H \leqslant G$, then $\langle X \rangle \leqslant H$. Of course, if $H \leqslant G$ then $\langle H \rangle = H$.

Note that $\langle \varnothing \rangle = 1$. If X is a non-empty finite subset of G, say $X = \{x_1, \ldots, x_n\}$, then we write $\langle x_1, \ldots, x_n \rangle$ rather than $\langle \{x_1, \ldots, x_n\} \rangle$ for the subgroup generated by X. In particular, if X consists of a single element x then $\langle X \rangle = \langle x \rangle$, the cyclic subgroup of G generated by x; the notation is consistent with that introduced in chapter 0.

It is easy to give an explicit description of the elements of $\langle X \rangle$ in terms of the elements of X.

2.28 Lemma. *Let* $\emptyset \neq X \subseteq G$. *Then*

$$\langle X \rangle = \{x_1^{n_1} x_2^{n_2} \dots x_r^{n_r} : r \text{ is a positive integer, each } x_i \in X \text{ and each } n_i \in \mathbf{Z}\}.$$

(In general an element of $\langle X \rangle$ may have many different expressions of the form on the right.)

Proof. Let H be the set on the right hand side above. Since $\langle X \rangle$ is closed under multiplication and the forming of inverses, $H \subseteq \langle X \rangle$. But also, by definition, $X \subseteq H$; and if $h_1, h_2 \in H$ then clearly $h_1 h_2^{-1} \in H$, so that $H \leqslant G$. Hence $\langle X \rangle \leqslant H$, and so $\langle X \rangle = H$.

2.29 Definition. If $X \subseteq G$ and $\langle X \rangle = G$, X is said to be *a set of generators* of G. This is to be compared with the notion of a set of vectors spanning a vector space; though there is in general in group theory no analogue for group elements of the notion of linear independence of vectors, and no basis theorem. (For abelian groups and for groups of finite prime power orders, something of the kind is possible; see 8.24, 11.12.) Obviously G itself and $G \setminus \{1\}$ are sets of generators of G. We are usually interested in 'small' sets of generators. If G has a set X of generators with $|X| \leqslant n$, where n is a positive integer, we say that G is an *n-generator group*. Thus the 1-generator groups are just the cyclic groups; and cyclic groups are *n*-generator for every positive integer n.

2.30 Examples. (i) We know that the symmetric group Σ_n is non-abelian whenever $n \geqslant 3$; then Σ_n is certainly not cyclic. However, Σ_n is a 2-generator group, for every n: for instance,

$$\Sigma_n = \langle (12), (123 \dots n) \rangle, \quad \text{whenever } n \geqslant 3.$$

To see this, let $\tau = (12), \sigma = (123 \dots n)$ and

$$H = \langle \sigma, \tau \rangle \leqslant \Sigma_n.$$

Then H contains $\sigma^{-1} \tau \sigma = (23)$, $\sigma^{-2} \tau \sigma^2 = \sigma^{-1}(\sigma^{-1} \tau \sigma)\sigma = (34), \dots,$ $\sigma^{-(n-2)} \tau \sigma^{n-2} = (n-1, n)$. Let m be an integer with $2 < m \leqslant n$. Then H also contains

$$(m-1, m)(m-2, m-1) \dots (34)(23)(12)(23)(34) \dots (m-1, m) = (1m).$$

Now let k, l be integers with $1 < k < l \leqslant n$. Then H contains $(1k)(1l)(1k) = (kl)$. Thus H contains all transpositions in Σ_n. Now it follows from **21** that $H = \Sigma_n$.

(ii) Let n be a positive integer and let V be a vector space of dimension

n over $\mathbf{Z}_p$. Then V^+ is an n-generator group, but is not an m-generator group for any integer $m < n$. To see that V^+ is an n-generator group, we need only consider a base $v_1, \ldots, v_n$ of V: then, since every element of V is expressible as $k_1 v_1 + \ldots + k_n v_n$, with $k_1, \ldots, k_n$ integers, it is clear that $\langle v_1, \ldots, v_n \rangle = V^+$. On the other hand, if m is a positive integer and $w_1, \ldots, w_m$ are elements of V such that $\langle w_1, \ldots, w_m \rangle = V^+$, then vectors $w_1, \ldots, w_m$ also span V as a vector space and hence $m \geqslant n$.

(iii) For every integer $n \geqslant 3$, the dihedral group D_{2n} is a 2-generator group: for, with the notation of 2.24, $D_{2n} = \langle \rho, \varepsilon \rangle$.

*68 Let $\emptyset \neq X \subseteq G$. Then every element of $\langle X \rangle$ is expressible in the form $x_1^{\varepsilon_1} x_2^{\varepsilon_2} \ldots x_r^{\varepsilon_r}$ with $x_1, x_2, \ldots, x_r \in X$ and $\varepsilon_i = \pm 1$ for each $i = 1, \ldots, r$. Moreover, if every element of X has finite order we can take every $\varepsilon_i = +1$.

*69 Let r be a positive integer and let $x_1, \ldots, x_r \in G$. If x_i and x_j commute for all i, j with $1 \leqslant i < j \leqslant r$ then $\langle x_1, \ldots, x_r \rangle$ is an abelian subgroup of G, and every element of $\langle x_1, \ldots, x_r \rangle$ is expressible in the form $x_1^{n_1} x_2^{n_2} \ldots x_r^{n_r}$, where $n_1, \ldots, n_r \in \mathbf{Z}$.

70 (i) Let $H, K \leqslant G$. Then

$$H \cup K \leqslant G \quad \text{if and only either} \quad K \leqslant H \quad \text{or} \quad H \leqslant K$$

(in which case $H \cup K$ is either H or K). In particular, G cannot be the union of two proper subgroups.

(ii) The dihedral group D_8 of order 8 is the union of three proper subgroups.

*71 Let $H, K \leqslant G$. Then $\langle H \cup K \rangle$ is usually denoted by $\langle H, K \rangle$ and called the *join* of H and K: it is the unique smallest subgroup of G containing both H and K.

(i) Every element of $\langle H, K \rangle$ is expressible (in general in many different ways) in the form $h_1 k_1 h_2 k_2 \ldots h_r k_r$, where r is a positive integer and $h_1, \ldots, h_r \in H, k_1, \ldots, k_r \in K$.

(ii) If H and K are finite subgroups of G, is $\langle H, K \rangle$ necessarily finite? (cf. Poincaré's theorem: **66**).

*72 Let X be a (non-empty) set of generators of G. Let $\varphi : G \to H$ and $\psi : G \to H$ be homomorphisms. If, for every $x \in X$,

$$x\varphi = x\psi$$

then

$$\varphi = \psi.$$

73 Suppose that $G = \langle x, y \rangle$ and that $x^{-1} yx = y^k$ for some $k \in \mathbf{Z}$. Show that every element of G is expressible in the form $x^m y^n$ with $m, n \in \mathbf{Z}$. Deduce that if x and y have finite orders then G is a finite group and $|G| \leqslant o(x)o(y)$.

74 Let $H < G$ and let $X = G \backslash H$. Then $\langle X \rangle = G$.

Next, from any groups H and K we construct another group in which both H and K can be embedded.

2.31 Definition. From any two sets X and Y (possibly equal) we can form another set $X \times Y$, called the *cartesian product* of X and Y: it is the set of all ordered pairs (x, y) with $x \in X$ and $y \in Y$. When X and Y are finite sets, $X \times Y$ is a finite set and

$$|X \times Y| = |X| . |Y|.$$

Now if H and K are any *groups* then the set $H \times K$ acquires the structure of a group when we define, for all $h, h' \in H$ and $k, k' \in K$,

$$(h, k)(h', k') = (hh', kk').$$

The group axioms are easy to verify: closure is immediate from the definition of multiplication; associativity follows from the associativity of multiplication in H and in K; the identity element of $H \times K$ is $(1, 1)$; and

$$(h, k)^{-1} = (h^{-1}, k^{-1}).$$

From now on, whenever H and K are groups, $H \times K$ will denote this *group*: it is called the *direct product* of H and K.

2.32 Examples. (i) $\qquad\qquad \mathbf{C}^+ \cong \mathbf{R}^+ \times \mathbf{R}^+ :$

for the map

$$a + ib \mapsto (a, b),$$

defined for all $a, b \in \mathbf{R}$, is an isomorphism of $\mathbf{C}^+$ onto $\mathbf{R}^+ \times \mathbf{R}^+$.

(ii) $\qquad\qquad \mathbf{Q}^\times \cong \mathbf{Q}^\times_{\text{pos}} \times C_2 :$

for the map

$$x \mapsto (|x|, \text{sign } x),$$

defined for all $x \in \mathbf{Q}^\times$, where

$$\text{sign } x = \begin{cases} 1 \text{ if } x > 0 \\ -1 \text{ if } x < 0, \end{cases}$$

is an isomorphism of $\mathbf{Q}^\times$ onto $\mathbf{Q}^\times_{\text{pos}} \times C_2$. Similarly

(iii) $\qquad\qquad \mathbf{R}^\times \cong \mathbf{R}^\times_{\text{pos}} \times C_2.$

(iv) Let $U = \{z \in \mathbf{C}^\times : |z| = 1\}$. It is clear that U is a subgroup of $\mathbf{C}^\times$; U is called the *circle group*. Then

$$\mathbf{C}^\times \cong \mathbf{R}^\times_{\text{pos}} \times U :$$

for the map

$$z \mapsto (|z|, z/|z|)$$

defined for all $z \in \mathbf{C}^\times$, is an isomorphism of $\mathbf{C}^\times$ onto $\mathbf{R}^\times_{\text{pos}} \times U$. Alternatively, if z is expressed in polar form $re^{i\theta}$, where $r, \theta \in \mathbf{R}$, this isomorphism is the map

$$re^{i\theta} \mapsto (r, e^{i\theta}).$$

2.33. *The group $H \times K$ has subgroups*

$$H \times 1 = \{(h,1) : h \in H\} \cong H$$

and $\qquad 1 \times K = \{(1,k) : k \in K\} \cong K.$

Every element of $H \times K$ is expressible as the product of an element of $H \times 1$ and an element of $1 \times K$. Furthermore, every element of $H \times 1$ commutes with every element of $1 \times K$, and

$$(H \times 1) \cap (1 \times K) = 1.$$

Proof. Let $\varphi : H \to H \times K$ be defined by

$$\varphi : h \mapsto (h,1) \quad \text{for all } h \in H.$$

Then, for $h_1, h_2 \in H$,

$$(h_1 h_2)\varphi = (h_1 h_2, 1) = (h_1, 1)(h_2, 1) = (h_1\varphi)(h_2\varphi).$$

Thus φ is a homomorphism, and clearly φ is injective.

Hence, by 2.10, $H \cong H\varphi = H \times 1$ and $H \times 1 \leqslant H \times K$.

Similarly $\qquad K \cong 1 \times K \leqslant H \times K.$

For every $h \in H$ and $k \in K$,

$$(h,1)(1,k) = (h,k) = (1,k)(h,1);$$

and $\qquad (H \times 1) \cap (1 \times K) = \{(1,1)\} = 1.$

75 $H \times K$ is abelian if and only if H and K are both abelian.

*76 If $H \cong J$ and $K \cong L$ then $(H \times K) \cong (J \times L)$.

*77 (i) Let $V = \{1, (12)(34), (13)(24), (14)(23)\} \subseteq \Sigma_4$. Show that $V \leqslant \Sigma_4$ and $V \cong C_2 \times C_2$. The group V is sometimes called *Klein's four-group*.
(ii) $C_2 \times C_2$ can be embedded in $GL_2(\mathbf{Q})$, but not in $F^\times$ for any field F.
(iii) $C_2 \times C_2$ and C_4 are non-isomorphic groups of order 4.
(iv) Any group of order 4 is isomorphic to either $C_2 \times C_2$ or C_4. Hence $\nu(4) = 2$ (cf. **60**).

*78 (i) If m and n are co-prime positive integers, then

$$C_m \times C_n \cong C_{mn}.$$

(ii) Let H and K be finite cyclic groups. Then $H \times K$ is cyclic if and only if the numbers $|H|$ and $|K|$ are co-prime.

Now we note a result converse to 2.33.

2.34 Lemma. *Suppose that G has subgroups H and K such that every element of G is expressible as a product hk with $h \in H$ and $k \in K$, every element of H commutes with every element of K, and $H \cap K = 1$. Then $G \cong H \times K$.*

Proof. First we note that each $g \in G$ is expressible *uniquely* as a product of an element of H and an element of K: for suppose that

$$g = hk = h'k'$$

with $h, h' \in H$ and $k, k' \in K$; then

$$(h')^{-1}h = k'k^{-1} \in H \cap K = 1,$$

by hypothesis, so that $h = h'$ and $k = k'$. Therefore we may define a map

$$\varphi : G \to H \times K$$

by $\qquad \varphi : hk \mapsto (h, k) \quad$ for all $h \in H$ and $k \in K$.

This map is well defined, by the remark above, and is clearly bijective. It is an isomorphism, for if $h_1, h_2 \in H$ and $k_1, k_2 \in K$ then in G,

$$(h_1 k_1)(h_2 k_2) = h_1 h_2 k_1 k_2,$$

since by hypothesis $k_1 h_2 = h_2 k_1$; hence

$$\begin{aligned}
((h_1 k_1)(h_2 k_2))\varphi &= (h_1 h_2, k_1 k_2) \\
&= (h_1, k_1)(h_2, k_2) \\
&= (h_1 k_1)\varphi.(h_2 k_2)\varphi, \text{ as required.}
\end{aligned}$$

This criterion applies directly to the examples in 2.32.

79 Let n be an integer, $n \geqslant 3$, and let $G = D_{2n}$, the dihedral group of order $2n$. Show that G has a cyclic subgroup H of order n and a subgroup K of order 2 such that every element of G is expressible as hk with $h \in H$ and $k \in K$, and $H \cap K = 1$. Is $G \cong H \times K$?

2.35. *For any groups H and K, $H \times K \cong K \times H$.*
Proof. The map $(h, k) \mapsto (k, h)$ (defined for all $h \in H, k \in K$) is an isomorphism of $H \times K$ onto $K \times H$.

2.36. The definition of the direct product of two groups can evidently be extended to the direct product of any finite number of groups. Let n be any positive integer, and let $G_1, G_2, \ldots, G_n$ be any n groups (not necessarily distinct). Then $G_1 \times G_2 \times \ldots \times G_n$ is the set of all ordered n-tuples $(g_1, g_2, \ldots, g_n)$ with $g_i \in G_i$ for $i = 1, \ldots, n$. This set is given the structure of a group, called the *direct product* of $G_1, G_2, \ldots, G_n$, by defining multiplication of n-tuples componentwise. (If $n = 1$, we naturally identify this 'direct product' with G_1.)

For instance, if V is a vector space of finite dimension $n > 0$ over a field F, then

$$V^+ \cong F^+ \times \ldots \times F^+,$$

the direct product of n copies of F^+ (cf. 2.15).

2.37. *For any groups* G, H *and* K,

$$G \times (H \times K) \cong G \times H \times K \cong (G \times H) \times K.$$

80 Let F be a field and n a positive integer. Show that the set H of all non-singular diagonal $n \times n$ matrices with entries in F forms a subgroup of $\mathrm{GL}_n(F)$, and that

$$H \cong F^\times \times \ldots \times F^\times,$$

the direct product of n copies of $F^\times$.

*81 A group G is said to be *decomposable* if it has *proper* subgroups H and K satisfying the hypotheses of 2.34; if not, G is said to be *indecomposable*.

(i) Let n be an integer, $n > 1$, and let the factorization of n as a product of primes be

$$n = p_1^{m_1} p_2^{m_2} \ldots p_s^{m_s},$$

where $s, m_1, \ldots, m_s$ are positive integers and $p_1, \ldots, p_s$ distinct primes. Then C_n is decomposable if $s > 1$, and

$$C_n \cong C_{q_1} \times C_{q_2} \times \ldots \times C_{q_s},$$

where $q_i = p_i^{m_i}$ for each $i = 1, \ldots, s$.

(ii) For each prime p and positive integer m, C_{p^m} is indecomposable.

82 Σ_3 is indecomposable (cf. **143**).

3

NORMAL SUBGROUPS, HOMOMORPHISMS AND QUOTIENTS

3.1 Definitions. Let $H \leqslant G$ and let A be a non-empty set of automorphisms of G. We say that H is an *A-invariant* subgroup of G if

$$h^\alpha \in H \qquad \text{for all } h \in H \text{ and } \alpha \in A.$$

For instance, if $A = 1$ then, trivially, every subgroup of G is A-invariant.

In two important special cases we use special terms. If H is Aut G-invariant, H is said to be *characteristic* in G. If H is Inn G-invariant, H is said to be *normal* (or *invariant* or *self-conjugate*) in G. The concept of normal subgroup dominates the whole of group theory, and a special notation is used. We write $H \trianglelefteq G$ to mean 'H is a normal subgroup of G', and $H \ntrianglelefteq G$ to mean 'H is not a normal subgroup of G'. Because of its fundamental importance, we restate explicitly the defining condition of normality:

3.2 Definition. Let $H \leqslant G$. Then $H \trianglelefteq G$ if and only if $g^{-1}hg \in H$ for all $h \in H$ and $g \in G$.

The alternative term 'self-conjugate' is easily explained. If $H \trianglelefteq G$ then, for any $g \in G$,

$$g^{-1}Hg = \{g^{-1}hg : h \in H\} \leqslant H. \tag{i}$$

Then also $gHg^{-1} = (g^{-1})^{-1}Hg^{-1} \leqslant H$, from which it is easy to see that

$$H \leqslant g^{-1}Hg. \tag{ii}$$

From (i) and (ii), $\qquad\qquad g^{-1}Hg = H.$

If, conversely, $g^{-1}Hg = H$ for every $g \in G$ then certainly $H \trianglelefteq G$. Hence we have

3.3. *Let $H \leqslant G$. Then $H \trianglelefteq G$ if and only if $g^{-1}Hg = H$ for every $g \in G$; that is, if and only if H coincides with all its conjugates in G.*

3.4. $1 \trianglelefteq G$ *and* $G \trianglelefteq G$.

3.5. *If G is an abelian group then every subgroup of G is normal in G.*
Proof. Let $H \leqslant G$. If $h \in H$ and $g \in G$ then $g^{-1}hg = hg^{-1}g = h \in H$.

Recall from chapter 1 that G is called *simple* if $G \neq 1$ and the only normal subgroups of G are 1 and G.

3.6 (cf. 1.11). *The only abelian simple groups are the groups of prime orders.*

Proof. If G is a group of prime order then, by Lagrange's theorem, the only subgroups of G are 1 and G, and so G is certainly simple.

Conversely, if G is an abelian simple group then, by 3.5, the only subgroups of G are 1 and G. Hence, by 29, G is finite and of prime order.

3.7 Example. *Find all the subgroups of Σ_3 and determine which of these are normal in Σ_3.*

First, $1 \trianglelefteq \Sigma_3$ and $\Sigma_3 \trianglelefteq \Sigma_3$. We have to find the non-trivial proper subgroups of Σ_3: by Lagrange's theorem, they can only have orders 2 and 3. By 1, any group of prime order is cyclic. Now

$$\Sigma_3 \text{ has } 1 \text{ element of order } 1 : 1,$$
$$3 \text{ elements of order } 2 : (12), (13), (23),$$
$$2 \text{ elements of order } 3 : (123), (132);$$

and $(132) = (123)^2$. Hence Σ_3 has 3 subgroups of order $2 : \{1, (12)\} = T$, $\{1, (13)\} = U$, $\{1, (23)\} = V$, say, and 1 subgroup of order 3: $\{1, (123), (132)\} = K$, say. Now $(13)^{-1}(12)(13) = (13)(12)(13) = (23) \notin T$, so $T \ntrianglelefteq \Sigma_3$. Similarly, $U \ntrianglelefteq \Sigma_3$ and $V \ntrianglelefteq \Sigma_3$. But for all $g \in \Sigma_3, g^{-1}Kg$ is a subgroup of Σ_3 of order 3, by 2.20, and therefore $g^{-1}Kg = K$. Hence $K \trianglelefteq \Sigma_3$.

The following remark is immediate from 2.26 and the definition of a normal subgroup.

3.8. *If $H \trianglelefteq G$ and $K \trianglelefteq G$ then $H \cap K \trianglelefteq G$. More generally, if $\{H_i : i \in I\}$ is any set of normal subgroups of G (indexed by a set I) then $\bigcap_{i \in I} H_i \trianglelefteq G$.*

83 If $\varnothing \subset A \subseteq \text{Aut } G$ and if $\{H_i : i \in I\}$ is any set of A-invariant subgroups of G then $\bigcap_{i \in I} H_i$ is an A-invariant subgroup of G.

84 Let $\varnothing \subset A \subseteq \text{Aut } G$, and let H be an A-invariant subgroup of $G, \alpha \in A$ and $H^\alpha = \{h^\alpha : h \in H\}$.
(i) Show that $H^\alpha \leqslant H$.
(ii) Show by an example that we may have $H^\alpha < H$. (Hint. For an example, consider $G = \mathbf{Q}^+, H = \mathbf{Z}^+, A = \{\lambda_2\}$, where λ_2 is the automorphism $x \mapsto 2x$ of $\mathbf{Q}^+$.)
(iii) Prove that if either H is finite or $A \leqslant \text{Aut } G$ then $H^\alpha = H$.

85 Let V be a vector space over a field F, and suppose that $0 \neq w \in V$. Let W be the 1-dimensional subspace of V spanned by w and let $\alpha \in \text{GL}(V)$. Then by **47**(i), $\alpha \in \text{Aut } V^+$.

Show that w is an eigenvector of α if and only if W^+ is an $\{\alpha\}$-invariant subgroup of V^+.

86 Let G be a finite group, of order n say, let m be a divisor of n, and suppose that G

has just one subgroup H of order m. Then H is characteristic in G.

*87 Suppose that $K \trianglelefteq J \leqslant G$. Let $\varphi : G \to H$ be a homomorphism, and let $\bar{K} = K\varphi$ and $\bar{J} = J\varphi$ (see 2.9). Then $\bar{K} \trianglelefteq \bar{J} \leqslant H$.

88 Suppose that $H \trianglelefteq G$. Show that if x and y are elements of G such that $xy \in H$ then $yx \in H$. Would this be true merely on the hypothesis that $H \leqslant G$?

*89 Consider the *direct square* $G \times G$ of G. Let

$$\hat{G} = \{(g,g) : g \in G\} \subseteq G \times G.$$

Show that $\hat{G}$ is a subgroup of $G \times G$ which is isomorphic to G ; $\hat{G}$ is called the *diagonal subgroup* of $G \times G$. Show also that $\hat{G} \trianglelefteq G \times G$ if and only if G is abelian.

*90 Let $H \leqslant G$ and define $H_G = \bigcap_{g \in G} (g^{-1}Hg)$.

Then $H_G \trianglelefteq G$ and, whenever $K \leqslant H \leqslant G$ with $K \trianglelefteq G, K \leqslant H_G$. Thus H_G is the unique largest normal subgroup of G contained in H : it is called the *core* (or *normal interior*) of H in G.

What is the core of the subgroup $\{1, (12)\}$ in Σ_3?

91 Let n be an integer, $n \geqslant 2$, and let F be a field.
(i) Consider the set of all $n \times n$ matrices with entries in F. Let e_{ij} denote the (singular) matrix with entry 1 in the ith row and jth column and all other entries $0 (1 \leqslant i \leqslant n, 1 \leqslant j \leqslant n)$. Verify that

$$e_{ij}e_{kl} = \delta_{jk}e_{il},$$

where δ_{jk} is the Kronecker delta. Show that if $i \neq j$ then, for any $a \in F$, the matrix $1 + ae_{ij}$ is non-singular, and find its inverse.
(ii) Let $G = \mathrm{GL}_n(F)$, and let H be the subgroup of G consisting of all diagonal matrices in G (see **80**). Prove that the core H_G of H in G consists of all scalar matrices in G (that is, matrices $a1$ with $0 \neq a \in F$) and that $H_G \cong F^\times$. (See **90**. Hint. Show that if $x \in H$, with the ith and jth diagonal entries of x not equal, then the conjugate of x by $1 + e_{ij}$ is not a diagonal matrix.)

*92 Let $g \in G$ and $\alpha \in \mathrm{Aut}\, G$. As in 2.19, let τ_g denote the inner automorphism of G induced by g. Show that $\alpha^{-1}\tau_g\alpha = \tau_{g\alpha}$. Deduce that for any group G, $\mathrm{Inn}\, G \trianglelefteq \mathrm{Aut}\, G$.

There is a fundamental connexion between homomorphisms and normal subgroups, which is the main theme of this chapter.

3.9 Theorem. *Let $\varphi : G \to H$ be a homomorphism, and let $K = \{g \in G : g\varphi = 1\}$. Then $K \trianglelefteq G$. We call K the* kernel *of φ and denote it by* $\mathrm{Ker}\, \varphi$. *(Compare with the kernel of a linear map.)*

Proof. By 2.9, $1 \in K$, so that $K \neq \emptyset$. If $k_1, k_2 \in K$ then, using 2.9, we have

$$(k_1 k_2^{-1})\varphi = (k_1\varphi)(k_2^{-1}\varphi) = (k_1\varphi)(k_2\varphi)^{-1} = 1.1^{-1} = 1.$$

Thus $k_1 k_2^{-1} \in K$. Hence $K \leqslant G$. Finally, if $k \in K$ and $g \in G$ then

$$(g^{-1}kg)\varphi = (g^{-1}\varphi)(k\varphi)(g\varphi) = (g\varphi)^{-1}.1.(g\varphi) = 1.$$

Thus $g^{-1}kg \in K$. Hence $K \trianglelefteq G$.

3.10. *Let $\varphi : G \to H$ be a homomorphism. Then φ is injective if and only if Ker $\varphi = 1$.* (Compare with the fact that a linear map θ between vector spaces is injective if and only if Ker $\theta = 0$.)
Proof. If φ is injective then, since, by 2.9, $1\varphi = 1$, it follows by definition that Ker $\varphi = 1$. Conversely, suppose that Ker $\varphi = 1$, and let $g_1, g_2 \in G$ with $g_1\varphi = g_2\varphi$. Then, by 2.9,

$$(g_1 g_2^{-1})\varphi = (g_1\varphi)(g_2\varphi)^{-1} = 1.$$

Hence $g_1 g_2^{-1} \in \text{Ker } \varphi = 1$, and so $g_1 = g_2$. Thus φ is injective.

We have seen that the kernel of every homomorphism $\varphi : G \to H$ is a normal subgroup of G. In 3.23 we shall show conversely that, for every $K \trianglelefteq G$, there is a homomorphism from G to a suitable group with kernel K. Before doing this, we note some other properties.

3.11. *Suppose that $G = H \times K$. Define maps $\pi_1 : G \to H$ and $\pi_2 : G \to K$ by*

$$\pi_1 : (h, k) \mapsto h \quad \text{and} \quad \pi_2 : (h, k) \mapsto k \quad \text{for all } (h, k) \in G.$$

Then π_1 and π_2 are surjective homomorphisms, called the projections *of G onto H and onto K, respectively. In the notation of 2.33, Ker $\pi_1 = 1 \times K$ and Ker $\pi_2 = H \times 1$.*
Proof. Immediate, from the definition of $H \times K$.

3.12. *Suppose that $G = H \times K$. Then $H \times 1 \trianglelefteq G$ and $1 \times K \trianglelefteq G$. But $H \times 1$ and $1 \times K$ need not be characteristic in G.*
Proof. The first assertion follows from 3.11 and 3.9. To see that $H \times 1$ and $1 \times K$ need not be characteristic in G, consider the situation when $H = K \neq 1$. Thus let H be any non-trivial group and let $G = H \times H$. Then the map

$$\alpha : G \to G$$

defined by $\qquad \alpha : (h_1, h_2) \mapsto (h_2, h_1) \quad \text{for all } (h_1, h_2) \in G$

is easily seen to be an automorphism of G. For $1 \neq h \in H$, $(h, 1)^\alpha = (1, h) \notin H \times 1$ and $(1, h)^\alpha = (h, 1) \notin 1 \times H$. Thus $H \times 1$ and $1 \times H$ are normal but not characteristic in G.

3.13. *If $K \trianglelefteq G$ and $K \leqslant H \leqslant G$ then $K \trianglelefteq H$; but H need not be normal in G.*
Proof. The first assertion is immediate from the definition of normality. To demonstrate the second assertion, it is enough, by 3.7, to choose $G = \Sigma_3, H = \{1, (12)\}, K = 1$.

Next we show that normality is not a transitive relation.

3.14. *It can happen that $K \trianglelefteq H \trianglelefteq G$ but $K \ntrianglelefteq G$.*

To see this, let

$$J = \Sigma_3 \quad \text{and} \quad G = J \times J.$$

By 3.7, J has a normal subgroup L of order 3: $L = \{1, (123), (132)\}$. Let $H = L \times L \trianglelefteq G$. It is easy to verify that $H \trianglelefteq G$ (see 111). Now H is an abelian group, so that every subgroup of H is normal in H. Let

$$K = \{(1, 1), ((123), (123)), ((132), (132))\},$$

a subgroup of H of order 3 (see 89). Then $K \trianglelefteq H$. But $K \ntrianglelefteq G$, since, for example,

$$((12), 1)^{-1} ((123), (123)) ((12), 1) = ((12)(123)(12), (123))$$
$$= ((132), (123)) \notin K.$$

In view of that negative fact, the following positive result is useful.

3.15 Lemma. *If $H \trianglelefteq G$ and K is a characteristic subgroup of H then $K \trianglelefteq G$.*

Proof. For each $g \in G$, $g^{-1}Hg = H$ (3.3). Therefore the inner automorphism τ_g of G induced by g maps H onto H. Hence, by restriction to H, τ_g determines an automorphism σ_g of H:

$$\sigma_g : h \mapsto g^{-1}hg \quad \text{for all } h \in H.$$

Since K is characteristic in H, K is σ_g-invariant; thus, for all $k \in K$,

$$k^{\sigma_g} \in K,$$

that is, $g^{-1}kg \in K.$

This is true for all $g \in G$, so that $K \trianglelefteq G$.

93 Let $K \leqslant H \leqslant G$. Prove that if $\varnothing \subset A \subseteq \text{Aut } G$, H is an A-invariant subgroup of G and K is a characteristic subgroup of H, then K is an A-invariant subgroup of G. (This generalizes 3.15.) In particular, if K is characteristic in H and H is characteristic in G then K is characteristic in G (cf. 3.14).

*94 Suppose that $G = H \times K$.
  (i) Then $(\text{Aut } H) \times (\text{Aut } K)$ can be embedded in Aut G.
  (ii) If $H \times 1$ and $1 \times K$ are characteristic in G then Aut $G \cong (\text{Aut } H) \times (\text{Aut } K)$.

3.16 Definition. For any two non-empty subsets X, Y of G we define the *product set*

$$XY = \{xy : x \in X, y \in Y\} \subseteq G.$$

If X consists of a single element x, we write xY instead of $\{x\}Y$; and similarly if $Y = \{y\}$, we write Xy for $X\{y\}$. This accords with the usual notation for *cosets* mentioned in chapter 0: when $H \leqslant G$ and $g \in G$, gH is a left coset of H in G and Hg is a right coset of H in G.

This multiplication of subsets of G is associative; for if X, Y, Z are non-empty subsets of G,

$$(XY)Z = \{xyz : x \in X . y \in Y, z \in Z\} = X(YZ).$$

Hence

3.17. *The set of all non-empty subsets of G forms a semigroup $\mathscr{Q}(G)$, when multiplication is defined as in 3.16.*

Of course, these definitions remain valid when G is replaced by any semigroup. Moreover, we can deal with the empty subset $\varnothing$ by defining $X\varnothing = \varnothing = \varnothing X$ for every $X \subseteq G$: then $\mathscr{Q}(G) \cup \{\varnothing\}$ becomes a semigroup $\mathscr{P}(G)$ in which $\varnothing$ plays the multiplicative rôle of zero. However, we shall discuss only $\mathscr{Q}(G)$ with G a group.

Note that the previous notation for a conjugate of a subgroup is consistent with the present definitions: if $K \leqslant G$ and $g \in G$ then the conjugate subgroup $g^{-1}Kg$ is the appropriate product of the sets $\{g^{-1}\}, K, \{g\}$.

*95 Let $H, K \leqslant G$. Then
   (i) Every subgroup of G containing both H and K contains the product sets HK and KH.
   (ii) $HK \leqslant G$ if and only if $HK = KH$.
(Note. The equation $HK = KH$ does *not* mean that every element of H commutes with every element of K. It means that for each $h \in H$ and $k \in K$, $hk = k'h'$ for some $k' \in K$ and $h' \in H$, and similarly kh can be expressed as an element of HK.)
   (iii) Find an example for which HK is not a subgroup of G (cf. **71**; see also **3.38**).

96 Let $H, K \leqslant G$ and $x, y \in G$ with $Hx = Ky$. Then $H = K$.

97 The semigroup $\mathscr{Q}(G)$ has an identity element, and the group of units of $\mathscr{Q}(G)$ is G.

*98 Let G be a finite group and let $H, K \leqslant G$. Then

$$|HK| = \frac{|H|\,|K|}{|H \cap K|} = |KH|.$$

(Note. HK and KH need not be subgroups of G: see **95**.)

*99 Let G be a finite group and let $H, K \leqslant G$.
   (i) Then $|\langle H, K \rangle : K| \geqslant |H : H \cap K|$ (see **71**).
   (ii) If $|H : H \cap K| > \frac{1}{2}|G : K|$ then $\langle H, K \rangle = G$.

*100 Let G be a finite group and let $H, K \leqslant G$ with $(|G : H|, |G : K|) = 1$. Then $|G : H \cap K| = |G : H|\,|G : K|$ and $G = HK$.
(Hint. Use **11**, **98** and **99**(i). Later we shall be able to remove the condition of finiteness of G: see **197**.)

Any subgroup of the semigroup $\mathscr{Q}(G)$ must contain an identity element, hence a subset X of G such that $X^2 = X$ (that is, an idempotent subset: see **15**).

3.18. (i) *If $H \leqslant G$ then $H^2 = H$.*

(ii) *Let $X \in \mathcal{Q}(G)$. If $X^2 = X$ and X is finite then $X \leqslant G$. This statement fails, however, if we delete the condition that X is finite.*

Proof. (i) Since $H \leqslant G, H = 1H \subseteq H^2$. But also, because $H \leqslant G, H^2 \subseteq H$. Hence $H^2 = H$.

(ii) Let $x \in X$. Then $xX \subseteq X^2 = X$. Since X is a finite set and clearly $|xX| = |X|$, it follows that $xX = X$. Therefore $x \in xX$ and so $x = xe$ for some $e \in X$. But then, in $G, 1 = x^{-1}x = e \in X$. Now $1 \in xX$ and so $1 = xy$ for some $y \in X$. Then, in $G, x^{-1} = y \in X$. Since also $X^2 \subseteq X$, this shows that $X \leqslant G$.

For an example to show that we need X to be finite here, let $G = \mathbf{Z}^+$ and let X be the set of all non-negative integers. The sum of any two non-negative integers is a non-negative integer, and, conversely, any non-negative integer n is the sum of two non-negative integers, for $n = 0 + n$. Thus $X^2 = X$. But X is not a subgroup of $\mathbf{Z}^+$.

By 3.18, for each $H \leqslant G, \{H\}$ is a subgroup of the semigroup $\mathcal{Q}(G)$ of order 1 (and when G is a finite group, every subgroup of $\mathcal{Q}(G)$ of order 1 is $\{H\}$ for some $H \leqslant G$). We are going to define some more subgroups of $\mathcal{Q}(G)$ associated to *normal* subgroups of G. Later we shall be able to give a simple characterization of all the subgroups of $\mathcal{Q}(G)$ when G is a finite group: see 3.57.

First we reformulate the statement of 3.3:

3.19. *Let $H \leqslant G$. Then $H \trianglelefteq G$ if and only if $Hg = gH$ for every $g \in G$.*

101 (G. Horrocks) Let X be a non-empty finite subset of G; say $X = \{x_1, \ldots, x_n\}$, where n is a positive integer. Suppose that $x_i x_j \in X$ whenever $1 \leqslant i \leqslant j \leqslant n$.
   (i) Prove by induction on m that $x_i^m \in X$ for every positive integer m and for every $i = 1, \ldots, n$.
   (ii) Deduce that, for every $i = 1, \ldots, n, x_i$ has finite order and $x_i^{-1} \in X$.
   (iii) Deduce that whenever $x_i X = X$ then also $x_i^{-1} X = X$.
   (iv) Hence prove by induction on j that, for every $j = 1, \ldots, n, x_j X = X$.
   (v) By means of 3.18, conclude that $X \leqslant G$.

*102 If $H \leqslant G$ and $|G : H| = 2$ then $H \trianglelefteq G$.

103 Let $X = E^2$, the euclidean plane, $G = \text{Isom } X$, $T = \text{Tr } X$ and $H = T \cup \bigcup_{s \in X} \text{Rot}(X; s)$: then, according to 2.23, $|G : H| = 2$.
   (i) Let $\tau \in H$. Then τ fixes no point of X if and only if $1 \neq \tau \in T$.
   (ii) Let $\sigma, \tau \in G$. If $\sigma^{-1}\tau\sigma$ fixes some point of X then so also does τ fix some point of X.
   (iii) $T \trianglelefteq G$. (Hint. Use **102**.)

3.20. *Let $K \trianglelefteq G$. Let G/K denote the set of all cosets of K in G. (By 3.19, it is unnecessary to qualify 'cosets' by left or right.) Then G/K is a subgroup of the semigroup $\mathcal{Q}(G)$, called the* quotient group (*or* factor group) *of G by K. In G/K,*

$$(xK)(yK) = xyK,$$

the identity element of G/K is K, and

$$(xK)^{-1} = x^{-1}K \quad \text{(for } x, y \in G\text{)}.$$

If G is abelian then G/K is also abelian.

Note. G/K is not a subgroup of G: it is a group formed by the multiplication of certain subsets of G.

Proof. Let $x, y \in G$. Then in $\mathscr{Q}(G)$,

$$(xK)(yK) = xy(y^{-1}Ky)K = xyKK = xyK,$$

by 3.3 and 3.18. Thus the set G/K is closed under the multiplication defined in $\mathscr{Q}(G)$. In particular,

$$(xK)K = xK = K(xK),$$

so that G/K has identity element K. Moreover,

$$(xK)(x^{-1}K) = K = (x^{-1}K)(xK),$$

so that xK has the element $x^{-1}K$ as inverse in G/K.

Immediately from the definition we have

3.21. $G/1 \cong G$ and $G/G \cong 1$.

Let $K \trianglelefteq G$. Then, by definition of G/K,

$$|G/K| = |G : K|,$$

the index of K in G. Hence we have

3.22 (cf. 1.7). *If $K \trianglelefteq G$ and G is a finite group then $|G/K| = |G|/|K|$.*

104 Let $K \leqslant G$. If the set of all left cosets of K in G forms a subgroup of $\mathscr{Q}(G)$ then $K \trianglelefteq G$. (This is a converse to 3.20.)

*105 Suppose that $K \trianglelefteq G$ with $|G/K| = n < \infty$.
   (i) Then $g^n \in K$ for every $g \in G$.
   (ii) If $g \in G$ and $g^m \in K$ for some integer m such that $(m, n) = 1$ then $g \in K$ (cf. **12**).

*106 Suppose that $K \trianglelefteq G$ with $|K| = m < \infty$. Let $x \in G$ and let n be a positive integer such that $(m, n) = 1$.
   (i) If $o(x) = n$ then $o(xK) = n$ (where xK is viewed as an element of the group G/K).
   (ii) If $o(xK) = n$ then there is an element $y \in G$ such that $o(y) = n$ and $xK = yK$.

*107 Prove by induction on $|G|$ that if G is a finite abelian group such that p divides $|G|$ then G has an element of order p. (Hint. If G has a proper non-trivial subgroup K then, since p is prime, either p divides $|K|$ or p does not divide $|K|$ and p divides $|G/K|$. In the latter case use **106**(ii).)

*108 Let $K \trianglelefteq G$ and let $\bar{G} = G/K$. For each subset X of G, let $\bar{X} = \{xK : x \in X\} \subseteq \bar{G}$.
   (i) If X is a set of generators of G then $\bar{X}$ is a set of generators of $\bar{G}$. In particular, if G is an n-generator group, where n is a positive integer, then $\bar{G}$ is an n-generator group.

(ii) If X is a subset of G such that $\bar{X}$ is a set of generators of $\bar{G}$, and if Y is a set of generators of K, then $X \cup Y$ is a set of generators of G.

In particular, if K is an m-generator group and G/K is an n-generator group, where m and n are positive integers, then G is an $(n + m)$-generator group. (Hint. Apply 2.28.)

*109 Let $H \trianglelefteq G$ and $K \trianglelefteq G$. Then (3.8) $H \cap K \trianglelefteq G$. Show that we can define a map $\psi : G/(H \cap K) \to (G/H) \times (G/K)$ by $\psi : g(H \cap K) \mapsto (gH, gK)$ (for all $g \in G$), and that ψ is an injective homomorphism. Thus $G/(H \cap K)$ can be embedded in $(G/H) \times (G/K)$. Deduce that if G/H and G/K are both abelian then $G/(H \cap K)$ is abelian.

110 Let X be a non-empty set. For each $\sigma \in \Sigma_X$, we define a subset $s(\sigma)$ of X by

$$s(\sigma) = \{x \in X : x\sigma \neq x\}.$$

(a) Let $\sigma, \tau \in \Sigma_X$. Then
  (i) $s(\sigma^{-1}) = s(\sigma)$,
  (ii) $s(\sigma\tau) \subseteq s(\sigma) \cup s(\tau)$,
  (iii) $s(\sigma^{-1}\tau\sigma) = \{x\sigma : x \in s(\tau)\}$.
  (iv) If $s(\sigma) \cap s(\tau) = \emptyset$ then $\sigma\tau = \tau\sigma$.
(b) Let $\Sigma_{(X)} = \{\sigma \in \Sigma_X : |s(\sigma)| < \infty\}$.
Then $\Sigma_{(X)} \trianglelefteq \Sigma_X$. The group $\Sigma_{(X)}$ is called the *restricted symmetric group on X*. Furthermore, $\Sigma_{(X)} = \Sigma_X$ if and only if $|X| < \infty$; and if X is infinite then $\Sigma_{(X)}$ is an infinite group in which every element has finite order, and the quotient group $\Sigma_X/\Sigma_{(X)}$ is infinite.
(Hint. It may be assumed that if X is infinite then there is an injective map of $\mathbf{Z}$ into X.)

The following observation is fundamental.

3.23. *Let $K \trianglelefteq G$. Then the map $v : G \to G/K$ defined by*

$$v : g \mapsto gK \quad \text{for all } g \in G$$

is a surjective homomorphism, and Ker $v = K$. *The map v is called the* natural (*or* canonical) *homomorphism of G onto G/K.*
Proof. That v is a surjective homomorphism follows at once from 3.20. Let $g \in G$. Then $g \in$ Ker v if and only if $gK = K$ (the identity element of G/K), that is, if and only if $g \in K$. Hence Ker $v = K$.

This establishes the fact, mentioned earlier, that the normal subgroups of G are precisely the kernels of homomorphisms of G to other groups. We now complete the description of the links connecting normal subgroups, quotient groups and homomorphisms by relating an arbitrary homomorphism to a suitable natural homomorphism.

3.24 Fundamental theorem on homomorphisms. *Let $\varphi : G \to H$ be a homomorphism and let $K = $ Ker $\varphi \trianglelefteq G$, by 3.9. Let v be the natural homomorphism of G onto G/K. Then there is an injective homomorphism $\psi : G/K \to H$ such that $\varphi = v\psi$. In particular,* Im $\varphi \cong G/$Ker φ.
Proof. We define the map

$$\psi : G/K \to H$$

by

$$\psi : gK \mapsto g\varphi \quad \text{(for all } g \in G).$$

We must check that ψ is well defined: that is, that if $x, y \in G$ and $xK = yK$ then $x\varphi = y\varphi$. Now $xK = yK$ if and only if $x^{-1}y \in K$, that is, if and only if $(x^{-1}y)\varphi = 1$. By 2.9, this is true if and only if $(x\varphi)^{-1}(y\varphi) = 1$; that is, if and only if $x\varphi = y\varphi$. This shows that ψ is well defined and also that ψ is injective. Moreover, ψ is a homomorphism, for if $x, y \in G$ then

$$((xK)(yK))\psi = (xyK)\psi = (xy)\varphi = (x\varphi)(y\varphi).$$

Now
$$xv\psi = (xK)\psi = x\varphi.$$

This is true for all $x \in G$, and so

$$v\psi = \varphi.$$

Finally, by 2.10, $G/K \cong \operatorname{Im} \psi = \operatorname{Im} \varphi$.

We now illustrate the fundamental theorem with some examples.

3.25. Recall from 2.14 the surjective homomorphism

$$v_n : \mathbf{Z}^+ \to C_n$$

defined by $v_n : x \mapsto e^{2\pi x i/n}$ for all $x \in \mathbf{Z}^+$, where n is a positive integer. Now $\operatorname{Ker} v_n = \{nx : x \in \mathbf{Z}^+\}$. We denote this subgroup of $\mathbf{Z}^+$ by $n\mathbf{Z}^+$: it consists of all the integral multiples of n. By the fundamental theorem we have

$$C_n = \operatorname{Im} v_n \cong \mathbf{Z}^+ / \operatorname{Ker} v_n = \mathbf{Z}^+ / n\mathbf{Z}^+.$$

Thus $\mathbf{Z}^+$ has a quotient group of order n for every positive integer n.

Now we can classify all subgroups and all quotient groups of $\mathbf{Z}^+$. Remember that, since $\mathbf{Z}^+$ is abelian, every subgroup is normal in $\mathbf{Z}^+$ and therefore has a corresponding quotient group. Let

$$0 < H \leqslant \mathbf{Z}^+.$$

Then H contains *positive* integers, since if $h \in H$ then also $-h \in H$. Let n be the least positive integer belonging to H. (Here we use the well ordering principle for the positive integers.) Then H contains with n every integral multiple of n: that is, $n\mathbf{Z}^+ \leqslant H$. Let $h \in H$. By the division algorithm for integers, there are integers q and r such that $h = nq + r$ and $0 \leqslant r < n$. Since $nq \in n\mathbf{Z}^+ \leqslant H, r = h - nq \in H$. By choice of n, it follows that $r = 0$. Hence $h \in n\mathbf{Z}^+$. This proves that $H = n\mathbf{Z}^+$.

Thus the only non-trivial subgroups of $\mathbf{Z}^+$ are the subgroups $n\mathbf{Z}^+$, one for each positive integer n. Moreover, in the notation of 2.11, $n\mathbf{Z}^+ = \operatorname{Im} \lambda_n \cong \mathbf{Z}^+$. We have shown that $\mathbf{Z}^+ / n\mathbf{Z}^+ \cong C_n$; and of course $\mathbf{Z}^+ / 0 \cong \mathbf{Z}^+$. Note that every subgroup and every quotient group of $\mathbf{Z}^+$

is cyclic; and also that each subgroup is either finite, in which case it is 0, or has finite quotient group.

We may contrast this with the (abelian) group $\mathbf{Q}^+$. Here $\mathbf{Z}^+$ is an infinite subgroup of $\mathbf{Q}^+$, and also the quotient $\mathbf{Q}^+/\mathbf{Z}^+$ is an infinite group: for if a, b, c, d are integers with $b \neq 0 \neq d$ then in $\mathbf{Q}^+/\mathbf{Z}^+$,

$$\frac{a}{b} + \mathbf{Z}^+ = \frac{c}{d} + \mathbf{Z}^+ \quad \text{if and only if} \quad \frac{a}{b} - \frac{c}{d} \text{ is an integer.}$$

(Cosets are written additively in this case, since the group operation in $\mathbf{Q}^+$ is addition.) Thus, for instance, $\mathbf{Z}^+, \frac{1}{2} + \mathbf{Z}^+, \frac{1}{3} + \mathbf{Z}^+, \frac{1}{4} + \mathbf{Z}^+, \ldots$ are distinct elements of $\mathbf{Q}^+/\mathbf{Z}^+$. The group $\mathbf{Q}^+/\mathbf{Z}^+$ is called the *additive group of rationals* mod 1.

3.26. Let U denote the circle group (see 2.32), and let

$$\eta : \mathbf{R}^+ \to U$$

be defined by $\qquad \eta : x \mapsto e^{2\pi x i} \quad$ for all $x \in \mathbf{R}^+$.

Clearly η is a surjective homomorphism, and Ker $\eta = \mathbf{Z}^+$. Hence, by the fundamental theorem,

$$U \cong \mathbf{R}^+/\mathbf{Z}^+.$$

3.27. Suppose that G has subgroups H and K such that $G = HK$, every element of H commutes with every element of K, and $H \cap K = 1$. Then, by 2.34, there is an isomorphism $\varphi : G \to H \times K$ with $\varphi : hk \mapsto (h, k)$ for all $h \in H$ and all $k \in K$. Let π_1, π_2 denote the projections of $H \times K$ onto H, K respectively: see 3.11. Then, by 3.11 and **19**, $\varphi\pi_1 : G \to H$ is a surjective homomorphism, and Ker $(\varphi\pi_1) = \{hk \in G : h = 1, k \in K\} = K$. Hence, by the fundamental theorem,

$$K \trianglelefteq G \quad \text{and} \quad G/K \cong H.$$

Similarly, from $\varphi\pi_2 : G \to K$,

$$H \trianglelefteq G \quad \text{and} \quad G/H \cong K.$$

Thus, for example, from 2.32 we get

$$\mathbf{C}^+/\mathbf{R}^+ \cong \mathbf{R}^+, \qquad \mathbf{C}^\times/\mathbf{R}_{\text{pos}}^\times \cong U,$$

$$\mathbf{C}^\times/U \cong \mathbf{R}_{\text{pos}}^\times \cong \mathbf{R}^\times/C_2,$$

$$\mathbf{R}^\times/\mathbf{R}_{\text{pos}}^\times \cong C_2 \cong \mathbf{Q}^\times/\mathbf{Q}_{\text{pos}}^\times,$$

and $\qquad \mathbf{Q}^\times/C_2 \cong \mathbf{Q}_{\text{pos}}^\times.$

3.28. For F a field and n a positive integer, consider again the surjective homomorphism

$$\det : GL_n(F) \to F^\times \quad \text{(see 2.17)}.$$

Here $$\text{Ker det} = \{x \in GL_n(F) : \det x = 1\}.$$

This normal subgroup of $GL_n(F)$ is called the *special linear group of degree n over F* and denoted by $SL_n(F)$. By the fundamental theorem,

$$GL_n(F)/SL_n(F) \cong F^\times.$$

*111 Let $J \trianglelefteq H$ and $L \trianglelefteq K$. Then $(J \times L) \trianglelefteq (H \times K)$ and

$$(H \times K)/(J \times L) \cong (H/J) \times (K/L).$$

112 Let $s \in E^2$, the euclidean plane. Then $\text{Rot}(E^2 ; s) \cong U$, the circle group.

113 Let F be a field and n a positive integer. Suppose that for each $a \in F$ there is a unique $b \in F$ such that $b^n = a$. Then $GL_n(F) \cong F^\times \times SL_n(F)$. (In particular, this is true when $F = \mathbf{R}$ and n is odd.)

*114 Let $K \trianglelefteq G$. Then the following two statements are equivalent :
   (i) There is a homomorphism φ of G onto H with $\text{Ker } \varphi = K$.
   (ii) $G/K \cong H$.

115 (i) Let $K \trianglelefteq G$ and $\alpha \in \text{Aut } G$. Write $K^\alpha = \{k^\alpha : k \in K\}$. Then $K \cong K^\alpha \trianglelefteq G$ and $G/K \cong G/K^\alpha$.
   (ii) A group G can have normal subgroups K and L such that $K \cong L$ but $G/K \not\cong G/L$, and normal subgroups H and J such that $G/H \cong G/J$ but $H \not\cong J$. (Hint. Consider $G = C_2 \times C_4$.)

116 Find non-isomorphic groups G_1 and G_2 with $K_1 \trianglelefteq G_1$ and $K_2 \trianglelefteq G_2$ such that $K_1 \cong K_2$ and $G_1/K_1 \cong G_2/K_2$. (Hint. Consider groups of order 4.)

*117 The *centre* of G is defined to be

$$Z(G) = \{g \in G : gx = xg \text{ for all } x \in G\}.$$

Show that $Z(G) = \text{Ker } \tau$, where $\tau : G \to \Sigma_G$ is the homomorphism defined in 2.21. Deduce that

$$Z(G) \trianglelefteq G \quad \text{and} \quad \text{Inn } G \cong G/Z(G).$$

*118 $Z(G)$ is an abelian characteristic subgroup of G, and every subgroup H of $Z(G)$ is normal in G. Need such a subgroup H be characteristic in G?

*119 If $K \trianglelefteq G$ and $|K| = 2$ then $K \leqslant Z(G)$.

*120 Let U be the set of all matrices of the form

$$\begin{pmatrix} 1 & a & b \\ 0 & 1 & c \\ 0 & 0 & 1 \end{pmatrix}$$

where a, b, c are arbitrary elements of a field F, and 0 and 1 denote respectively the zero and identity elements of F. Prove that

$$U \leqslant GL_3(F), \qquad Z(U) \cong F^+ \quad \text{and} \quad U/Z(U) \cong F^+ \times F^+.$$

*121 Prove that if $K \trianglelefteq G$ then $Z(K) \trianglelefteq G$. Show by an example that $Z(K)$ need not be contained in $Z(G)$.

*122 For $H \leqslant G$, we define the *centralizer of H in G* to be

$$C_G(H) = \{g \in G : gh = hg \text{ for all } h \in H\}.$$

Then $Z(G) \leqslant C_G(H) \leqslant G$.

123 Let n be an integer and F a field, with $n \geqslant 2$ and $F \neq \mathbf{Z}_2$. Let $G = \mathrm{GL}_n(F)$ and let H be the subgroup of G consisting of all diagonal matrices in G (see **80**).
(i) Prove that $C_G(H) = H$. Deduce, by means of **91**, that $Z(G)$ consists of all scalar matrices in G.
(ii) Suppose further that either $n > 2$ or $F \neq \mathbf{Z}_3$. Let $S = \mathrm{SL}_n(F)$. Prove that $C_G(H \cap S) = H$ and deduce that $Z(S) = S \cap Z(G)$.

124 Let n be an integer, $n \geqslant 3$, and let $G = D_{2n}$, the dihedral group of order $2n$. Prove that if n is odd, $Z(G) = 1$, while if n is even, $|Z(G)| = 2$; and in the latter case that $G/Z(G) \cong D_n$ for $n \geqslant 6$, while for $n = 4$, $G/Z(G) \cong C_2 \times C_2$.

*125 If $G/Z(G)$ is cyclic then G is abelian (and so $Z(G) = G$). (Hint. Let $G/Z(G) = \langle gZ(G) \rangle$, where $g \in G$. Show that every element of G is expressible in the form $g^r z$, where r is an integer and $z \in Z(G)$.)

126 Let $x, y \in G$ and let $xy = z$. If $z \in Z(G)$ then x and y commute.

127 Let $K \trianglelefteq G$ and let $v : G \rightarrow G/K$ be the natural homomorphism. Then the surjective homomorphisms $G \rightarrow G/K$ with kernel K are precisely the maps $v\beta$ with $\beta \in \mathrm{Aut}\,(G/K)$. Deduce that if $\alpha \in \mathrm{Aut}\,G$ and K is mapped onto itself by α then $\alpha v = v\beta$ for some $\beta \in \mathrm{Aut}\,(G/K)$.

128 $\mathbf{C}^\times \cong \mathbf{R}^+ \times (\mathbf{R}^+/\mathbf{Z}^+)$ (cf. $\mathbf{C}^+ \cong \mathbf{R}^+ \times \mathbf{R}^+$).

129 $\mathbf{C}^\times \cong \mathbf{C}^+/\mathbf{Z}^+$.

130 $\mathbf{C}^\times/\mathbf{R}^\times \cong U/C_2$, where U is the circle group.

131 Let $V = \{z \in \mathbf{C}^\times : \text{there is a positive integer } n \text{ such that } z^n = 1\}$, the multiplicative group of all complex roots of 1. Then $V < U$, the circle group. Show that

$$V \cong \mathbf{Q}^+/\mathbf{Z}^+ \text{ and } U/V \cong \mathbf{R}^+/\mathbf{Q}^+.$$

(Remark. By vector space theory it can be shown that $\mathbf{R}^+/\mathbf{Q}^+ \cong \mathbf{R}^+$; and we know from 3.26 that $\mathbf{R}^+/\mathbf{Z}^+ \cong U$. Thus U and $\mathbf{R}^+$ are non-isomorphic groups—why?—each of which has a quotient group isomorphic to the other.)

*132 $\mathbf{Z}^+$ is indecomposable (see **81**).

*133 (i) If H is any proper subgroup of $\mathbf{Q}^+$ then $\mathbf{Q}^+/H$ is infinite. (Use **105**. Compare this with the fact that if K is any non-trivial subgroup of $\mathbf{Z}^+$, $\mathbf{Z}^+/K$ is finite.)
(ii) Any two non-trivial subgroups of $\mathbf{Q}^+$ have non-trivial intersection. Hence $\mathbf{Q}^+$ is indecomposable (see **81**).
(iii) There is no proper subgroup H of $\mathbf{Q}^+$ such that $\mathbf{Q}^+/H$ is cyclic. (Hint. If $H \leqslant \mathbf{Q}^+$ with $\mathbf{Q}^+/H$ cyclic then, by (ii), $H \cap \mathbf{Z}^+ \neq 0$. Then use (i).)

Next, we consider the relation which exists between the subgroups of a homomorphic image of G and the subgroups of G.

3.29 Theorem. *Let $\varphi : G \rightarrow \bar{G}$ be a surjective homomorphism. Let $\mathscr{S}$ be the set of all subgroups of G which contain $\mathrm{Ker}\,\varphi$ and let $\bar{\mathscr{S}}$ be the set of all subgroups of $\bar{G}$. Then there is a bijective map*

$$\hat{\varphi} : \mathscr{S} \to \bar{\mathscr{S}}$$

defined by $\qquad \hat{\varphi} : H \mapsto H\varphi = \{h\varphi : h \in H\} = \bar{H}$, *say.*

Moreover, for $H \in \mathscr{S}$, $\bar{H} \trianglelefteq \bar{G}$ if and only if $H \trianglelefteq G$; and, if so, $\bar{G}/\bar{H} \cong G/H$ (cf. **87**).

Proof. Certainly $\hat{\varphi}$ is a well-defined map: for, by 2.9, if $H \leqslant G$ then $H\varphi \leqslant \bar{G}$. To show that $\hat{\varphi}$ is bijective, it is enough to show that there is an inverse map

$$\hat{\psi} : \bar{\mathscr{S}} \to \mathscr{S}.$$

We define $\hat{\psi}$ as follows. For each $J \in \bar{\mathscr{S}}$, let

$$\hat{\psi} : J \mapsto \{g \in G : g\varphi \in J\} = J^*, \text{ say.}$$

(J^* is often called the *inverse image of J under φ*.)

Then $J^* \in \mathscr{S}$. To see this, first note that if $g \in \text{Ker } \varphi$ then $g\varphi = 1 \in J$, and so $g \in J^*$. Thus Ker $\varphi \subseteq J^*$. Also, if $g_1, g_2 \in J^*$ then, by 2.9, $(g_1 g_2^{-1})\varphi = (g_1\varphi)(g_2\varphi)^{-1} \in J$, since $g_1\varphi, g_2\varphi \in J$ and $J \leqslant \bar{G}$. Hence $g_1 g_2^{-1} \in J^*$. Therefore Ker $\varphi \leqslant J^* \leqslant G$. This shows that $\hat{\psi}$ is well defined.

Let $H \in \mathscr{S}$ and $J \in \bar{\mathscr{S}}$. Then
$$\begin{aligned}
H\hat{\varphi}\hat{\psi} &= \{g \in G : g\varphi \in H\hat{\varphi}\} \\
&= \{g \in G : g\varphi = h\varphi \text{ for some } h \in H\} \\
&= \{g \in G : h^{-1}g \in \text{Ker } \varphi \text{ for some } h \in H\} \\
&= \{g \in G : g \in H\} \quad (\text{since Ker } \varphi \leqslant H) \\
&= H,
\end{aligned}$$

and
$$\begin{aligned}
J\hat{\psi}\hat{\varphi} &= \{x\varphi : x \in J\hat{\psi}\} \\
&= \{x\varphi : x \in G \text{ and } x\varphi \in J\} \\
&= J \quad (\text{since } J \leqslant \bar{G} = G\varphi).
\end{aligned}$$

Thus $\hat{\varphi}\hat{\psi} = $ identity on $\mathscr{S}$, and $\hat{\psi}\hat{\varphi} = $ identity on $\bar{\mathscr{S}}$. Hence $\hat{\psi}$ is inverse to $\hat{\varphi}$, and so $\hat{\varphi}$ is bijective.

Suppose that Ker $\varphi \leqslant H \trianglelefteq G$. Then $\bar{H} \leqslant \bar{G}$ and, for all $g\varphi \in \bar{G}$ and $h\varphi \in \bar{H}$ (where $g \in G, h \in H$), $(g\varphi)^{-1}(h\varphi)(g\varphi) = (g^{-1}hg)\varphi \in \bar{H}$, by 2.9 and since $g^{-1}hg \in H$. Hence $\bar{H} \trianglelefteq \bar{G}$.

Conversely, suppose that $\bar{H} \trianglelefteq \bar{G}$, and let ν be the natural homomorphism $\bar{G} \to \bar{G}/\bar{H}$. By what we have proved above, $\bar{H} = H\varphi$ for some $H \in \mathscr{S}$. Now $\varphi\nu$ is a homomorphism of G onto $\bar{G}/\bar{H}$, and
$$\begin{aligned}
\text{Ker } \varphi\nu &= \{g \in G : g\varphi \in \text{Ker } \nu\} \\
&= \{g \in G : g\varphi \in \bar{H}\} \\
&= \bar{H}\hat{\psi} \\
&= H\hat{\varphi}\hat{\psi} \\
&= H.
\end{aligned}$$

Hence $H \trianglelefteq G$, and, by the fundamental theorem,
$$\bar{G}/\bar{H} = \text{Im } \varphi\nu \cong G/H.$$

134 With the notation and hypotheses of 3.29, let $H, K \in \mathscr{S}$.
Then
   (i) $\overline{H \cap K} = \bar{H} \cap \bar{K}$,
   (ii) $\bar{K} \leqslant \bar{H}$ if and only if $K \leqslant H$,
   (iii) $\overline{\langle H, K \rangle} = \langle \bar{H}, \bar{K} \rangle$. (Hint. Use **71**.)

An important special case of 3.29 occurs when $K \trianglelefteq G$ and we choose for φ the natural homomorphism $v : G \to G/K = \bar{G}$. Then $K = \mathrm{Ker}\ \varphi$, and if $K \leqslant H \leqslant G, \bar{H} = \{hK : h \in H\} = H/K$. So we have

3.30 An isomorphism theorem. *Let* $K \trianglelefteq G$. *Then every subgroup of* G/K *is of the form* H/K, *where* $K \leqslant H \leqslant G$. *Moreover,* $H/K \trianglelefteq G/K$ *if and only if* $H \trianglelefteq G$, *and if so,* $G/K \big/ H/K \cong G/H$.

The fundamental theorem, the result of 3.30 and a result to be proved soon (3.40) are called by some authors the first, second, and third isomorphism theorems. Since there is a lack of unanimity in the assignment of these numbers, we prefer to refer simply to 'an isomorphism theorem' and 'another isomorphism theorem'.

We illustrate 3.30 by classifying all subgroups and all quotient groups of cyclic groups.

3.31. We know (**29**) that any infinite cyclic group is isomorphic to $\mathbf{Z}^+$, and we have already classified all subgroups and quotients of $\mathbf{Z}^+$ in 3.25. So we need only consider finite cyclic groups. Any cyclic group of finite order n is isomorphic to C_n (**2**), and, by 3.25,

$$C_n \cong \mathbf{Z}^+ / n\mathbf{Z}^+.$$

By 3.30, every subgroup of $\mathbf{Z}^+/n\mathbf{Z}^+$ is of the form $H/n\mathbf{Z}^+$, where $n\mathbf{Z}^+ \leqslant H \leqslant \mathbf{Z}^+$. Also, by 3.25, every non-trivial subgroup of $\mathbf{Z}^+$ is of the form $m\mathbf{Z}^+$, where m is a positive integer. It is easy to see that $n\mathbf{Z}^+ \leqslant m\mathbf{Z}^+$ if and only if m divides n. Hence the subgroups of $\mathbf{Z}^+/n\mathbf{Z}^+$ are just the subgroups $m\mathbf{Z}^+/n\mathbf{Z}^+$, one for each divisor m of n.

By 3.30, when m is a divisor of n, we have

$$\mathbf{Z}^+/n\mathbf{Z}^+ \big/ m\mathbf{Z}^+/n\mathbf{Z}^+ \cong \mathbf{Z}^+/m\mathbf{Z}^+ \cong C_m.$$

Moreover, since $|\mathbf{Z}^+/n\mathbf{Z}^+| = n$ and $|\mathbf{Z}^+/m\mathbf{Z}^+| = m$, by 3.22, $|m\mathbf{Z}^+/n\mathbf{Z}^+| = n/m$. Also $m\mathbf{Z}^+ \cong \mathbf{Z}^+$, by 3.25 and 2.11, and since every quotient of $\mathbf{Z}^+$ is cyclic, $m\mathbf{Z}^+/n\mathbf{Z}^+$ is cyclic. Thus every subgroup of $\mathbf{Z}^+/n\mathbf{Z}^+$ is cyclic. In summary we have

3.32. *All subgroups and all quotient groups of any cyclic group are cyclic. If* G *is a cyclic group of finite order* n *then* G *has just one subgroup* H *of order* s *for each divisor* s *of* n, H *is cyclic and* G/H *is cyclic of order* n/s.

135 Let G be an abelian group of finite order n. Prove by induction on n that for every divisor m of n, G has a subgroup of order m. (cf. **185**, 5.32, **674**, **675**, **676**. Hint.

If $m > 1$, let p be a prime divisor of m, use **107** to show that G has a subgroup K of order p, and consider G/K.)

*136 Let K be a characteristic subgroup of G.

    (i) Let $\alpha \in \text{Aut } G$. Let $\bar{\alpha}$ be the map of G/K into itself defined (for all $g \in G$) by

$$\bar{\alpha} : gK \mapsto g^{\alpha}K.$$

Then $\bar{\alpha}$ is well defined and is an automorphism of G/K.

    Moreover, the map $\alpha \mapsto \bar{\alpha}$ is a homomorphism of Aut G into Aut (G/K).

    (ii) Let $K \leqslant H \leqslant G$. If H/K is characteristic in G/K then H is characteristic in G. (Remark. The converse of (ii) is false : see **137**; cf. 3.30.)

137 Let $G = D_8$, the dihedral group of order 8, and let $K = Z(G)$. Then $|K| = 2$ and $G/K \cong C_2 \times C_2$: see **124**. Let H be the unique cyclic subgroup of G of order 4 (**59**).

    Show that $K < H$ and that H and K are characteristic subgroups of G, but that H/K is not characteristic in G/K (cf. **136**).

*138 (i) Every subgroup of a finite cyclic group G is characteristic in G.

    (ii) Let n be an integer, $n > 1$, and let the factorization of n as a product of primes be $n = p_1^{m_1} p_2^{m_2} \ldots p_s^{m_s}$, where $s, m_1, \ldots, m_s$ are positive integers and $p_1, \ldots, p_s$ distinct primes. For each $i = 1, \ldots, s$, let $q_i = p_i^{m_i}$. We know (**81**) that

$$\mathbf{Z}_n^+ \cong \mathbf{Z}_{q_1}^+ \times \mathbf{Z}_{q_2}^+ \times \ldots \times \mathbf{Z}_{q_s}^+.$$

Prove that

$$\mathbf{Z}_n^\times \cong \mathbf{Z}_{q_1}^\times \times \mathbf{Z}_{q_2}^\times \times \ldots \times \mathbf{Z}_{q_s}^\times.$$

(Hint. Use **46** and **94**.)

*139 Let n be a positive integer.

    (i) Let G be a cyclic group of order n. For each divisor s of n, let G_s be the unique subgroup of G of order s (see 3.32). Then

$$G_s = \{x \in G : x^s = 1\}.$$

    (ii) $\sum_s \varphi(s) = n$, where the summation is over all divisors s of n and φ is Euler's function (see **5**).

    (iii) Let G be a group of order n such that, for each divisor s of n, G has at most one subgroup of order s. Then G is cyclic. (See also 9.15(i). Hint. Use (ii) and **5** to show that G must have an element of order n.)

*140 Let G be a non-trivial group. A proper subgroup M of G is said to be a *maximal subgroup of G* if there is no subgroup L such that $M < L < G$.

    (i) Let $K \leqslant M < G$, with $K \trianglelefteq G$. Then M/K is a maximal subgroup of G/K if and only if M is a maximal subgroup of G.

    (ii) If G is finite then every proper subgroup of G is contained in a maximal subgroup of G.

    (iii) Every proper subgroup of $\mathbf{Z}^+$ is contained in a maximal subgroup of $\mathbf{Z}^+$. (Hint. See 3.25 and use (i) and (ii).)

    (iv) $\mathbf{Q}^+$ has no maximal subgroup. (Hint. Use (i), **29**(i) and **133**(i).)

    (v) If $M < G$ and $|G : M| = p$ for some prime p, then M is a maximal subgroup of G (see **11**).

    (vi) Suppose that G is finite. Then G has a unique maximal subgroup if and only if G is cyclic and of order p^m for some prime p and positive integer m.

141 (i) If G is a non-abelian group, Aut G cannot be cyclic. (Hint. See **117** and **125**.)

(ii) There is no finite group G for which Aut G is cyclic of odd order greater than 1. (Hint. See **42** and **52**. Remark. The condition of finiteness of G is actually superfluous here.)

142 Let $G = D_\infty$, the infinite dihedral group (see **57**).

(i) G has just one cyclic subgroup H of index 2 in G, and every element of $G\backslash H$ has order 2 (cf. **59**).

(ii) Let $1 < K \leqslant H$. We know, by 3.25, that H/K is finite, say of order n. Then $K \trianglelefteq G$ and $G/K \cong D_{2n}$ if $n \geqslant 3$, $C_2 \times C_2$ if $n = 2$, C_2 if $n = 1$.

(iii) Let $J \leqslant G$ with $J \not\leqslant H$. Then $|J : H \cap J| = 2$; and if $H \cap J \neq 1$ then $J \cong D_\infty$.

(iv) Say $H = \langle h \rangle$. By 3.25, the non-trivial subgroups of H are just the subgroups $H_n = \langle h^n \rangle$, one for each positive integer n. Then there are just n distinct subgroups J of G such that $J \not\leqslant H$ and $H \cap J = H_n$. Also $J \trianglelefteq G$ if and only if $n \leqslant 2$.

(v) The proper normal subgroups of G are just the subgroups of H and two subgroups of index 2 in G, both isomorphic to D_∞.

(vi) Every non-trivial subgroup of G is isomorphic to D_∞ or C_∞ or C_2, and every proper non-trivial quotient of G is isomorphic to D_{2n} for some integer $n \geqslant 3$ or to $C_2 \times C_2$ or to C_2.

(vii) By means of 3.30, or otherwise, classify, for each integer $n \geqslant 3$, all subgroups, normal subgroups and quotient groups of D_{2n}.

143 Let n be an integer, $n \geqslant 3$.

(i) If $n = 2m$ for some odd integer m then $D_{2n} \cong D_n \times C_2$.

(ii) If n is not twice an odd integer then D_{2n} is indecomposable (see **81**).

*144 For each prime p, let

$$C_{p^\infty} = \{ z \in \mathbf{C}^\times : \text{there is a positive integer } n \text{ such that } z^{p^n} = 1 \}.$$

Then $C_{p^\infty} < V$, the multiplicative group of all complex roots of 1, and

$$1 < C_p < C_{p^2} < C_{p^3} < \ldots < C_{p^\infty} = \bigcup_{n=0}^{\infty} C_{p^n}.$$

Prove that every proper subgroup of C_{p^∞} is C_{p^n} for some non-negative integer n, and that $C_{p^\infty}/C_{p^n} \cong C_{p^\infty}$. Thus C_{p^∞} is an infinite group every proper subgroup of which is finite and every non-trivial quotient of which is isomorphic to C_{p^∞}. (Compare with the properties of $\mathbf{Z}^+$ in 3.25.) Show that C_{p^∞} has no maximal subgroup (see **140**).

The groups C_{p^∞}, one for each p, are called *quasi-cyclic* (or *Prüfer*) groups.

We have seen that any subgroup of a 1-generator group is a 1-generator group (3.32) and also that for any positive integer n, any quotient group of an n-generator group is an n-generator group (**108**). We shall show that in general subgroups of n-generator groups need not be n-generator groups. A group is said to be *finitely generated* if it has a finite set of generators. We shall construct a 2-generator group with a subgroup which is not even finitely generated.

Before doing this we note a curious property of $\mathbf{Q}^+$.

3.33. *Every finitely generated subgroup of* $\mathbf{Q}^+$ *is cyclic. Hence* $\mathbf{Q}^+$ *is not finitely generated.*

Proof. Let X be a non-empty finite subset of $\mathbf{Q}^+$: say

$$X = \left\{ \frac{a_1}{b_1}, \frac{a_2}{b_2}, \dots, \frac{a_r}{b_r} \right\},$$

where $a_1, \dots, a_r, b_1, \dots, b_r$ are integers, and we may suppose that $b_1, \dots, b_r$ are positive. By 2.28 and the fact that $\mathbf{Q}^+$ is abelian (see **69**),

$$\langle X \rangle = \left\{ n_1 \frac{a_1}{b_1} + n_2 \frac{a_2}{b_2} + \dots + n_r \frac{a_r}{b_r} : n_1, \dots, n_r \in \mathbf{Z} \right\}.$$

Hence every element of $\langle X \rangle$ is a rational number of the form $a/b_1 b_2 \dots b_r$, where $a \in \mathbf{Z}$. Thus $\langle X \rangle \leqslant \langle 1/b_1 b_2 \dots b_r \rangle$, a cyclic subgroup of $\mathbf{Q}^+$. Hence, by 3.32, $\langle X \rangle$ is a cyclic subgroup of $\mathbf{Q}^+$. Since $\mathbf{Q}^+ \not\cong \mathbf{Z}^+$ (**32**), $\mathbf{Q}^+$ is not cyclic (**29**) and hence not finitely generated.

3.34. *Let* $H_1, H_2, H_3, \dots$ *be a sequence of subgroups of G such that* $H_1 \leqslant H_2 \leqslant H_3 \leqslant \dots$. *This is called an* ascending sequence *of subgroups. Let* $H = \bigcup\limits_{i=1}^{\infty} H_i$. *Then* $H \leqslant G$.

Proof. Certainly $H \neq \emptyset$. Let $h_1, h_2 \in H$. Then there are positive integers i_1, i_2 such that $h_1 \in H_{i_1}, h_2 \in H_{i_2}$. Let $j = \max \{i_1, i_2\}$. Then $H_{i_1} \leqslant H_j$ and $H_{i_2} \leqslant H_j$, and so $h_1, h_2 \in H_j$. Since $H_j \leqslant G$, $h_1 h_2^{-1} \in H_j$, so that $h_1 h_2^{-1} \in H$. Hence $H \leqslant G$.

3.35. We may note that $\mathbf{Q}^+$ is the union of an ascending sequence of cyclic subgroups (necessarily proper subgroups, since $\mathbf{Q}^+$ is not cyclic). To see this, for each positive integer n let $H_n = \langle 1/n! \rangle \leqslant \mathbf{Q}^+$. Then $\mathbf{Z}^+ = H_1 < H_2 < H_3 < \dots$ and $\bigcup\limits_{n=1}^{\infty} H_n = \mathbf{Q}^+$, since any rational number is expressible as a/b with a, b integers and $b > 0$, and then

$$\frac{a}{b} = (b-1)! \, a \cdot \frac{1}{b!} \in H_b.$$

For another example, note that the group C_{p^∞} defined in **144** is the union of an ascending sequence of cyclic proper subgroups.

Clearly any group which is the union of an ascending sequence of proper subgroups must be infinite. In fact, we can prove

3.36. *A finitely generated group cannot be the union of an ascending sequence of proper subgroups.*

Proof. Suppose that $G = \langle X \rangle$, where X is a non-empty finite subset of G. Assume that $H_1 \leqslant H_2 \leqslant H_3 \leqslant \dots$ is an ascending sequence of subgroups of G such that $H_i < G$ for every positive integer i and $\bigcup\limits_{i=1}^{\infty} H_i = G$. Then, for each $x \in X$, there is a positive integer $i(x)$ such that $x \in H_{i(x)}$. Let $j =$

$\max\{i(x) : x \in X\}$. Then $x \in H_j$ for all $x \in X$, and so

$$G = \langle X \rangle \leqslant H_j < G,$$

a contradiction.

This provides another proof that $\mathbf{Q}^+$ is not finitely generated. Now we show

3.37. *A 2-generator group can have a subgroup which is not finitely generated.*

Proof. The 2-generator group G will be defined as a subgroup of the symmetric group $\Sigma_{\mathbf{R}}$. Let maps $x, y : \mathbf{R} \to \mathbf{R}$ be defined by

$$x : a \mapsto a + 1, \quad y : a \mapsto 2a, \quad \text{for all } a \in \mathbf{R}.$$

Then x and y are clearly permutations of $\mathbf{R}$. Let

$$G = \langle x, y \rangle \leqslant \Sigma_{\mathbf{R}}.$$

For each positive integer n,

$$y^n : a \mapsto 2^n a \quad \text{and} \quad y^{-n} : a \mapsto 2^{-n}a, \quad \text{for all } a \in \mathbf{R}.$$

Let $x_n = y^n x y^{-n} \in G$. Then one verifies that

$$x_n : a \mapsto a + 2^{-n}, \quad \text{for all } a \in \mathbf{R},$$

hence that for $n > 1$,

$$x_n^2 = x_{n-1}.$$

Now let $H_n = \langle x_n \rangle \leqslant G$. Then whenever $n > 1, H_{n-1} \leqslant H_n$, and in fact $H_{n-1} < H_n$, because $x_n \notin H_{n-1}$. Thus $H_1 < H_2 < H_3 < \dots \leqslant G$. Now let $H = \bigcup_{n=1}^{\infty} H_n \leqslant G$, by 3.34. Each H_n is a *proper* subgroup of H, because $H_n < H_{n+1} \leqslant H$. Hence, by 3.36, H is not finitely generated, although G is a 2-generator group.

Remark. For the reader familiar with infinite cardinals, we mention that it is easy to prove that every finitely generated group is countable although, since $\mathbf{Q}^+$ is countable (or by 3.37), not every countable group is finitely generated. G. Higman, B. H. Neumann and H. Neumann [a58] proved in 1949 that any countable group can be embedded in a 2-generator group. A proof of this result is contained in Rotman [b34] p. 275.

145 G is said to be *locally cyclic* if every finitely generated subgroup of G is cyclic. Thus, by 3.33, $\mathbf{Q}^+$ is locally cyclic.

(i) A locally cyclic group is abelian.

(ii) A group which is the union of an ascending sequence of locally cyclic subgroups is locally cyclic.

(Note. In fact, a group is locally cyclic if and only if it is isomorphic to a subgroup of a quotient group of $\mathbf{Q}^+$. A proof of this result appears in Schenkman [b35] §II.2.)

146 Let G and H be defined as in 3.37. Then $H \trianglelefteq G, H$ is locally cyclic (see **145**), and G/H is cyclic. (Hints. Note that in order to prove that $H \trianglelefteq G$, it is enough by 2.28 to show that for each $h \in H$, the elements $x^{-1}hx, xhx^{-1}, y^{-1}hy, yhy^{-1}$ all belong to H. To prove that G/H is cyclic, apply **108**.)

147 Suppose that $H_1, H_2, H_3, \ldots$ is an ascending sequence of subgroups of G such that, for every positive integer $n, H_n \cong C_{p^n}$. Then $\bigcup\limits_{n=1}^{\infty} H_n \cong C_{p^\infty}$. (See **144**. Hint. By induction on n, show that for each positive integer n there is an element $x_n \in H_n$ such that $H_n = \langle x_n \rangle$ and $x_n = x_{n+1}^p$.)

148 Let N denote the set of all positive integers and let $G = \Sigma_{(N)}$, the restricted symmetric group on N (see **110**). For each $n \in N$, let

$$G_n = \{\sigma \in G : j\sigma = j \text{ for every } j \in N \text{ with } j > n\}.$$

Then

(i) For every $n \in N, G_n < G$ and $G_n \cong \Sigma_n$.

(ii) $G_1, G_2, G_3, \ldots$ is an ascending sequence of subgroups of G and $\bigcup\limits_{n=1}^{\infty} G_n = G$.

(iii) G contains cyclic subgroups of order n for every $n \in N$; but G does not contain a subgroup isomorphic to C_{p^∞} for any prime p. (cf. **147**. Hint. Suppose to the contrary that G has a subgroup H isomorphic to C_{p^∞}. Use **22** and **23** to derive a contradiction by showing that for every positive integer n, the unique subgroup of H of order p contains an element σ such that $|s(\sigma)| \geqslant p^n$.)

149 Let G be a finitely generated group. Then every subgroup of finite index in G is finitely generated. (cf. 3.37. Hints. Let $\{x_1, \ldots, x_n\}$ be a set of generators of G and let $H \leqslant G$ with $|G : H| = m$, where n and m are positive integers. For each $j = 1, \ldots, n$, let $x_{n+j} = x_j^{-1}$; and let $g_1, \ldots, g_m$ be elements of G such that $G = \bigcup\limits_{i=1}^{m} Hg_i$, with $g_1 = 1$. Then, for each ordered pair (i,j) with $i \in \{1, \ldots, m\}$ and $j \in \{1, 2, \ldots, 2n\}$, there is a unique element $h_{ij} \in H$ and a unique integer $k \in \{1, \ldots, m\}$ such that $g_i x_j = h_{ij}g_k$. Let $h \in H$, and note that $h = g_1 h$. Use **68** to show that $h \in \langle h_{ij} : i = 1, \ldots, m; j = 1, \ldots, 2n \rangle$. Remark. It follows from a deeper result of O. Schreier [a87] that if H is a subgroup of index m in an n-generator group G, where m and n are positive integers, then H is a $(1 + m(n-1))$-generator group; and this bound on the number of generators of H is the best possible.)

We have observed (**95**) that if H and K are subgroups of G, HK need not be a subgroup of G. Now we note

3.38. If $H \leqslant G$ and $K \trianglelefteq G$ then $HK \leqslant G$.
Proof. By definition, $HK = \{hk : h \in H, k \in K\} \neq \emptyset$.
Let $h_1, h_2 \in H$ and $k_1, k_2 \in K$. Then

$$(h_1 k_1)(h_2 k_2)^{-1} = h_1 k_1 k_2^{-1} h_2^{-1} = (h_1 h_2^{-1})(h_2 k_1 k_2^{-1} h_2^{-1}) \in HK,$$

since $h_1 h_2^{-1} \in H, k_1 k_2^{-1} \in K, h_2 = (h_2^{-1})^{-1}$ and $K \trianglelefteq G$. Hence $HK \leqslant G$.

3.39 Corollary. If $H \trianglelefteq G$ and $K \trianglelefteq G$ then $HK \trianglelefteq G$.
Proof. By 3.38, $HK \leqslant G$. Let $g \in G, h \in H, k \in K$.

Then $g^{-1}hkg = (g^{-1}hg)(g^{-1}kg) \in HK$. Hence $HK \trianglelefteq G$.

The following result is very useful.

3.40 Another Isomorphism Theorem. *Let* $H \leqslant G$ *and* $K \trianglelefteq G$. *Then* $H \cap K \trianglelefteq H$ *and* $H/H \cap K \cong HK/K$.

Proof. Let ν be the natural homomorphism $G \to G/K$, and let ν_1 be the restriction of ν to H. Then $\nu_1 : H \to G/K$ is a homomorphism, with $\operatorname{Ker} \nu_1 = \{h \in H : h \in \operatorname{Ker} \nu\} = H \cap K$. Hence, by the fundamental theorem,

$$H \cap K \trianglelefteq H \quad \text{and} \quad H/H \cap K \cong \operatorname{Im} \nu_1.$$

By 3.38, $K \trianglelefteq HK \leqslant G$. For each $h \in H, h\nu_1 = hK \in HK/K$. Moreover, each element of HK/K is of the form $hkK = hK = h\nu_1$, with $h \in H, k \in K$. Thus $\operatorname{Im} \nu_1 = HK/K$.

150 Suppose that $K \trianglelefteq G$. Let $\bar{G} = G/K$ and $\overline{Z(G)} = Z(G)K/K$. Show that $\overline{Z(G)} \leqslant Z(\bar{G})$. Show by an example that we can have $Z(G) < Z(\bar{G})$.

*151 Let H, J and K be normal subgroups of G such that $J \leqslant H$. Prove that if $H/J \leqslant Z(G/J)$ then $HK/JK \leqslant Z(G/JK)$ (see 3.39).

*152 G is called *metacyclic* if it has a cyclic normal subgroup L such that G/L is cyclic. For instance, every dihedral group is metacyclic: see **79, 102, 142**. Prove that every subgroup and every quotient group of a metacyclic group is also metacyclic.

153 (Hölder) Let $K \trianglelefteq G$ with G finite and K simple. If $|K|^2$ does not divide $|G|$ then K is the only subgroup of G which is isomorphic to K.

154 There is no proper subgroup H of $\mathbf{Q}^+$ such that $\mathbf{Q}^+ = H + \mathbf{Z}^+$. (Hint. Use 3.40 with 3.25 and 133. Here, since the group operation in $\mathbf{Q}^+$ is addition, we also use additive notation for the corresponding semigroup $\mathscr{P}(\mathbf{Q}^+)$ of non-empty subsets of $\mathbf{Q}^+$.)

155 Let $H_1, H_2, H_3, \ldots$ be an ascending sequence of subgroups of G, and let $H = \bigcup_{i=1}^{\infty} \leqslant G$ (see 3.34). If H_i is simple for infinitely many distinct positive integers i, then H is simple. (Hint. Show that if $K \trianglelefteq H$ then either $H_i \cap K = 1$ whenever H_i is simple or $H_i \leqslant K$ whenever H_i is simple.)

3.41 Definitions. Recall our convention that ϖ always denotes a set of prime numbers.

(i) A positive integer n is said to be a ϖ-*number* if every prime divisor of n belongs to ϖ. Note that we do not require that every prime in ϖ actually divides n: so, for instance, 6 is a $\{2, 3, 5\}$-number. By convention, 1 is a ϖ-number for every set ϖ of primes (and if $\varpi = \varnothing$, 1 is the only ϖ-*number*). If $\varpi = \{p\}$, the ϖ-numbers are just the powers of $p : 1, p, p^2, p^3, \ldots$

(ii) Let G be a finite group. We say that G is a ϖ-*group* if $|G|$ is a ϖ-number. Thus, for example, Σ_3 and Σ_4 are $\{2, 3\}$-groups but Σ_5 is not a $\{2, 3\}$-group. However, Σ_3, Σ_4 and Σ_5 are all $\{2, 3, 5\}$-groups. If $\varpi = \{p\}$, a ϖ-group is called a *p-group* (rather than a $\{p\}$-group).

3.42. *If G is a finite ϖ-group then all subgroups and all quotient groups of G are ϖ-groups.*

Proof. The orders of all subgroups and all quotient groups of G divide $|G|$.

3.43. *Let G be a finite group. Then G has a unique largest normal ϖ-subgroup, which is denoted by $O_\varpi(G)$ and called the ϖ-radical of G.* (Here $O_\varpi(G)$ is 'largest' in the sense that it contains every normal ϖ-subgroup of G.)

Proof. Among all the normal subgroups of G, choose a normal ϖ-subgroup K of largest order. (Possibly $K = 1$.) Then let H be any normal ϖ-subgroup of G. By 3.39, $HK \trianglelefteq G$, and, by 3.40, $|HK/K| = |H/H \cap K|$, which, by 3.42, is a ϖ-number. Hence $|HK| = |HK/K||K|$, the product of two ϖ-numbers and therefore a ϖ-number. Thus HK is a normal ϖ-subgroup of G, and so, by choice of K and since $K \leqslant HK$, $HK = K$. Hence $H \leqslant K$. Thus K has the required property of $O_\varpi(G)$.

For example, by 3.7,

$$O_2(\Sigma_3) = 1 \quad \text{and} \quad O_3(\Sigma_3) = \{1, (123), (132)\}.$$

3.44. *Let G be a finite group. Then G has a unique smallest normal subgroup K such that G/K is a ϖ-group. We write $K = O^\varpi(G)$ and call $G/O^\varpi(G)$ the ϖ-residual of G.* (Here $O^\varpi(G)$ is 'smallest' in the sense that whenever $H \trianglelefteq G$ and G/H is a ϖ-group then $O^\varpi(G) \leqslant H$.)

Proof. Among all the normal subgroups of G, choose K of smallest order such that G/K is a ϖ-group. (Possibly $K = G$.) Then let $H \trianglelefteq G$ with G/H a ϖ-group. By 3.8, $H \cap K \trianglelefteq G$, and, by 3.40, $|H/H \cap K| = |HK/K|$, which, by 3.42, is a ϖ-number, since $HK/K \leqslant G/K$. Hence, by 3.30, $|G/H \cap K| = |G/H||H/H \cap K|$, the product of two ϖ-numbers and therefore a ϖ-number. Since $H \cap K \leqslant K$ and by choice of K, it follows that $H \cap K = K$. Hence $K \leqslant H$. Thus K has the required property of $O^\varpi(G)$.

For example, by 3.7,

$$O^2(\Sigma_3) = \{1, (123), (132)\} \quad \text{and} \quad O^3(\Sigma_3) = \Sigma_3.$$

3.45 Definitions. It is often convenient to refer to the *class* of all groups possessing a particular property. For example, we have the class of abelian groups, the class of finite groups, the class of finite ϖ-groups, etc. When we speak of a class $\mathfrak{X}$ of groups we always understand that if $G \cong H \in \mathfrak{X}$ then also $G \in \mathfrak{X}$: in words, that if $\mathfrak{X}$ contains a particular group, $\mathfrak{X}$ also contains all groups of the same type. We suppose also that $\mathfrak{X}$ contains the trivial group (of order 1).

For a particular group G and class $\mathfrak{X}$ of groups, we may ask whether G has an $\mathfrak{X}$-*radical* and an $\mathfrak{X}$-*residual*: that is, whether G has a unique largest normal $\mathfrak{X}$-subgroup H (in which case H is the $\mathfrak{X}$-radical of G) and whether G has a unique smallest normal subgroup K such that

G/K is an $\mathfrak{X}$-group (in which case G/K is the $\mathfrak{X}$-residual of G). Here, as usual in group theory, 'largest' and 'smallest' are meant in the sense of containment. Thus 3.43 and 3.44 assert the existence for every finite group of an $\mathfrak{X}$-radical and an $\mathfrak{X}$-residual when $\mathfrak{X}$ is the class of finite ϖ-groups.

*156 Let G be a finite group. Then $O_\varpi(G)$ and $O^\varpi(G)$ are characteristic subgroups of G and $G/O_\varpi(G)$ has trivial ϖ-radical and $O^\varpi(G)$ has trivial ϖ-residual.

*157 Let $H \leqslant G$, a finite group.
  (i) Prove that $H \cap O_\varpi(G) \leqslant O_\varpi(H)$; and that if $H \trianglelefteq G$ then $H \cap O_\varpi(G) = O_\varpi(H)$.
  Show by an example that if H is not normal in G, we can have $H \cap O_\varpi(G) < O_\varpi(H)$.
  (ii) Prove that $O^\varpi(H) \leqslant H \cap O^\varpi(G)$.
  Show by an example that even if H is normal in G, we can have $O^\varpi(H) < H \cap O^\varpi(G)$.

158 Let $K \trianglelefteq G$, a finite group. If G/K is a ϖ-group then $O^\varpi(K) = O^\varpi(G)$ (cf. **157**(ii)).

159 Let H and K be normal subgroups of a finite group G.
  (i) Prove that $O_\varpi(H)O_\varpi(K) \leqslant O_\varpi(HK)$.
  Show by an example that equality need not hold. (Hint. Consider $G = \Sigma_3 \times C_2$ and $\varpi = \{2\}$: show that G has two distinct normal subgroups isomorphic to Σ_3 and choose these for H and K.)
  (ii) Prove that $O^\varpi(H)O^\varpi(K) = O^\varpi(HK)$.

*160 Let $\mathfrak{X}$ be a class of groups.
  (i) If G has an $\mathfrak{X}$-radical H then H is a characteristic subgroup of G.
  (ii) If G has an $\mathfrak{X}$-residual G/K then K is a characteristic subgroup of G (cf. **156**).

161 Let $\mathfrak{X}$ and $\mathfrak{Y}$ be classes of groups. We say that G is an $\mathfrak{X}$-by-$\mathfrak{Y}$ group if G has a normal subgroup L such that $L \in \mathfrak{X}$ and $G/L \in \mathfrak{Y}$. Prove that if every subgroup and every quotient group of an $\mathfrak{X}$-group is an $\mathfrak{X}$-group and if every subgroup and every quotient group of a $\mathfrak{Y}$-group is a $\mathfrak{Y}$-group then every subgroup and every quotient group of an $\mathfrak{X}$-by-$\mathfrak{Y}$ group is an $\mathfrak{X}$-by-$\mathfrak{Y}$ group. (This generalizes **152**.)

We shall show next that *every* group has an abelian residual (though in general not an abelian radical – see **171**). With this aim in mind, we make the following

3.46 Definition. The *commutator* of an ordered pair g_1, g_2 of elements of G is the element
$$[g_1, g_2] = g_1^{-1}g_2^{-1}g_1g_2 \in G.$$

Immediately from the definition we note

3.47. Let $g_1, g_2 \in G$. Then
  (i) $[g_2, g_1] = [g_1, g_2]^{-1}$, and
  (ii) $[g_1, g_2] = 1$ if and only if g_1 and g_2 commute.

3.48 Definitions. Let $H, K \leqslant G$. Then the corresponding *commutator subgroup* is
$$[H, K] = \langle [h, k] : h \in H, k \in K \rangle \leqslant G.$$

We emphasize that $[H, K]$ is the subgroup *generated* by all the commutators $[h, k]$ with $h \in H$ and $k \in K$: it may happen that the product of two commutators cannot itself be expressed as a commutator. The particular subgroup $[G, G]$, generated by all commutators in G, is usually denoted by G' and called the *derived group* (or *commutator subgroup*) of G.

3.49. *Let $H, K \leqslant G$. Then $[H, K] = [K, H]$.*
Proof. By 2.28, if X is a non-empty subset of G and $Y = \{x^{-1} : x \in X\}$ then $\langle X \rangle = \langle Y \rangle$. The assertion now follows from 3.47 (i).

3.50. *Let X be a non-empty subset of G and let A be a non-empty subset of* Aut G. *Suppose that $x^{\alpha} \in \langle X \rangle$ for all $x \in X$ and $\alpha \in A$. Then $\langle X \rangle$ is an A-invariant subgroup of G.*
Proof. Let $y \in \langle X \rangle$. By 2.28, y can be expressed in the form

$$y = x_1^{n_1} x_2^{n_2} \ldots x_r^{n_r},$$

where r is a positive integer, and each $x_i \in X, n_i \in \mathbf{Z}$. Let $\alpha \in A$. Then

$$y^{\alpha} = (x_1^{\alpha})^{n_1} (x_2^{\alpha})^{n_2} \ldots (x_r^{\alpha})^{n_r}.$$

Since, by hypothesis, each $x_i^{\alpha} \in \langle X \rangle \leqslant G$, it follows that $y^{\alpha} \in \langle X \rangle$. Hence $\langle X \rangle$ is A-invariant.

From this we deduce

3.51. *Let A be a non-empty subset of* Aut G *and let H and K be A-invariant subgroups of G. Then $[H, K]$ is an A-invariant subgroup of G. In particular, the derived group G' is a characteristic subgroup of G.*
Proof. Let $h \in H, k \in K, \alpha \in A$. Then

$$[h, k]^{\alpha} = (h^{-1})^{\alpha}(k^{-1})^{\alpha}h^{\alpha}k^{\alpha} = (h^{\alpha})^{-1}(k^{\alpha})^{-1}h^{\alpha}k^{\alpha} = [h^{\alpha}, k^{\alpha}].$$

Since, by hypothesis, $h^{\alpha} \in H$ and $k^{\alpha} \in K, [h, k]^{\alpha} \in [H, K]$. The result now follows from 3.50.

3.52 Theorem. *For any group G, the derived group G' is the unique smallest normal subgroup K of G such that G/K is abelian. (Thus G/G' is the abelian residual of G, sometimes called G made abelian.)*
Proof. Let $K \trianglelefteq G$. Then G/K is abelian if and only if $(xK)(yK) = (yK)(xK)$ for all $x, y \in G$; that is, if and only if $xyK = yxK$, or, equivalently, $x^{-1}y^{-1}xyK = K$; that is, if and only if $[x, y] \in K$ for all $x, y \in G$. Thus G/K is abelian if and only if $G' \leqslant K$. Since, by 3.51, $G' \trianglelefteq G$, this completes the proof.

3.53 Lemma. *Let $H \trianglelefteq G$ and $K \trianglelefteq G$. Then $[H, K] \leqslant H \cap K$. In particular, if $H \cap K = 1$ then every element of H commutes with every element of K.*
Proof. Let $h \in H$ and $k \in K$. Then

$$[h,k] = h^{-1}(k^{-1}hk) \in H, \quad \text{since } H \trianglelefteq G,$$

and
$$[h,k] = (h^{-1}k^{-1}h)k \in K, \quad \text{since } K \trianglelefteq G.$$

Thus
$$[h,k] \in H \cap K \text{ for all } h \in H, k \in K.$$

Hence
$$[H,K] \leqslant H \cap K.$$

The second assertion now follows from 3.47(ii).

We now note a useful reformulation of 2.34 (and its converse).

3.54 Theorem. $G \cong H \times K$ *if and only if G has normal subgroups H_1, K_2 such that $H_1 \cong H, K_2 \cong K, G = H_1 K_2$ and $H_1 \cap K_2 = 1$.*
Proof. If φ is an isomorphism of $H \times K$ onto G, let $H_1 = (H \times 1)\varphi, K_2 = (1 \times K)\varphi$ and apply 2.33, 3.12 and 3.29. For the converse, suppose that $H \cong H_1 \trianglelefteq G, K \cong K_2 \trianglelefteq G, G = H_1 K_2$ and $H_1 \cap K_2 = 1$. By 3.53, every element of H_1 commutes with every element of K_2. Then, by 2.34 and 76, $G \cong H_1 \times K_2 \cong H \times K$.

*162 Let $K \leqslant H \leqslant G$ with $K \trianglelefteq G$. Then
  (i) $H \trianglelefteq G$ if and only if $[H,G] \leqslant H$,
  (ii) $H/K \leqslant Z(G/K)$ if and only if $[H,G] \leqslant K$.

163 Every subgroup of G which contains G' is normal in G.

*164 Let $K \trianglelefteq G$.
  (i) If $x, y \in G$ then, in G/K,
$$[xK, yK] = [x,y]K.$$
  (ii) If $H, J \leqslant G$ then
$$[HK/K, JK/K] = [H,J]K/K.$$
In particular, $(G/K)' = G'K/K$ (cf. 150).

*165 Let $G = H \times K$.
  (i) If $h_1, h_2 \in H$ and $k_1, k_2 \in K$ then
$$[(h_1,k_1),(h_2,k_2)] = ([h_1,h_2],[k_1,k_2]).$$
  (ii) If H_1, H_2 are subgroups of H and K_1, K_2 are subgroups of K then
$$[H_1 \times K_1, H_2 \times K_2] = [H_1, H_2] \times [K_1, K_2].$$
In particular,
$$G' = H' \times K'.$$

166 Let n be an integer, $n \geqslant 3$, and let $G = D_{2n}$, the dihedral group of order $2n$. Then $|G/G'|$ is either 4 or 2, according as n is even or odd.

167 Let A be an abelian group.
  (i) For any homomorphism $\varphi : G \to A, G' \leqslant \operatorname{Ker} \varphi$.
  (ii) $\operatorname{Hom}(G,A) \cong \operatorname{Hom}(G/G', A)$ (see 33).

*168 A group is said to be *perfect* if it coincides with its derived group, or, equivalently, if it has no non-trivial abelian quotient group. Prove that every perfect subgroup H of an arbitrary group G is contained in G'.

*169 (i) Let $x, y, z \in G$. Then

$$[xy, z] = y^{-1}[x, z]y[y, z].$$

(ii) Let H, J, K be normal subgroups of G. Then

$$[HJ, K] = [H, K][J, K].$$

170 (i) Let $Z_1 = Z(G)$ and define $Z_2 \trianglelefteq G$ by $Z_2/Z_1 = Z(G/Z_1)$. Let $z \in Z_2$, and let θ_z be the map of G into itself defined, for all $x \in G$, by

$$\theta_z : x \mapsto [x, z].$$

Prove that θ_z is a homomorphism, that $\operatorname{Im} \theta_z \leqslant Z_1$, and that $G' \leqslant \operatorname{Ker} \theta_z$. (Hint. See **162** and **169**.)

(ii) Suppose that G is a perfect group (**168**). Prove that $G/Z(G)$ has trivial centre. (Remarks. This result is known as Grün's lemma. A perfect group need not itself have trivial centre. For instance, it can be shown that for any field F with more than 3 elements, the group $SL_2(F)$ defined in 3.28 is perfect; and if $1 + 1 \neq 0$ in F then the centre of $SL_2(F)$ has order 2: see **123**.)

*171 A group need not have an abelian radical. Demonstrate this by considering $G = D_8$, the dihedral group of order 8. Find abelian normal subgroups H and K of G such that $HK = G$.

We now show that to each subgroup H of G there is associated a unique largest subgroup L of G such that $H \trianglelefteq L \leqslant G$.

3.55. Let $H \leqslant G$ and define $N_G(H) = \{g \in G : g^{-1}Hg = H\}$. Then $H \trianglelefteq N_G(H) \leqslant G$, and, whenever $H \trianglelefteq J \leqslant G, J \leqslant N_G(H)$. We call $N_G(H)$ the normalizer of H in G.

Proof. Let $L = N_G(H)$, as defined above. Certainly $H \subseteq L$, so that $L \neq \emptyset$. Let $x, y \in L$. Since $y^{-1}Hy = H$, it follows that $H = yHy^{-1}$, and so

$$(xy^{-1})^{-1}Hxy^{-1} = y(x^{-1}Hx)y^{-1} = yHy^{-1} = H.$$

Hence $xy^{-1} \in L$. Therefore $L \leqslant G$. Now $H \leqslant L$, and, immediately from the definition of $L, H \trianglelefteq L$. Finally, if $H \trianglelefteq J \leqslant G$ and $x \in J$ then $x \in G$ and, by 3.3, $x^{-1}Hx = H$. Hence, by definition of $L, x \in L$. Thus $J \leqslant L$.

Note that by 3.3 we have

3.56. Let $H \leqslant G$. Then $H \trianglelefteq G$ if and only if $N_G(H) = G$.

We can now solve a problem mentioned earlier in this chapter.

3.57. Let G be a finite group. Then the subgroups of the semigroup $\mathcal{Q}(G)$ are precisely the groups H/K, where $K \trianglelefteq H \leqslant G$.

Proof. Whenever $K \trianglelefteq H \leqslant G$, the quotient H/K is a group whose elements are non-empty subsets of G and with the same multiplication as in $\mathcal{Q}(G)$; that is, H/K is a subgroup of $\mathcal{Q}(G)$.

Now let $\mathcal{G}$ be any subgroup of $\mathcal{Q}(G)$. Then the identity element of $\mathcal{G}$ is a non-empty subset K of G such that $K^2 = K$. Since G is finite, it follows, by 3.18, that $K \leqslant G$. Now if $X \in \mathcal{G}$ then there is a $Y \in \mathcal{G}$ such that $XY = K$. Also $XK = X = KX$. Hence $|X| \leqslant |K|$ and $|K| \leqslant |X|$. Therefore $|X| = |K|$. Let $x \in X$. Then $xK \subseteq XK = X$. Since $|xK| = |K| = |X| < \infty$, it follows that $xK = X$. Similarly, $Kx = X$.

Thus every element of $\mathcal{G}$ is a coset xK of K in G such that $xK = Kx$, hence such that $K = x^{-1}Kx$; that is, such that $x \in N_G(K)$. Thus the elements of $\mathcal{G}$ are elements of $N_G(K)/K$, and the multiplication in $\mathcal{G}$ is the same as in $N_G(K)/K$; that is, $\mathcal{G}$ is a subgroup of $N_G(K)/K$. Hence, by 3.30, $\mathcal{G} = H/K$, where $K \leqslant H \leqslant N_G(K)$; equivalently, by 3.55, where $K \trianglelefteq H \leqslant G$.

172 Let $G = \Sigma_3$ and $H = \{1, (12)\}$. What is $N_G(H)$?

173 Let $H \leqslant G$ and $g \in G$. Then $N_G(g^{-1}Hg) = g^{-1}N_G(H)g$.

*174 Let $H \leqslant G$. Prove that if H is finite then

$$N_G(H) = \{g \in G : g^{-1}hg \in H \text{ for all } h \in H\}.$$

Show by an example that this may fail if H is infinite. (Hints. See **84**. For an example, let $G = \langle x, y \rangle$ be defined as in 3.37, and let $H_0 = \langle x \rangle$. Then $y^{-1}hy \in H_0$ for all $h \in H_0$ but $y \notin N_G(H_0)$.)

*175 Let $H, J \leqslant G$. If $x^{-1}Hx \leqslant H$ for every $x \in J$, then $J \leqslant N_G(H)$. (cf. **174**. Hint. $\{x^{-1} : x \in J\} = J$.)

176 Let $\emptyset \subset A \subseteq \operatorname{Aut} G$. If H is a finite A-invariant subgroup of G then $C_G(H)$ and $N_G(H)$ are A-invariant subgroups of G (see **84, 122, 174**).

177 Let $\hat{G} = \{(g, g) : g \in G\}$, the diagonal subgroup of $G \times G$ (see **89**). Then

$$N_{G \times G}(\hat{G}) = \hat{G} \quad \text{if and only if } Z(G) = 1.$$

*178 Let $H \leqslant G$, and suppose that X and Y are non-empty subsets of G such that $\langle X \rangle = G$ and $\langle Y \rangle = H$.
 (i) If $x^{-1}Hx = H$ for all $x \in X$ then $H \trianglelefteq G$.
(Hint. Use the fact that $N_G(H) \leqslant G$.)
 (ii) If $g^{-1}yg \in H$ for all $g \in G$ and $y \in Y$, then $H \trianglelefteq G$.
(Apply 3.50.)
 (iii) If $x^{-1}yx \in H$ for all $x \in X$ and $y \in Y$ then $x^{-1}Hx \leqslant H$ for all $x \in X$. Deduce that if H is finite then $H \trianglelefteq G$.
 (iv) The second assertion of (iii) may fail without the condition of finiteness of H. (To see this, consider the group $G = \langle x, y \rangle$ of 3.37. Let $H_0 = \langle x \rangle$. Then $x^{-1}xx = x \in H_0$ and $y^{-1}xy = x^2 \in H_0$, but $H_0 \ntrianglelefteq G$ since $yxy^{-1} \notin H_0$.)
 (v) If $x^{-1}yx \in H$ and $xyx^{-1} \in H$ for all $x \in X$ and $y \in Y$, then $H \trianglelefteq G$.

*179 (i) Let $H, K \leqslant G$. Then

$$[H, K] \trianglelefteq \langle H, K \rangle.$$

(See **71**. Hint. Apply **169**(i), 3.47(i) and **178**(v).)
 (ii) Let G be a non-abelian simple group, and suppose that H and K are proper subgroups of G such that $G = \langle H, K \rangle$. Then

$$G = [H, K].$$

 (iii) Suppose that G is a non-abelian simple group in which there are elements

t and x such that $o(t) = 2$, $o(x) = 3$ and $G = \langle x, t \rangle$. Then also

$$G = \langle x, t^{-1}xt \rangle.$$

(Remark. We shall see later that such a group G does exist: cf. 5.24, **315**. Hint. Use (ii) to show that $G = \langle [x,t], [x^{-1},t] \rangle$ and consider $[x^{-1},t][x,t]^{-1}$.)

*180 Let $H \leqslant G$. The *normal closure of H in G* is defined to be the intersection of all normal subgroups of G which contain H, and is denoted by H^G. Then
  (i) H^G is the unique smallest normal subgroup of G containing H.
  (ii) $H^G = \langle g^{-1}hg : g \in G, h \in H \rangle$. (Hint. Use **178**(ii).)
  (iii) $H^G = H[H,G]$. (Note that by **179**(i), $[H,G] \trianglelefteq G$.)

181 Let $G = \langle j, k \rangle \leqslant GL_2(\mathbb{C})$, where

$$j = \begin{pmatrix} i & 0 \\ 0 & -i \end{pmatrix} \quad \text{and} \quad k = \begin{pmatrix} 0 & 1 \\ -1 & 0 \end{pmatrix}.$$

Let $J = \langle j \rangle$ and $K = \langle k \rangle$. Show that $|J| = |K| = 4$ and that $|J \cap K| = 2$. By means of **178**(iii), show that $J \trianglelefteq G$ and $K \trianglelefteq G$. Deduce that $|G| = 8$. Show that the only element of order 2 in G is $j^2 = k^2$. Deduce that every subgroup of G is normal in G, although G is non-abelian. G is another example of a group which does not have an abelian. radical (cf. **171**).

This group G is denoted by Q_8 and called the *quaternion group*. If we identify the complex number i with the matrix

$$jk = \begin{pmatrix} 0 & i \\ i & 0 \end{pmatrix},$$

then the group elements i, j, k satisfy the relations

$$i^2 = j^2 = k^2 = -1, \quad i = jk = -kj, \quad j = ki = -ik, \quad k = ij = -ji,$$

discovered by Sir William Rowan Hamilton (1805–65); and

$$Q_8 = \{1, i, j, k, -1, -i, -j, -k\}.$$

182 Prove that $Q_8 \ncong D_8$.

The first examples of non-abelian simple groups were discovered by Evariste Galois (1811–32). We shall now introduce these groups: simplicity will be established in chapter 5.

3.58. Let n be an integer greater than 1, and consider the group Σ_n of all permutations of the set $\{1, 2, \ldots, n\} = X$, say. Consider the $\frac{1}{2}n(n-1)$ unordered pairs $\{i, j\}$ with $i, j \in X$ and $i \neq j$, and let

$$N = \prod_{1 \leqslant i < j \leqslant n} (j - i), \text{ a positive integer.}$$

For each $\sigma \in \Sigma_n$ let

$$N_\sigma = \prod_{1 \leqslant i < j \leqslant n} (j\sigma - i\sigma).$$

Thus $N_1 = N$. Moreover, for any σ, $N_\sigma = \pm N$: for, as $\{i, j\}$ runs through all the $\frac{1}{2}n(n-1)$ 2-element subsets of X, so does $\{i\sigma, j\sigma\}$ (since if $\{i\sigma, j\sigma\} =$

$\{k\sigma, l\sigma\}$ then either $i\sigma = k\sigma$ and $j\sigma = l\sigma$ or $i\sigma = l\sigma$ and $j\sigma = k\sigma$, hence either $i = k$ and $j = l$ or $i = l$ and $j = k$, and so $\{i,j\} = \{k,l\}$). We write

$$N_\sigma = \varepsilon_\sigma N:$$

thus

$$\varepsilon_\sigma = \begin{cases} 1 & \text{if } N_\sigma = N, \\ -1 & \text{if } N_\sigma = -N; \end{cases}$$

then ε_σ is called the *sign* of σ.

Note that $\varepsilon_\sigma = (-1)^{t_\sigma}$, where t_σ is the number of ordered pairs (i,j) such that $i,j \in X, i < j$ and $i\sigma > j\sigma$. For instance, if σ is a transposition (rs) then $\varepsilon_\sigma = -1$: for we may suppose that $r < s$ and then, for each $i \in X$,

$$i\sigma = \begin{cases} i \text{ whenever } i \text{ is distinct from } r, s, \\ s \text{ when } i = r, \\ r \text{ when } i = s; \end{cases}$$

hence, for $i,j \in X$ with $i < j$, $i\sigma > j\sigma$ if and only if either $i = r$ and $r < j \leqslant s$ or $r < i < s$ and $j = s$; thus $t_\sigma = 2(s - r) - 1$, an odd integer.

Now let $\sigma, \tau \in \Sigma_n$ and, for each $i, j \in X$ with $i < j$, write $i\sigma = i', j\sigma = j'$. Then

$$N_{\sigma\tau} = \prod_{1 \leqslant i < j \leqslant n} (j\sigma\tau - i\sigma\tau) = \prod_{1 \leqslant i < j \leqslant n} (j'\tau - i'\tau).$$

In the product on the right we replace any factor $j'\tau - i'\tau$ for which $i' > j'$ by $-(i'\tau - j'\tau)$. Then, by definition of ε_σ, we have

$$N_{\sigma\tau} = \varepsilon_\sigma \prod_{1 \leqslant i < j \leqslant n} (j\tau - i\tau)$$

$$= \varepsilon_\sigma \varepsilon_\tau N.$$

However,

$$N_{\sigma\tau} = \varepsilon_{\sigma\tau} N.$$

Since these are equations between non-zero numbers, it follows that

$$\varepsilon_{\sigma\tau} = \varepsilon_\sigma \varepsilon_\tau. \tag{i}$$

This shows that the map

$$\varepsilon : \Sigma_n \to C_2 = \{1, -1\}, \quad \text{defined by } \varepsilon : \sigma \mapsto \varepsilon_\sigma,$$

is a homomorphism; it is called the *alternating character* on Σ_n. Moreover, it is surjective, for we have pointed out that if σ is a transposition then $\varepsilon_\sigma = -1$. (Note that since $n \geqslant 2$, there are transpositions in Σ_n.) By the fundamental theorem on homomorphisms,

$$\text{Ker } \varepsilon \trianglelefteq \Sigma_n \quad \text{and} \quad \Sigma_n/\text{Ker } \varepsilon \cong \text{Im } \varepsilon = C_2.$$

This means that Ker ε is a subgroup of index 2 in Σ_n. We call Ker ε the *alternating group of degree n* and denote it by A_n. The elements of A_n are called *even* permutations of X and the elements of $\Sigma_n \backslash A_n$ are called *odd* permutations of X. Note that

$$|A_n| = n!/2 = |\Sigma_n \backslash A_n|.$$

Note also that even and odd permutations multiply as even and odd integers *add*:

(even) (even) = even = (odd) (odd), (even) (odd) = odd = (odd) (even).

How can we decide whether a given permutation is even or odd?

3.59. *Let n be an integer greater than 1, and let $\sigma \in \Sigma_n$. Let the expression of σ as a product of cycles on disjoint sets be $\sigma = v_1 v_2 \ldots v_r$, and suppose that the cycle v_j has length $m_j (j = 1, \ldots, r)$. Then the permutation σ is even or odd according as the integer $\sum\limits_{j=1}^{r} (m_j - 1)$ is even or odd.*

Proof. Consider first a cycle v, say of length m, where $2 \leqslant m \leqslant n$: say

$$v = (a_1 a_2 \ldots a_m)$$
$$= (a_1 a_2)(a_1 a_3) \ldots (a_1 a_m) \quad \text{(see 21).}$$

Then $\qquad \varepsilon_v = \varepsilon_{(a_1 a_2)} \varepsilon_{(a_1 a_3)} \ldots \varepsilon_{(a_1 a_m)}$, by equation (i) of 3.58,
$$= (-1)^{m-1},$$

since the sign of any transposition is -1.

Now if, as in the statement, $\sigma = v_1 v_2 \ldots v_r$,

by equation (i) of 3.58, $\qquad \varepsilon_\sigma = \varepsilon_{v_1} \varepsilon_{v_2} \ldots \varepsilon_{v_r}$,
$$= (-1)^{m_1 - 1}(-1)^{m_2 - 1} \ldots (-1)^{m_r - 1}$$
$$= (-1)^k, \quad \text{where } k = \sum_{j=1}^{r} (m_j - 1).$$

This establishes the result.

For example, in Σ_5,

$$\begin{pmatrix} 1 & 2 & 3 & 4 & 5 \\ 4 & 5 & 2 & 1 & 3 \end{pmatrix} = (14)(253),$$

which is odd since $(2 - 1) + (3 - 1) = 3$, an odd integer.

Galois proved

3.60. *Whenever $n \geqslant 5$, A_n is a non-abelian simple group.*

We shall establish this in 5.28. The occurrence of the number 5 in this result is intimately connected with the unsolvability of the general quintic equation. For an explanation of this, the reader may consult the references to Galois theory mentioned in 2.25. The group A_5 has order 60, and, as we shall show in chapter 5, there is no non-abelian simple group of smaller order.

183 Which of the following permutations in Σ_6 are even and which are odd:

$$(123456), (12345), (123)(45), \begin{pmatrix} 123456 \\ 246135 \end{pmatrix}?$$

*184 Let n be an integer, $n > 1$, and let $H \leqslant \Sigma_n$. Prove that if H includes an odd permutation then H has a subgroup of index 2. Deduce that if H is simple and $|H| > 2$ then $H \leqslant A_n$.

*185 (i) Write down the 12 elements of A_4 and note their orders.

(ii) Prove that $Z(A_4) = 1$. (Hint. Use **6** to show that if $Z(A_4) \neq 1$ then A_4 would have an element of order 6.)

(iii) Show that A_4 has a unique subgroup V of order 4, and deduce that $V \trianglelefteq A_4$.

(iv) Prove that A_4 has no subgroup of order 6. (Hint. Assume that A_4 has a subgroup H of order 6. Consider $H \cap V$, and derive a contradiction by means of **102**, 3.40, **119** and (ii).)

3.61. We end this chapter with some remarks on simple groups. We have mentioned one family of non-abelian finite simple groups containing infinitely many different types, namely the alternating groups of degrees 5 and greater. We mention some other examples.

Let n be an integer greater than 1 and let F be any field. Let $S = \mathrm{SL}_n(F)$, the group of all $n \times n$ matrices with entries in F and determinant 1 (3.28). Then it can be shown that $Z(S)$ consists of all the scalar matrices in S; that is, the matrices $a1$ with $a \in F$ and $a^n = 1$. (See **123** for the case when $F \neq \mathbf{Z}_2$ and either $n > 2$ or $F \neq \mathbf{Z}_3$.) The group $S/Z(S)$ is called the *projective special linear group of degree n over F* and denoted by $\mathrm{PSL}_n(F)$. It can be proved that $\mathrm{PSL}_n(F)$ is a non-abelian simple group, except when $n = 2$ and F is either $\mathbf{Z}_2$ or $\mathbf{Z}_3$. (See Artin [b2] or Huppert [b21] for the proof.) In particular, this provides examples of infinite simple groups, for when the field F is infinite then the group $\mathrm{PSL}_n(F)$ is infinite. When F is finite, $\mathrm{PSL}_n(F)$ is finite.

There are other families of simple groups defined in a similar way from groups of matrices: the so-called *orthogonal*, *symplectic* and *unitary* groups. For instance, let n be an even positive integer, say $n = 2m$, where m is a positive integer, and let F be any field. Let y be a fixed non-singular skew-symmetric $n \times n$ matrix with entries in F: there is such a matrix because n is even. For each $x \in \mathrm{GL}_{2m}(F)$ let x' denote the transpose of x. Then the set

$$\{ x \in \mathrm{GL}_{2m}(F) : x' y x = y \}$$

forms a subgroup of $\mathrm{GL}_{2m}(F)$ which is called the *symplectic group of degree 2m over F* and denoted by $\mathrm{Sp}_{2m}(F)$. The type of the group does not depend on the choice of the skew-symmetric matrix y. It can be shown that every matrix in $\mathrm{Sp}_{2m}(F)$ has determinant 1, so that $\mathrm{Sp}_{2m}(F) \leqslant \mathrm{SL}_{2m}(F)$. Let $Y = \mathrm{Sp}_{2m}(F)$. Then $Z(Y)$ consists of the scalar matrices in Y, and in fact $|Z(Y)| = 2$ if in F, $1 + 1 \neq 0$, while $Z(Y) = 1$ if in F, $1 + 1 = 0$. The group $Y/Z(Y)$ is called the *projective symplectic group of degree 2m over F*, denoted by $\mathrm{PSp}_{2m}(F)$, and $\mathrm{PSp}_{2m}(F)$ is a non-abelian simple group, except when $2m = 2$ and F is either $\mathbf{Z}_2$ or $\mathbf{Z}_3$, and when $2m = 4$ and $F = \mathbf{Z}_2$. (For details of these groups and the orthogonal and unitary groups, see Artin [b2], Dieudonné [b9] and [b10], and Huppert [b21].)

These families of simple groups defined by means of groups of matrices were discovered by C. Jordan (1838–1922), and discussed in his book [b23]. They are usually called the *classical* simple groups. A further family was discovered in 1905 by L. E. Dickson (1874–1954), who also made a detailed study of the classical groups in his book [b8]. Earlier E. Mathieu (1835–1890) had found five individual simple groups which have never been identified as belonging to an infinite family; they are subgroups of the alternating groups $A_{11}, A_{12}, A_{22}, A_{23}, A_{24}$. Such simple groups which do not appear as members of an infinite family are now referred to as *sporadic* simple groups.

No further discoveries of finite simple groups were made until 1955. In that year, C. Chevalley described a method which provided a construction of the groups of Jordan and Dickson, together with new infinite families of simple groups. Further families were then discovered by several other authors by varying Chevalley's method. For details, we refer to an article and book of R. W. Carter, [a14] and [b4].

At this point, it was widely expected that the list of finite simple groups would prove to be complete. However, in 1965 a new simple group was discovered by Z. Janko: a group of 7×7 matrices with entries in the field $\mathbf{Z}_{11}$ and of order $2^3.3.5.7.11.19$. Since 1965, around two dozen more sporadic finite simple groups have been found, and the state of our understanding of finite simple groups remains unstable. We do not even know as yet whether our present list of simple groups of orders less than 1 000 000 is complete. For further information we refer to the survey articles of W. Feit [a21] and D. Gorenstein [a44], chapters 16 and 17 of Gorenstein's book [b13], and the volume edited by M. B. Powell and G. Higman [b33].

4

GROUP ACTIONS ON SETS

We have discussed in chapter 2 the significance of groups occurring as symmetry groups of mathematical systems. Let X be some system, that is to say a set with a certain distinguished structure, it may be algebraic or geometric, and let G be a subgroup of the symmetry group of X (the group of all structure-preserving permutations of X). Then each $g \in G$ moves each $x \in X$ to some element of X (where in using the word 'moves' we allow the possibility that g fixes x). Let us write xg for this element of X to which g moves x. In this way we think of the group G as *acting* on the system X. The action is determined when for each $g \in G$ and $x \in X$ the corresponding element $xg \in X$ is specified. This simple notion of a group action has proved very fruitful. We shall find it profitable to build from a definition which generalizes the initial idea. In this chapter we define and develop the notion of a group action on a set (without further structure).

4.1 Definition. We say that G *acts on the non-empty set* X (or that G *permutes* X) if to each $g \in G$ and each $x \in X$ there corresponds a unique element $xg \in X$ such that, for all $x \in X$ and $g_1, g_2 \in G$,

$$(xg_1)g_2 = x(g_1 g_2)$$

and
$$x1 = x.$$

To be explicit, we say under these conditions that G acts on X *on the right*. We can define in a similar way what is meant by the action of a group on a set *on the left*. Later on (in chapter 10) we shall need to discuss right and left actions together and so preserve the distinction between them. Till then, if we speak of a group action without qualification, we shall mean an action on the right.

4.2 Examples. (i) Let X be any non-empty set and let $G \leqslant \Sigma_X$. Then G acts on X. In this case each $g \in G$ is a map $X \to X$ and, for $x \in X$, xg is the image of x under the map g. The condition $(xg_1)g_2 = x(g_1 g_2)$ of 4.1 is satisfied by definition of composition of maps, and the condition $x1 = x$ by definition of the identity element 1 of Σ_X. This action is called the *natural action* of G on X.

(ii) Let V be a vector space $\neq 0$ over a field F. Then, with the usual vector space notation, to each $a \in F$ and $v \in V$ there corresponds an element $av \in V$. By this correspondence, the multiplicative group $F^\times$ acts on the left on V (regarded as a set): for if $a_1, a_2 \in F$ and $v \in V$ then, by vector space axioms, $a_1(a_2 v) = (a_1 a_2)v$ and $1v = v$. But the additive group F^+ does *not* in this way act on the left on V: for if it did, we should have $a_1(a_2 v) = (a_1 + a_2)v$ and $0v = v$, and these equations are both false unless either $v = 0$ or $a_1 \neq 1$ and $a_2 = a_1(a_1 - 1)^{-1}$.

186 Suppose that G acts (on the right) on the set X. Then we get a left action of G on X by defining, for all $g \in G$ and $x \in X$, $gx = xg^{-1}$. Why will it not do in general to define $gx = xg$?

We note at once the relation between group actions on a set X and the symmetry group of X, that is, Σ_X.

4.3 Theorem. *Let G act on the set X. Then to each $g \in G$ there corresponds a map $\rho_g : X \to X$ defined by $\rho_g : x \mapsto xg$, and this is a permutation of X. Moreover, the map $\rho : G \to \Sigma_X$ defined by $\rho : g \mapsto \rho_g$ is a homomorphism;* it is called the *permutation representation* of G corresponding to the group action.

Proof. Let $g \in G$. By definition, ρ_g is a map of X into itself. For $g_1, g_2 \in G$ and $x \in X$, using the first axiom of 4.1 we have

$$x\rho_{g_1 g_2} = x(g_1 g_2) = (xg_1)g_2 = (x\rho_{g_1})\rho_{g_2} = x(\rho_{g_1}\rho_{g_2}),$$

so that

$$\rho_{g_1 g_2} = \rho_{g_1}\rho_{g_2}. \tag{i}$$

Moreover, using the second axiom of 4.1 we have

$$x\rho_1 = x1 = x,$$

so that

$$\rho_1 = 1 \in \Sigma_X. \tag{ii}$$

By (i) and (ii),

$$\rho_g \rho_{g^{-1}} = 1 = \rho_{g^{-1}}\rho_g.$$

Thus ρ_g is an invertible map of X to itself, that is, a permutation of X. Then (i) shows that ρ is a homomorphism of G into Σ_X.

4.4 Theorem. *Let σ be a homomorphism of G into Σ_X, where X is a non-empty set. Then G acts on X when we define, for each $g \in G$ and $x \in X$,*

$$xg = x(g\sigma);$$

and the permutation representation of G corresponding to this action is σ.

Proof. For $g_1, g_2 \in G$ and $x \in X$, by definition of composition of maps,

$$(x(g_1\sigma))(g_2\sigma) = x((g_1\sigma)(g_2\sigma))$$
$$= x((g_1g_2)\sigma),$$

since σ is a homomorphism;
and $x(1\sigma) = x$, since, by 2.9, σ must map $1 \in G$ to $1 \in \Sigma_X$. Hence, by setting

$$xg = x(g\sigma),$$

we do define an action of G on X. Let the corresponding permutation representation of G be ρ. Then

$$x\rho_g = xg = x(g\sigma),$$

hence $$\rho_g = g\sigma \quad \text{for all } g \in G,$$

and so $$\rho = \sigma.$$

Thus, in considering group actions on a set X, we now look not merely at subgroups of Σ_X but at homomorphisms of groups into Σ_X.

4.5 Definition. Let G act on the set X. We say that the action is *faithful* if the corresponding permutation representation of G is injective.

In 4.2(i), the permutation representation is just the inclusion map $G \to \Sigma_X$. This is certainly injective, so that the action is faithful. The left action of $F^\times$ on V in 4.2(ii) is also faithful.

4.6 Lemma. *Let G act on the set X. We define a relation $\sim$ on X by setting $x_1 \sim x_2$ if and only if $x_1, x_2 \in X$ and there is an element $g \in G$ such that $x_1g = x_2$. Then $\sim$ is an equivalence relation on X.*
Proof. For any $x \in X, x1 = x$, so that $x \sim x$. If $x_1 \sim x_2$ then $x_1g = x_2$ for some $g \in G$, hence $x_2g^{-1} = (x_1g)g^{-1} = x_11 = x_1$, and so $x_2 \sim x_1$. If $x_1 \sim x_2$ and $x_2 \sim x_3$ then $x_1g_1 = x_2$ and $x_2g_2 = x_3$ for some $g_1, g_2 \in G$, hence $x_1(g_1g_2) = (x_1g_1)g_2 = x_3$, and so $x_1 \sim x_3$.
The following definition is of fundamental importance.

4.7 Definition. Let G act on the set X. Then X is partitioned into disjoint equivalence classes with respect to the equivalence relation $\sim$ of 4.6. These equivalence classes are called the *orbits* or *transitivity classes* of the action. For each $x \in X$, the orbit containing x is called the *orbit of x*: it is the subset $\{xg : g \in G\}$ of X.

4.8. *Let G act on the set X, and let $x \in X$. Set $\text{Stab}_G(x) = \{g \in G : xg = x\}$. Then $\text{Stab}_G(x)$ is a subgroup of G, called the* stabilizer *of x in G. (In the literature, this subgroup is often denoted by G_x and called the* isotropy group *of x in G.)*
Proof. By 4.1, $1 \in \text{Stab}_G(x)$, so that $\text{Stab}_G(x) \neq \emptyset$. Let $g_1, g_2 \in \text{Stab}_G(x)$. Then $xg_1 = x = xg_2$, hence

$$x(g_1 g_2^{-1}) = (xg_1)g_2^{-1} = (xg_2)g_2^{-1} = x1 = x,$$

and so $g_1 g_2^{-1} \in \mathrm{Stab}_G(x)$. Hence $\mathrm{Stab}_G(x) \leqslant G$.

The following fact is an immediate consequence of the definitions already given.

4.9. *Let G act on the set X, and let the corresponding permutation representation of G be ρ. Then*

$$\mathrm{Ker}\ \rho = \bigcap_{x \in X} \mathrm{Stab}_G(x).$$

*187 Suppose that G acts on the set X. For each $g \in G$ and each non-empty subset Y of X, define

$$Yg = \{yg : y \in Y\} \subseteq X.$$

Define also

$$G_Y = \{h \in G : yh = y \text{ for all } y \in Y\} = \bigcap_{y \in Y} \mathrm{Stab}_G(y),$$
$$G_Y^* = \{h \in G : Yh = Y\}. \text{ Then}$$

(i) $G_{Yg} = g^{-1} G_Y g$ and $G_{Yg}^* = g^{-1} G_Y^* g$.
In particular, for all $x \in X$, $\mathrm{Stab}_G(xg) = g^{-1}\,\mathrm{Stab}_G(x)g$.
(ii) $G_Y \trianglelefteq G_Y^* \leqslant G$ (cf. **62**(i)).
(iii) If Y is an orbit of the action of G on X then $G_Y \trianglelefteq G$, G acts on Y, and G/G_Y is isomorphic to the image of the corresponding permutation representation of G on Y.

188 Suppose that the finite group G acts faithfully on the finite set X. Let the orbits of the action be $X_1, \ldots, X_s$, where s is a positive integer, and let $|X_i| = n_i$ for $i = 1, \ldots, s$ (where $n_1 + \ldots + n_s = |X|$). Then G can be embedded in the group $\Sigma_{n_1} \times \ldots \times \Sigma_{n_s}$. (Hint. Use **187**(iii) and **109**.)

4.10 Examples. (i) Let n be a positive integer, $\sigma \in \Sigma_n$ and $G = \langle \sigma \rangle$. Suppose that σ is expressed as a product of disjoint cycles as

$$\sigma = (a_{11} a_{12} \ldots a_{1n_1})(a_{21} \ldots a_{2n_2}) \ldots (a_{s1} \ldots a_{sn_s}),$$

where $s, n_1, \ldots, n_s$ are positive integers such that $n_1 + \ldots + n_s = n$. Then the orbits of the natural action of G on the set $\{1, 2, \ldots, n\}$ are the s disjoint subsets $\{a_{11}, \ldots, a_{1n_1}\}, \{a_{21}, \ldots, a_{2n_2}\}, \ldots, \{a_{s1}, \ldots, a_{sn_s}\}$.

For instance, for $n = 5$ and $\sigma = (123)(45)$, there are just two orbits $\{1, 2, 3\}$ and $\{4, 5\}$. Note that then $\mathrm{Stab}_G(1) = \mathrm{Stab}_G(2) = \mathrm{Stab}_G(3) = \langle \sigma^3 \rangle$ and $\mathrm{Stab}_G(4) = \mathrm{Stab}_G(5) = \langle \sigma^2 \rangle$. Since $o(\sigma) = 6$, we see that for each $x \in \{1, 2, 3, 4, 5\}$, the number of elements in the orbit of x is equal to $|G : \mathrm{Stab}_G(x)|$. We shall show in 4.11 that this is not coincidental but an instance of a general result.

(ii) Let $H \leqslant G$. Then H acts on G (regarded as a set) *by right multiplication* in G; that is, when to each $h \in H$ and each $g \in G$ there corresponds the element $gh \in G$. That this does define an action of H on G follows from the associative law for multiplication in G and the defining property of the identity element. Now, for $g \in G$,

$$\text{Stab}_H(g) = \{h \in H : gh = g\} = 1.$$

In particular, by 4.9, it follows that the action is faithful. Also, the orbit of g is the set $\{gh : h \in H\} = gH$, the left coset of H in G containing g. Thus from 4.6 we can deduce that distinct left cosets of H in G are disjoint, and hence derive Lagrange's theorem.

In a similar way, left multiplication of the elements of G by the elements of H defines a left action of H on G, the orbits of which are the right cosets of H in G.

The following result on lengths of orbits is a key fact needed for many applications.

4.11 Lemma. *Let G act on the set X, and let $x \in X$. Then*

$$|\text{the orbit of } x| = |G : \text{Stab}_G(x)|.$$

Proof. Let X_1 denote the orbit of x, let $H = \text{Stab}_G(x)$ and let Y denote the set of right cosets of H in G. Thus

$$X_1 = \{xg : g \in G\}.$$

We define a map

$$\mu : X_1 \to Y$$

by

$$\mu : xg \mapsto Hg \quad \text{(for all } g \in G).$$

We must check that this is well defined. Let $g_1, g_2 \in G$. We need to be sure that if $xg_1 = xg_2$ then also $Hg_1 = Hg_2$. Using the axioms of 4.1, we see that if $xg_1 = xg_2$ then

$$x(g_1 g_2^{-1}) = (xg_1)g_2^{-1} = (xg_2)g_2^{-1} = x1 = x,$$

so that

$$g_1 g_2^{-1} \in \text{Stab}_G(x) = H,$$

from which it follows that $Hg_1 = Hg_2$, as required.

Furthermore, we see conversely that if $Hg_1 = Hg_2$ then $g_1 g_2^{-1} \in H$, so that

$$x(g_1 g_2^{-1}) = x;$$

then

$$xg_1 = x((g_1 g_2^{-1})g_2) = xg_2.$$

This shows that μ is an injective map. It is clear from its definition that μ is surjective, so that μ is in fact a bijective map. Hence

$$|X_1| = |Y|,$$

as asserted (where this means in particular that $|X_1| = \infty$ if and only if $|Y| = \infty$).

189 Let $G \leqslant \Sigma_4$, and consider the natural action of G on the set $\{1, 2, 3, 4\}$. For each of the following choices of G, write down the orbits of the action and find the stabilizer of each point. Verify the result of 4.11 in each case:

   (i) $G = \langle (123) \rangle$,
   (ii) $G = \langle (1234) \rangle$,
   (iii) $G = \{1, (12)(34), (13)(24), (14)(23)\}$,
   (iv) $G = \{1, (12), (12)(34), (34)\}$,
   (v) $G = A_4$.

190 Let n be a positive integer, F a field, and V a vector space of dimension n over F. Let $F^\times$ act on the left on V, as in 4.2(ii). Find the orbits of this action and the stabilizer in $F^\times$ of each $v \in V$, and verify that the result of 4.11 is true in this case. How many orbits are there if $|F| = q < \infty$?

We shall discuss two special group actions which are of great importance: the action by right multiplication of a group on the set of right cosets of a subgroup, and the action by conjugation of a group on its subsets. We shall use information about these actions to obtain fundamental results on abstract groups.

4.12 Definition. Let G act on the set X. The action is said to be *transitive* if it has just one orbit. An action which is not transitive is called *intransitive*.

For instance, let n be a positive integer and let $X = \{1, 2, \ldots, n\}$. Then the natural action of Σ_n on X is transitive; so also is the natural action on X of the cyclic subgroup $\langle (12 \ldots n) \rangle$ of Σ_n. The natural action of A_n on X is transitive if $n \geqslant 3$: for if $n \geqslant 3$ and i, j are any two distinct points of X, there is a point $k \in X$ which is distinct from both i and j, and then $(ijk) \in A_n$ (by 3.59) and (ijk) moves i to j.

4.13. Let $H \leqslant G$ and let X be the set of right cosets of H in G. Then G acts on X *by right multiplication:* to each $g \in G$ and each $Hx \in X$ (where $x \in G$) there corresponds the coset $Hxg \in X$.

This does define an action of G on X, for if $x, g_1, g_2 \in G$ then $(Hxg_1)g_2 = Hxg_1g_2$ and $Hx1 = Hx$. The action is transitive, for any two right cosets of H in G are equivalent under the action: if $x_1, x_2 \in G$ then $x_1^{-1}x_2 \in G$ and $(Hx_1)x_1^{-1}x_2 = Hx_2$.

Next we note that

$$\begin{aligned} \mathrm{Stab}_G(Hx) &= \{g \in G : Hxg = Hx\} \\ &= \{g \in G : xgx^{-1} \in H\} \\ &= x^{-1}Hx, \end{aligned}$$

the conjugate of H by x (2.19). Note that, by 4.11, for every $x \in G$,

$$|G : x^{-1}Hx| = |X| = |G : H|.$$

Of course, this is clear from Lagrange's theorem if G is finite (see also **27**).

Let ρ^H denote the permutation representation of G corresponding to the action. Then, by 4.9,

$$\text{Ker } \rho^H = \bigcap_{x \in G} x^{-1}Hx$$

$$= H_G,$$

the core of H in G (**90**).

When $|G:H| < \infty$ we may, by 2.7, identify Σ_X with $\Sigma_{|G:H|}$. Then ρ^H is a homomorphism of G into $\Sigma_{|G:H|}$ and the fundamental theorem on homomorphisms yields the following simple but important fact.

4.14 Theorem. *If H is a subgroup of finite index in G then G/H_G can be embedded in $\Sigma_{|G:H|}$.*

The following is an immediate consequence.

4.15 Corollary. *If H is a subgroup of finite index in an infinite group G then there is a normal subgroup K of G such that $K \leqslant H$ and G/K is finite.*

We now derive two less obvious consequences, due to R. Baer [a5], for G a finitely generated infinite group.

4.16 Corollary. *Let G be a finitely generated infinite group. Then, for each positive integer n, G has only finitely many subgroups of index n.*
Proof. If H is any subgroup of index n in G then, by 4.14, there is a homomorphism of G into Σ_n with kernel H_G. Suppose that $G = \langle x_1, \ldots, x_m \rangle$, where m is a positive integer. Any homomorphism $\varphi : G \to \Sigma_n$ is determined (by 2.28) as soon as $x_1\varphi, \ldots, x_m\varphi$ are specified. Hence, since Σ_n is a finite group, there are only finitely many homomorphisms of G into Σ_n, and therefore there are only finitely many normal subgroups of G eligible to be cores in G of subgroups of index n. Moreover, any such normal subgroup K of G can be the core in G of only finitely many subgroups of G of index n, since G/K is a finite group. Hence there are in G only finitely many subgroups of index n.

4.17 Corollary. *Let H be a subgroup of finite index in a finitely generated infinite group G. Then there is a characteristic subgroup K of G such that $K \leqslant H$ and G/K is finite.*
Proof. Let $|G:H| = n$. By 4.16, G has only finitely many subgroups of index n: let them be $H = H_1, H_2, \ldots, H_s$. Then let

$$K = \bigcap_{i=1}^{s} H_i.$$

Any automorphism of G maps a subgroup of index n to a subgroup of index n (**27**), and therefore permutes $H_1, \ldots, H_s$ among themselves, hence maps K to itself. Hence K is characteristic in G. Finally, by Poincaré's theorem (66), G/K is finite.

In 4.16 and 4.17, the condition that G is finitely generated cannot be omitted, as can be seen by considering the additive group of an infinite-dimensional vector space over $\mathbf{Z}_p$.

The statement in 102 is another immediate consequence of 4.14, since $|\Sigma_2| = 2$. This can be generalized when G is a finite group.

4.18 Corollary. *Suppose that G is finite and that p is the smallest prime divisor of $|G|$. If H is a subgroup of index p in G then $H \trianglelefteq G$.*
Proof. Suppose that $H \leqslant G$ with $|G : H| = p$. Then $|G/H_G| = p|H : H_G|$. Suppose that $|H : H_G| > 1$ and let q be a prime divisor of $|H : H_G|$. Then q divides $|G|$, and so, by hypothesis, $q \geqslant p$. On the other hand, by 4.14, $|G/H_G|$ divides $p!$, hence pq divides $(p-1)!p$, and so q divides $(p-1)!$ Since q is prime, it follows that $q < p$, a contradiction. Therefore we conclude that $|H : H_G| = 1$. Hence $H = H_G \trianglelefteq G$.

191 Let $H \leqslant G$ with $|G : H| = n < \infty$. Let ϖ be the set of all primes not exceeding n. Then G/H_G is a finite ϖ-group.

192 Let G be a finite simple group with a subgroup H of prime index p. Then p is the largest prime divisor of $|G|$ and p^2 does not divide $|G|$.

193 Let $G = \mathrm{GL}_2(\mathbf{Z}_3)$ and $K = Z(G)$. By 2.16 and 2.17, $|G| = 48$, and, by 123, $|K| = 2$.

    (i) Let $H = \left\{ \begin{pmatrix} a & b \\ 0 & c \end{pmatrix} : a, b, c \in \mathbf{Z}_3 \text{ with } ac \neq 0 \right\}$.

Show that $K \leqslant H \leqslant G$, and that $|H| = 12$.
    (ii) Prove that $H_G = K$.
    (iii) By means of 4.14, deduce that $G/K \cong \Sigma_4$.

194 An infinite simple group cannot have a proper subgroup of finite index.

*195 A group is said to be *periodic* if every element has finite order. Thus every finite group is periodic. There are also infinite periodic groups, for example $\mathbf{Q}^+/\mathbf{Z}^+$: see 3.25.
    (i) All subgroups and all quotient groups of a periodic group are periodic.
    (ii) If $K \trianglelefteq G$ and the groups K and G/K are both periodic, then G is periodic.
    (iii) A group is periodic if it has a periodic subgroup of finite index.

196 G is said to be *locally finite* if every finitely generated subgroup of G is finite.
    (i) Every locally finite group is periodic (195). (Remark. The converse is false: see 8.29.)
    (ii) Every periodic abelian group is locally finite. (Hint. Apply 69.)
    (iii) (O. J. Schmidt [a86]). If $K \trianglelefteq G$ and the groups K and G/K are both locally finite, then G is locally finite. (Hint. Apply 3.40, 108 and 149.)
    (iv) A group is locally finite if it has a locally finite subgroup of finite index.

197 Let $H, K \leqslant G$ with $|G : H|, |G : K|$ finite and co-prime. Then $|G : H \cap K| = |G : H||G : K|$ and $G = HK$. (This generalizes 100. Apply 4.15, 66 and 100.)

Next we shall show that every transitive group action is equivalent, in the sense of the following definition, to an action of the kind described in 4.13.

4.19 Definition. Let groups G_1, G_2 act on sets X_1, X_2, respectively. We say that the actions are *equivalent* if there is an isomorphism $\varphi : G_1 \to G_2$ and a bijective map $\mu : X_1 \to X_2$ such that, for all $g_1 \in G_1$ and $x_1 \in X_1$,

$$(x_1 g_1)\mu = (x_1\mu)(g_1\varphi).$$

This defines an equivalence relation on group actions.

4.20 Theorem. *Let G act transitively on the set X. Let $x \in X$ and let $H = \mathrm{Stab}_G(x)$. Then the action of G on X is equivalent to the action by right multiplication of G on the set of right cosets of H in G.*
Proof. Since the action of G on X is transitive,

$$X = \{xg : g \in G\}.$$

We define a map μ from X to the set of right cosets of H in G by

$$\mu : xg \mapsto Hg \quad (\text{for all } g \in G).$$

This is well defined, for if $xg_1 = xg_2$, with $g_1, g_2 \in G$, then $g_1 g_2^{-1} \in \mathrm{Stab}_G(x) = H$ and so $Hg_1 = Hg_2$. This argument works also in reverse to show that if $Hg_1 = Hg_2$ then $xg_1 = xg_2$. Then clearly μ is a bijective map. So far, the argument is just the same as in 4.11. For all $g, g_1 \in G$,

$$((xg)g_1)\mu = (x(gg_1))\mu = Hgg_1 = ((xg)\mu)g_1.$$

This establishes the stated equivalence of group actions, when we choose as the appropriate isomorphism the identity automorphism of G.

4.21. From 4.13 and 4.20, it follows that if $H \leqslant G$ and $x \in G$ then the action by right multiplication of G on the set of right cosets of H in G is equivalent to the action by right multiplication of G on the set of right cosets of $x^{-1}Hx$ in G.

*198 Let X and Y be sets with a bijective map $\mu : X \to Y$, and suppose that G acts faithfully on the set X. Then the action is equivalent to the natural action of a suitable subgroup of Σ_Y on Y. (Hint. See 2.7.)

*199 Let groups G_1, G_2 act transitively on sets X_1, X_2, respectively. If these actions are equivalent and $x_1 \in X_1, x_2 \in X_2$ then $\mathrm{Stab}_{G_1}(x_1) \cong \mathrm{Stab}_{G_2}(x_2)$.

200 Let groups G, H act on the same set X, with corresponding permutation representations ρ, σ, respectively. Prove that the actions are equivalent if and only if there is an isomorphism $\varphi : G \to H$ and an element $\mu \in \Sigma_X$ such that $\varphi\sigma = \rho\tau_\mu$, where τ_μ is the inner automorphism of Σ_X induced by μ.
   Deduce that if $G \leqslant \Sigma_X, H \leqslant \Sigma_X$ and the actions of G, H on X are the natural ones, then the actions are equivalent if and only if G, H are conjugate subgroups of Σ_X.

201 Let G be a finite group and let $H, J \leqslant G$. Consider the actions of G by right multiplication on the set of right cosets of H in G and on the set of right cosets of J in G. These actions are equivalent if and only if there is an automorphism α of G such that $H^\alpha = J$.

202 Consider the actions of a group G on sets. A set X together with an action of G on X will be called a *G-set*. If X is a G-set and $Y \subseteq X$, Y is a *G-subset* of X if the action of G on X restricts to an action of G on Y; that is, if $yg \in Y$ for all $y \in Y$ and $g \in G$. The empty set $\emptyset$ is regarded as a G-subset of every G-set. A non-empty G-set X is called *irreducible* if the only G-subsets of X are $\emptyset$ and X. A *G-map* of a G-set X to a G-set Y is a map $\varphi : X \to Y$ such that, for all $x \in X$ and $g \in G$, $(xg)\varphi = (x\varphi)g$.

(i) Let X be a non-empty G-set. Then the orbits of the action of G on X are irreducible G-subsets of X, and they are the only irreducible G-subsets of X. In particular, X is irreducible if and only if the action of G on X is transitive.

(ii) Let X be a non-empty G-set and let $\{X_r : r \in R\}$ be the set of all irreducible G-subsets of X. For every non-empty G-subset Y of X there is a non-empty subset S of R such that $Y = \bigcup_{s \in S} X_s$.

(iii) Let φ be a G-map of a G-set X to a G-set Y. Then $\operatorname{Im} \varphi = \{x\varphi : x \in X\}$ is a G-subset of Y; and for each G-subset W of Y, $\{x \in X : x\varphi \in W\}$ is a G-subset of X.

(iv) A G-map of a non-empty G-set to an irreducible G-set is necessarily surjective.

(v) Let φ be a G-map of an irreducible G-set X to a G-set Y. Then, for each G-subset W of Y, either $\operatorname{Im} \varphi \subseteq W$ or $W \cap \operatorname{Im} \varphi = \emptyset$.

(vi) If X is a non-empty G-set then the set of all bijective G-maps $X \to X$ is a subgroup Σ_X^G of Σ_X.

(vii) If X is an irreducible G-set then $|\Sigma_X^G| \leqslant |X|$.

(viii) If $H \leqslant G$ and X is the set of right cosets of H in G, with action of G on X by right multiplication as in 4.13, then X is an irreducible G-set and $\Sigma_X^G \cong N_G(H)/H$.

4.22 Definition. An action of G on a set X is said to be *regular* if it is transitive and $\operatorname{Stab}_G(x) = 1$ for each $x \in X$.

It follows from the definition and 4.9 that a regular action is faithful.

4.23. We obtain a regular action of G by choosing $H = 1$ in 4.13. Then $X = G$, and G acts *on itself* by right multiplication. The corresponding permutation representation ρ^1 of G is called *the right regular permutation representation of G*: ρ^1 maps each $g \in G$ to the permutation of G obtained by multiplying all elements of G on the right by g. The use of the qualifying of G is equivalent, in the sense of 4.19, to the action of G on itself by right multiplication: this follows from 4.20.

When 2.10 is applied to ρ^1, we get the following well known result, the proof of which derives from ideas in paper [a16] of Cayley in 1854.

4.24 Theorem (A. Cayley). *G can be embedded in Σ_G.*

This provides another proof of the fact, established in 1.2, that for each positive integer n, there are only finitely many distinct types of groups of order n: for, by Cayley's theorem, every group of order n can be embedded in Σ_n and Σ_n has, as a finite group, only finitely many subgroups.

203 Let G act on the set X.

(i) The action is *regular* if it is transitive and $\mathrm{Stab}_G(x) = 1$ for some $x \in X$. (Hint. See **187**(i).)

(ii) If G is abelian and the action is faithful and transitive then it is regular.

204 Let G be a finite group which acts transitively on a finite set X, with $|X| = n$. Then $|G|$ is a multiple of n; and $|G| = n$ if and only if the action is regular.

*205 Suppose that $|G| = 2r$, where r is an odd integer with $r > 1$. By 1.13, there is in G an element t with $o(t) = 2$. Show that in the right regular permutation representation of G, t corresponds to an odd permutation. By means of **184**, deduce that G is not a simple group.

206 Let N denote the set of all positive integers. Every finite group can be embedded in $\Sigma_{(N)}$, the restricted symmetric group on N (see **110**, **148**).

We now consider another important group action.

4.25. G acts on itself *by conjugation*. In this case, for each $g \in G$ and each $x \in G$ we write x^g for the element of G to which g moves x, so that by definition

$$x^g = g^{-1}xg, \quad \text{the } conjugate \text{ of } x \text{ by } g \text{ (2.19)}.$$

This does define an action of G on itself, for if $x, g_1, g_2 \in G$ then

$$(x^{g_1})^{g_2} = g_2^{-1}(g_1^{-1}xg_1)g_2 = (g_1g_2)^{-1}x(g_1g_2) = x^{g_1g_2}$$

and $\qquad\qquad x^1 = 1^{-1}x1 = x.$

Now the orbit of x is the set

$$\{g^{-1}xg : g \in G\},$$

the *conjugacy class of x in G* (see **49**); and

$$\begin{aligned}\mathrm{Stab}_G(x) &= \{g \in G : g^{-1}xg = x\} \\ &= \{g \in G : xg = gx\} \\ &= C_G(x),\end{aligned}$$

the *centralizer of x in G* (see chapter 1). The corresponding permutation representation of G is the map $\tau : G \to \Sigma_G$ defined in 2.21. By 4.9,

$$\mathrm{Ker}\ \tau = \bigcap_{x \in G} C_G(x) = Z(G),$$

as in **117**.

When we apply 4.11 and the definition of orbit for this action, we get the following important results.

4.26 Corollary. *For each $x \in G$,*

$$\big| \text{the conjugacy class of } x \text{ in } G \big| = |G : C_G(x)|.$$

4.27 Corollary (The class equation). *If G is a finite group with k distinct* conjugacy classes of elements, and if $x_1, \ldots, x_k$ are elements of G, one from each of these k classes, then

$$|G| = \sum_{i=1}^{k} |G : C_G(x_i)|.$$

The positive integer k is called the *class number* of G, which we denote by $k(G)$.

We note two applications of the class equation.

4.28 Theorem. *If* $|G| = p^n$, *where* n *is a positive integer, then* $Z(G) \neq 1$.

We shall generalize this result in 5.8, but the statement given here is sufficiently important to be recorded separately.

Proof. Let $x \in G$. By 4.26, the conjugacy class of x contains just one element if and only if $C_G(x) = G$, that is, if and only if $x \in Z(G)$. Hence if $Z(G) = 1$, the class equation gives

$$p^n = 1 + m_2 + m_3 + \ldots + m_k,$$

where each of the positive integers $m_2, \ldots, m_k$ is a divisor of p^n and is greater than 1. (In the notation of 4.27, $m_i = |G : C_G(x_i)|$.) But then, since p is prime, each of $m_2, \ldots, m_k$ is a power of p to a positive exponent, and so $m_2 + m_3 + \ldots + m_k$ is divisible by p. The equation above then implies that 1 is divisible by p, a contradiction. We conclude that $Z(G) \neq 1$.

This property is fundamental for the investigation of groups of prime power orders. The property is not in general shared by finite groups whose orders involve two or more prime numbers: for instance, the group Σ_3, of order 6, has trivial centre.

From 4.28 we make two deductions.

4.29 Corollary. *If* G *is a finite non-abelian simple group then* $|G|$ *is divisible by at least two distinct primes.*

As a matter of fact, the order of a finite non-abelian simple group is divisible by at least *three* distinct primes. This is an important result of William Burnside (1852–1927), to which we shall refer again, but which will not be proved in this book.

4.30 Corollary. *Every group of order* p^2 *is abelian.*

Proof. Suppose to the contrary that $|G| = p^2$ and G is non-abelian. Then $Z(G) < G$ and so, by 4.28 and Lagrange's theorem, $|Z(G)| = p$. Hence $|G/Z(G)| = p$, and so $G/Z(G)$ is cyclic. But then it follows (**125**) that G is abelian, a contradiction.

For every prime p there is a non-abelian group of order p^3: see **221**.

207 Let G be a finite non-abelian simple group and p the largest prime divisor of $|G|$.
 (i) If $H < G$ then $|G : H| \geq p$.
 (ii) If X is a conjugacy class of non-trivial elements of G then $|X| \geq p$.

208 Let X be a conjugacy class of elements of G. Then $\langle X \rangle \trianglelefteq G$. (Hint. Use 3.50.)

209 Let X be a conjugacy class of non-trivial elements of G.

(i) Let $\alpha \in \operatorname{Aut} G$ and $X^\alpha = \{ x^\alpha : x \in X \}$. Then X^α is a conjugacy class of elements of G.

(ii) Suppose that G is a finite non-abelian simple group and that, for every conjugacy class Y of elements of G distinct from X, either $|X| \neq |Y|$ or the elements of X and the elements of Y have different orders. Let $|X| = n$. Then $\operatorname{Aut} G$ can be embedded in Σ_n. (Hint. For (ii), show that $\operatorname{Aut} G$ acts on X and that $\langle X \rangle = G$. Use **208** and 2.28.)

210 Let G be a finite group and let x, y be conjugate elements of G. Then the number of distinct elements $g \in G$ such that $x^g = y$ is equal to $|C_G(x)|$.

211 Let $x \in G$, a finite group. Then $|C_G(x)| \geqslant |G/G'|$ (where G' denotes the derived group of G: see 3.48).

212 Let n be a positive integer and F a field such that $|F| > n$. Let $G = \operatorname{GL}_n(F)$, and let x be a diagonal matrix in G whose n diagonal entries are distinct elements of F. Prove that $C_G(x)$ is the subgroup of G consisting of all diagonal matrices in G (cf. **123**).

213 Let n be a positive integer and F a field in which $1 + 1 \neq 0$. Let $G = \operatorname{GL}_n(F)$ and, for each $i = 0, 1, \ldots, n-1$, let t_i be the diagonal matrix in G whose first i diagonal entries are equal to 1 and whose other diagonal entries are equal to -1.

Prove that every element of order 2 in G is conjugate in G to one of the n elements $t_0, t_1, \ldots, t_{n-1}$. Prove also that no two of the elements $t_0, t_1, \ldots, t_{n-1}$ are conjugate in G.

Hence G has just n conjugacy classes of elements of order 2. (Hints. Let V be a vector space of dimension n over F and choose a base of V. With respect to this base, any element of G of order 2 represents an element θ of $\operatorname{GL}(V)$ of order 2. Note that for every $v \in V, v = \frac{1}{2}(v + v\theta) + \frac{1}{2}(v - v\theta)$. Hence show that there is a base of V with respect to which θ is represented by one of the matrices $t_0, t_1, \ldots, t_{n-1}$. Note that two elements of G are conjugate in G if and only if they represent the same element of $\operatorname{GL}(V)$ with respect to suitable bases of V.)

214 Let H be a subgroup of index 2 in the finite group G. Assume that for every $h \in H$ with $h \neq 1, C_G(h) \leqslant H$ (that is, by 4.26, the G-conjugacy class of h splits into two H-conjugacy classes). Then $G \backslash H$ forms a single conjugacy class of elements in G. (Hint. Let $g \in G \backslash H$. Show that the map $h \mapsto g^h$, defined for all $h \in H$, is an injective map of H into G.)

215 Prove that every group of order 15 is abelian. Deduce by means of **107** and **6** that every group of order 15 is cyclic (and therefore that $\nu(15) = 1$). (Hint. If G were a non-abelian group of order 15 then, by **125**, $Z(G) = 1$. Then use the class equation to show that G would have just 1 conjugacy class with 5 elements in it and that this would consist of all the elements of G of order 3, in contradiction to **67**. Remark. This result will be proved in another way in 5.18.)

216 Let G be a non-trivial finite group and let p be the least prime divisor of $|G|$. If $k(G) > |G|/p$ then $Z(G) \neq 1$.

217 If G is a finite non-abelian group then $k(G) > |Z(G)| + 1$.

218 Let G be a non-abelian group.

(i) For every $x \in G, Z(G) < C_G(x)$.

(ii) If $|G| = p^3$ then $|Z(G)| = p$ and $k(G) = p^2 + p - 1$.

219 (i) Let $H \leqslant G$. Then $H \trianglelefteq G$ if and only if H is a union of G-conjugacy classes of elements.

(ii) Let $H \trianglelefteq G$, a finite group. Then

$$k(G/H) \leqslant k(G) - j + 1,$$

where j is the number of G-conjugacy classes of elements in H.

(iii) If G is a finite non-abelian group such that $G/Z(G)$ is abelian then $k(G) \geqslant |G/Z(G)| + |Z(G)| - 1$.

*220 Use the class equation to prove the following theorem of Cauchy: if G is a finite group and p divides $|G|$ then G has an element of order p. (Hint. Use induction on $|G|$ and the fact that by **107** the result is true if G is abelian: cf. 1.13. Later we shall prove Cauchy's theorem in another way: see 5.11.)

*221 Let U be the subgroup of $GL_3(F)$ defined in **120**, and let $F = \mathbf{Z}_p$.

(i) Then U is a non-abelian group of order p^3.

(ii) If $p > 2$ then $x^p = 1$ for every $x \in U$ (cf. 3).

222 Every group of order p^2 is isomorphic to either C_{p^2} or $C_p \times C_p$. Hence $\nu(p^2) = 2$ (cf. 77).

It is obvious that if G is a finite group then $k(G) \leqslant |G|$ (with equality if and only if G is abelian). As a second application of the class equation we shall prove the less obvious fact that $|G|$ is bounded above by a function of $k(G)$. We follow the formulation of the proof in Scott [b36].

4.31 Theorem (E. Landau [a69], 1903). *For each positive integer k, there is a positive integer $N(k)$ such that, for every finite group G with class number k, $|G| \leqslant N(k)$.*

Proof. If G is a finite group with class number k then the class equation for G gives, with the notation of 4.27,

$$|G| = \sum_{i=1}^{k} |G : C_G(x_i)|. \qquad \text{(i)}$$

Let $n_i = |C_G(x_i)|$, for each $i = 1, \ldots, k$. Without loss of generality, we may suppose $x_1, \ldots, x_k$ labelled so that

$$n_1 \geqslant n_2 \geqslant \ldots \geqslant n_k. \qquad \text{(ii)}$$

Note that then $n_1 = |G|$, since $C_G(1) = G$. Division of equation (i) by $|G|$ gives

$$1 = \sum_{i=1}^{k} \frac{1}{n_i}. \qquad \text{(iii)}$$

In order to complete the proof it will therefore suffice to show that there are only finitely many sequences $(n_1, n_2, \ldots, n_k)$ of positive integers satisfying (ii) and (iii): for then we may take for $N(k)$ the largest value of n_1 among all such sequences.

To achieve this we shall prove, by induction on k, that for each positive integer k and real number A, if $\mathscr{S}(k, A)$ denotes the set of all sequences $(n_1, n_2, \ldots, n_k)$ of positive integers satisfying (ii) and

$$\sum_{i=1}^{k} \frac{1}{n_i} = A, \tag{iv}$$

then $\mathscr{S}(k, A)$ is a finite set. (It may of course happen that $\mathscr{S}(k, A) = \emptyset$, as for instance if $A \leqslant 0$, but this does not matter.) This assertion is obvious if $k = 1$. Assume that $k > 1$ and, inductively, that $\mathscr{S}(k - 1, B)$ is a finite set, for every real number B. We may also assume that $A > 0$. If $(n_1, n_2, \ldots, n_k)$ is a sequence in $\mathscr{S}(k, A)$ then

$$A = \sum_{i=1}^{k} \frac{1}{n_i} \leqslant \frac{k}{n_k},$$

so that

$$n_k \leqslant \frac{k}{A}.$$

Thus there are only finitely many possible choices for n_k.

Now $\mathscr{S}(k, A)$ is a subset of the set $\mathscr{T}(k, A)$ of all sequences $(n_1, n_2, \ldots, n_k)$ of positive integers such that $n_1 \geqslant n_2 \geqslant \ldots \geqslant n_{k-1}, n_k \leqslant \frac{k}{A}$ and

$$\sum_{i=1}^{k-1} \frac{1}{n_i} = A - \frac{1}{n_k}. \tag{v}$$

Since, for each choice of n_k, $\mathscr{S}(k - 1, A - (1/n_k))$ is a finite set, by the inductive assumption, and since there are only finitely many choices for n_k, it follows that $\mathscr{T}(k, A)$ is a finite set. Hence also $\mathscr{S}(k, A)$ is a finite set, and the induction argument goes through.

There also exist infinite groups with only finitely many distinct conjugacy classes. Indeed, G. Higman, B. H. Neumann and H. Neumann [a58] proved that any infinite group in which no non-trivial element has finite order can be embedded in a group in which all non-trivial elements form a single conjugacy class.

223 Let G be a finite group. Then
   (i) $k(G) = 2$ if and only if $G \cong C_2$,
   (ii) $k(G) = 3$ if and only if either $G \cong C_3$ or $G \cong \Sigma_3$.
(Hint. Use the class equation as in the proof of 4.31. In case (ii), show that the possible orders for G are 3, 4, 6 and examine the groups of these orders.)

224 (i) Let X be a conjugacy class of elements in G, and let $X^* = \{x^{-1} : x \in X\}$. Show that X^* is a conjugacy class of elements in G.
   (ii) Suppose that G is finite. Prove that if $|G|$ is odd then $\{1\}$ is the only conjugacy class X such that $X = X^*$, but that if $|G|$ is even then there is at least one conjugacy class X other than $\{1\}$ such that $X = X^*$.
   (iii) Prove that if G is a finite group with $k(G)$ even then $|G|$ is even. Show by an example that the converse is false.
(Hint. For the first assertion in (ii), **12** may be helpful.)

We consider next an extension of the action of G on itself described in 4.25 to an action on the set $\mathcal{Q}(G)$ of all non-empty subsets of G.

4.32. G *acts on* $\mathcal{Q}(G)$ *by conjugation:* for each $g \in G$ and each non-empty subset U of G, g moves U to the set

$$U^g = g^{-1}Ug = \{g^{-1}ug : u \in U\},$$

which is called the *conjugate of* U *by* g. (When U consists of a single element or is a subgroup of G, this conforms with our previous terminology.) It is easy to check that this does define an action of G on $\mathcal{Q}(G)$; and the action in 4.25 is obtained by restricting this action to the subsets of G consisting of single elements.

For each $U \in \mathcal{Q}(G)$, the orbit of U is the set of all conjugates of U, that is, the set $\{g^{-1}Ug : g \in G\}$ of subsets of G: it is called the *conjugacy class of* U *in* G; and $\mathrm{Stab}_G(U) = \{g \in G : g^{-1}Ug = U\}$, called the *normalizer of* U *in* G and denoted by $N_G(U)$ (in accordance with the terminology and notation introduced in 3.55 when U is a subgroup of G).

We sometimes say that H *normalizes* U to mean that $H \leqslant N_G(U)$.

The 'exponential' notation U^g for the conjugate $g^{-1}Ug$ is a very convenient one, which will be used as standard in the rest of the book.

From 4.11, applied to 4.32, we get the following generalization of 4.26:

4.33 Corollary. *For each non-empty subset* U *of* G,

$$\left| \text{the conjugacy class of } U \text{ in } G \right| = \left| G : N_G(U) \right|$$

(that is, $\left| G : N_G(U) \right|$ is the number of distinct conjugates of U in G).

4.34. For any $H \leqslant G$, we have seen in 3.55 that $N_G(H)$ is the unique largest subgroup of G in which H is contained as a normal subgroup. But for a subset U of G which is not a subgroup, $N_G(U)$ need not even contain U: see **225**. Note that when $U = \{x\}$ for some $x \in G$, $N_G(U) = N_G(x) = C_G(x)$. But we can also define in a natural way a subgroup $C_G(U)$ of G for any $U \in \mathcal{Q}(G)$, and frequently $C_G(U) \neq N_G(U)$. In the notation of **187**, with G acting on itself by conjugation and U any non-empty subset of G, $C_G(U) = G_U$ and $N_G(U) = G_U^*$.

4.35 Definition. For any non-empty subset U of G, we define the *centralizer of* U *in* G to be

$$C_G(U) = \{g \in G : ug = gu \text{ for all } u \in U\}$$

$$= \bigcap_{u \in U} C_G(u) \leqslant G.$$

(This conforms with the definition in **122** when $U = H \leqslant G$.) Note that $C_G(U) = G$ if and only if $U \subseteq Z(G)$.

We sometimes say that H *centralizes* U to mean that $H \leqslant C_G(U)$.

It is easy to see that $C_G(U) \leqslant N_G(U)$ always, and in fact $C_G(U) \trianglelefteq N_G(U)$ (**233**). When U is a subgroup of G, a useful extra assertion can be made.

4.36 Lemma. *For every* $H \leqslant G$, $C_G(H) \trianglelefteq N_G(H)$ *and* $N_G(H)/C_G(H)$ *can be embedded in* Aut H.

Proof. Since (by 3.55) $H \trianglelefteq N_G(H)$, $h^g \in H$ for every $h \in H$ and $g \in N_G(H)$. Then it is clear that $N_G(H)$ acts on H by conjugation. Let the corresponding permutation representation of $N_G(H)$ be σ, so that for each $g \in N_G(H)$,

$$g\sigma : h \mapsto h^g \text{ for all } h \in H.$$

Then
$$\begin{aligned}
\operatorname{Ker} \sigma &= \{g \in N_G(H) : h^g = h \text{ for all } h \in H\} \\
&= \{g \in N_G(H) : hg = gh \text{ for all } h \in H\} \\
&= C_G(H), \text{ since } C_G(H) \leqslant N_G(H).
\end{aligned}$$

Hence, by the fundamental theorem on homomorphisms,

$$C_G(H) \trianglelefteq N_G(H) \text{ and } \operatorname{Im} \sigma \cong N_G(H)/C_G(H).$$

For each $g \in N_G(H)$, $g\sigma$ is a permutation of H. In fact it is an automorphism of H, for if $h_1, h_2 \in H$ then

$$(h_1 h_2)^g = g^{-1} h_1 h_2 g = g^{-1} h_1 g g^{-1} h_2 g = h_1^g h_2^g.$$

Hence $\operatorname{Im} \sigma$ is a subgroup of Aut H, and so $N_G(H)/C_G(H)$ can be embedded in Aut H.

This lemma will be placed in a more general context in chapter 9. We end this chapter by noting some applications.

If H is a finite group then Aut H is also finite. Hence from 4.36 we deduce that in an infinite group every finite normal subgroup has a 'large' centralizer.

4.37 Corollary. *Let G be an infinite group. Then, for any finite normal subgroup H of G, $G/C_G(H)$ is finite. In particular, if G has no non-trivial finite quotient then every finite normal subgroup of G is abelian and contained in $Z(G)$.*

Note for example that $\mathbf{Q}^+/\mathbf{Z}^+$ and C_{p^∞} are infinite abelian groups with no non-trivial finite quotient but many finite subgroups: see **133** and **144**.

4.38 Lemma. (i) *For any cyclic group G,* Aut G *is abelian.*

(ii) *If $|G| = p$ then $|\text{Aut } G| = p - 1$.*

Proof. The statements follow from **46**, but we give direct proofs here.

Let $G = \langle g \rangle$. Each automorphism α of G is determined by its effect on g. Let $\alpha, \beta \in \text{Aut } G$ with, say, $g^\alpha = g^r, g^\beta = g^s$, where $r, s \in \mathbf{Z}$. Then

$$g^{\alpha\beta} = (g^r)^\beta = g^{sr} = g^{rs} = (g^s)^\alpha = g^{\beta\alpha}.$$

Since the automorphisms $\alpha\beta, \beta\alpha$ of G have the same effect on g, it follows that $\alpha\beta = \beta\alpha$. Thus Aut G is abelian.

Now let G be finite. Then for any particular integer r, there is an automorphism α of G such that $g^\alpha = g^r$ provided only that $o(g^r) = o(g)$. If $o(g) = p$ then there are $p - 1$ choices for g^r, so that $|\text{Aut } G| = p - 1$.

From 4.36 and 4.38 we deduce

4.39 Corollary. (i) *If G is a perfect group (see* **168**) *and K is a cyclic normal subgroup of G then $K \leqslant Z(G)$.*

(ii) *If G is a finite group, p the smallest prime divisor of $|G|$ and K a normal subgroup of G of order p then $K \leqslant Z(G)$ (cf.* **119** *and* **4.18**).

*225 Let $G = \Sigma_3$. Find a subset U of G such that $N_G(U) = 1$.

226 Let U be a non-empty subset of G. Then G acts transitively by conjugation on the conjugacy class of U in G, and this action is equivalent to the action by right multiplication of G on the set of right cosets of $N_G(U)$ in G.

227 Let H be a finite subgroup of G, and let $K = H^G$, the normal closure of H in G (see **180**). Then K is a finite normal subgroup of G if and only if $|G : N_G(H)| < \infty$. (cf. 4.14, 4.15. Hints. To prove that if $|G : N_G(H)| < \infty$ then $|K| < \infty$, let the distinct conjugates of H in G be $H_1, H_2, \ldots, H_n$: see 4.33. Let $k \in K$. Use **180**(ii) and 2.28 to show that k is expressible in the form $k = h_{i_1} h_{i_2} \ldots h_{i_r}$, where r is a positive integer and, for each $j = 1, \ldots, r$, $i_j \in \{1, 2, \ldots, n\}$ and $h_{i_j} \in H_{i_j}$. Choose such an expression for k with r as small as possible. Then observe that if $r > n$, there are integers j, l such that $1 \leqslant j < l \leqslant r$ and $i_j = i_l$, and then

$$h_{i_j} h_{i_{j+1}} \ldots h_{i_l} = (h_{i_j} h_{i_l})(h_{i_l}^{-1} h_{i_{j+1}} h_{i_l}) \ldots (h_{i_l}^{-1} h_{i_{l-1}} h_{i_l}).$$

Deduce that $r \leqslant n$. This is a special case of a result known as Ditsman's lemma.)

228 Let F be a field and let $G = \text{GL}_2(F)$.

(i) Prove that Σ_3 can be embedded in G. (Hint. Show that the elements $\begin{pmatrix} 0 & 1 \\ -1 & -1 \end{pmatrix}$ and $\begin{pmatrix} 0 & 1 \\ 1 & 0 \end{pmatrix}$ of G generate a non-abelian subgroup of order 6, and see **58**, **60**.)

(ii) Suppose that in $F, 1 + 1 \neq 0$. Prove that $C_2 \times C_2 \times C_2$ cannot be embedded in G. Deduce that the alternating group A_4 cannot be embedded in G. (Hints. Assume that $C_2 \times C_2 \times C_2$ can be embedded in G and derive a contradiction by applying **212** and **213**. Note that if L were a subgroup of G isomorphic to A_4 then $L \cap Z(G) = 1$. See **123**, **185** and 3.54.

Remarks. Since $\text{GL}_2(\mathbf{Z}_2) \cong \Sigma_3$ (**44**), A_4 cannot be embedded in $\text{GL}_2(\mathbf{Z}_2)$. However, there is a field F with $|F| = 4$, and it is known that for this $F, \text{SL}_2(F) \cong A_5$, so that then A_4 can be embedded in $\text{GL}_2(F)$.)

*229 (i) Let U be a non-empty subset of G and let $g \in G$. Then

$$C_G(U^g) = C_G(U)^g, \quad N_G(U^g) = N_G(U)^g \text{ and } \langle U^g \rangle = \langle U \rangle^g.$$

(ii) Let $H, K \leqslant G$ and let $g \in G$. Then

$$Z(H^g) = Z(H)^g, H^g \cap K^g = (H \cap K)^g \text{ and } \langle H^g, K^g \rangle = \langle H, K \rangle^g.$$

*230 Let $K \trianglelefteq G$.

  (i) If $H \leqslant G$ and $g \in G$, then $(HK/K)^{Kg} = H^g K/K$.

  (ii) If H_1 and H_2 are conjugate subgroups of G, then $H_1 K/K$ and $H_2 K/K$ are conjugate subgroups of G/K.

  (iii) If J_1/K and J_2/K are conjugate subgroups of G/K, then J_1 and J_2 are conjugate subgroups of G.

231 Suppose that G is an infinite simple group.

  (i) If U is a non-empty subset of G such that there are only finitely many distinct conjugates of U in G, then either $U = \{1\}$ or $\langle U \rangle = G$.

  (ii) If x is a non-trivial element of G then there are infinitely many distinct conjugates of x in G.

232 Let $x \in G$. Prove that if $C_G(x) \trianglelefteq G$ then x lies in an abelian normal subgroup of G. Show by an example that the converse is false. (Hint. If $C_G(x) \trianglelefteq G$, show that $\langle x \rangle^G$ is abelian: see **180**.)

*233 Show that for any non-empty subset U of G, $C_G(U) \trianglelefteq N_G(U)$ and $N_G(U)/C_G(U)$ can be embedded in Σ_U. For $G = \Sigma_3$, find a U with $C_G(U) \neq N_G(U)$.

*234 Let $x \in G$. The *extended centralizer of x in G* is defined to be the subgroup $C_G^*(x) = N_G(\{x, x^{-1}\})$. Show that $|C_G^*(x) : C_G(x)| \leqslant 2$. For $G = \Sigma_3$, find elements $x, y \in G$ such that $|C_G^*(x) : C_G(x)| = 2$, $C_G^*(y) = C_G(y)$

235 (i) Suppose that G is non-abelian and let $Z = Z(G)$. Then, for every $x \in G \setminus Z$, $\langle x \rangle Z$ is an abelian subgroup of G containing Z properly.

  (ii) Let A be an abelian subgroup of G. We say that A is a *maximal abelian subgroup of G* if there is no abelian subgroup of G which contains A properly. Then A is a maximal abelian subgroup of G if and only if $C_G(A) = A$.

236 Let $J \trianglelefteq H \leqslant G$, and let $K = C_G(J)$. Then $H \leqslant N_G(K)$. In particular, if $N_G(K) = K$ then $J \leqslant Z(H)$.

237 Let G be a finite group.

  (i) If $H \leqslant G$ then

$$\left| \bigcup_{g \in G} H^g \right| \leqslant 1 + |G| - |G : H|.$$

Hence the union of all the conjugates in G of a proper subgroup of G is a proper subset of G.

  (ii) If $K \leqslant G$ and K contains at least one element from each conjugacy class of elements of G then in fact $K = G$.

238 Let G be a non-trivial group.

  (i) If M is a maximal subgroup of G (see **140**) then, for every $g \in G$, M^g is a maximal subgroup of G.

  (ii) If G is finite and G has just one conjugacy class of maximal subgroups then G is cyclic of order p^m for some prime p and positive integer m. (cf. **140**(vi). Hint. Use **237**(i).)

239 Let V be a vector space of dimension n over a field F, where n is an integer greater than 1, and let $G = \mathrm{GL}(V)$ (2.16). Let $0 \neq v \in V$, and let H be the set of all elements of G for which v is an eigenvector. Then

  (i) $H < G$.

(ii) If $F = \mathbf{C}$ then $\bigcup_{g \in G} H^g = G$.

(cf. **237**. Hint. When $F = \mathbf{C}$, any $x \in G$ has an eigenvector, w say, and then there is an element $g \in G$ such that $vg = w$.)

240 Let G be an infinite group with $Z(G) = 1$. By 4.37, if G has no non-trivial finite quotient then G has no non-trivial finite normal subgroup. Show by an example that the converse is false. (Hint. Consider the infinite dihedral group D_∞.)

241 $G \cong D_\infty$ if and only if G has an infinite cyclic normal subgroup $H = \langle h \rangle$, say, such that $H < G$ and, for every positive integer n, $C_G(h^n) = H$. (Hint. Use **29**, **46** and 4.36.)

*242 Let p be an odd prime.
   (i) Then $\mathbf{Z}_p^\times$ has just one element of order 2.
   (ii) Let G be a group of order $2p$. Then G has a cyclic subgroup $\langle x \rangle$ of order p and a cyclic subgroup $\langle t \rangle$ of order 2, and x^t is either x or x^{-1}. Hence G is isomorphic to either C_{2p} or D_{2p}. (cf. **60**. Hint. Apply Cauchy's theorem **220** with **40** and **46**.)

*243 If G is a cyclic group of order p^n, where n is a positive integer, then $|\text{Aut } G| = p^n - p^{n-1}$ (cf. 4.38(ii), **40**, **46**).

244 Let $J \leqslant G$. Then $C_G(J) = 1$ if and only if $Z(H) = 1$ for every H such that $J \leqslant H \leqslant G$.

245 The elements of $C_{\text{Aut } G}(\text{Inn } G)$ are called *central automorphisms* of G. Let $\alpha \in \text{Aut } G$. Prove that α is a central automorphism of G if and only if $g^\alpha g^{-1} \in Z(G)$ for every $g \in G$. Deduce that if $Z(G) = 1$ then $Z(\text{Aut } G) = 1$. (Hint. See **92** and **117**.)

For further information about natural actions of groups on sets, see the books of Passman [b32] and Wielandt [b38].

5

FINITE p-GROUPS AND SYLOW'S THEOREM

In this chapter we make some further fundamental applications of the ideas developed in chapter 4 of group actions on sets. These applications will be concerned especially with finite p-groups and p-subgroups of finite groups. We shall obtain further information about finite simple groups and prove the simplicity of the alternating groups A_n for $n \geqslant 5$.

5.1 Definition. Let G act on the set X. Then the *fixed point subset* of X is defined to be

$$\text{Fix}_X(G) = \{x \in X : xg = x \text{ for all } g \in G\}$$
$$= \{x \in X : \text{Stab}_G(x) = G\}.$$

Thus $\text{Fix}_X(G)$ consists of those elements of X each of which forms an orbit by itself. Of course, it may happen that $\text{Fix}_X(G) = \emptyset$. In particular, if G acts transitively on X then $\text{Fix}_X(G) = \emptyset$ unless $|X| = 1$.

For instance, if $G \leqslant \Sigma_4$ and G acts naturally on the set $X = \{1, 2, 3, 4\}$ then, for $G = \langle (123) \rangle$, $\text{Fix}_X(G) = \{4\}$, while for $G = \langle (12)(34) \rangle$, $\text{Fix}_X(G) = \emptyset$.

The following simple application of 4.11 is very helpful. The proof is essentially the same as the proof of 4.28.

5.2 Lemma. *Let G be a finite p-group which acts on the finite set X. Then*

$$|\text{Fix}_X(G)| \equiv |X| \bmod p.$$

Proof. Let the orbits of the action be $X_1, \ldots, X_k$, where k is a positive integer. Now we count the elements of X:

$$|X| = \sum_{i=1}^{k} |X_i|.$$

By 4.11, each $|X_i|$ is a divisor of $|G|$ and hence, since p is prime, must be a power of p. If there are just j orbits consisting of single elements, where $0 \leqslant j \leqslant k$, then $|\text{Fix}_X(G)| = j$ and the equation above gives

$$|X| = j + \text{a sum of powers of } p \text{ to positive exponents}$$

(where the latter sum is empty if $j = k$). Hence

$$|\text{Fix}_X(G)| = j \equiv |X| \bmod p.$$

We shall make two important deductions from 5.2, using the group action of 4.13 and the following observation.

5.3. Let G act on the set X. Then each subgroup J of G acts on X *by restriction* of the action of G: that is, to each $j \in J$ and each $x \in X$ there corresponds the element $xj \in X$ determined by the action of G on X. This correspondence obviously satisfies the conditions for an action of J on X.

For each $x \in X$, $\text{Stab}_J(x) = \text{Stab}_G(x) \cap J$. Hence $x \in \text{Fix}_X(J)$ if and only if $J \leqslant \text{Stab}_G(x)$. If the permutation representation of G corresponding to the given action is ρ, then the permutation representation of J corresponding to the action of J is $\rho|_J$. If the action of G is faithful, then the action of J is also faithful. But the action of G may be transitive and the action of J intransitive.

For instance, let $H, J \leqslant G$ and let G act by right multiplication on the set X of right cosets of H in G, as in 4.13. Consider the action by restriction of J on X. Then, for each $g \in G$, $\text{Stab}_J(Hg) = H^g \cap J$, and $Hg \in \text{Fix}_X(J)$ if and only if $J \leqslant H^g$. The action of J on X is transitive if and only if $HJ = G$, in which case, by 4.20, this action is equivalent to the action of J by right multiplication on the set of right cosets of $H \cap J$ in J.

5.4 Theorem. *Let $H, J \leqslant G$. Suppose that $|G : H| = r < \infty$, that J is a finite p-group and that p does not divide r. Then $J \leqslant H^g$ for some $g \in G$.*
Proof. Let X be the set of right cosets of H in G. Then $|X| = |G : H| = r$. Let J act on X by restriction of the action of G on X by right multiplication. By 5.2, $|\text{Fix}_X(J)| \equiv r \bmod p$. Since, by hypothesis, p does not divide r, it follows that $|\text{Fix}_X(J)| \neq 0$, that is, that $\text{Fix}_X(J) \neq \emptyset$. Hence, by 5.3, $J \leqslant H^g$ for some $g \in G$.

5.5 Theorem. *Suppose that H is a p-subgroup of the finite group G and that p divides $|G : H|$. Then p divides $|N_G(H)/H|$.*
Proof. Let X be the set of right cosets of H in G, and let H act on X by restriction of the action of G on X by right multiplication (as in 5.3, with $J = H$). By 5.2, $|\text{Fix}_X(H)| \equiv |G : H| \bmod p$. Let $g \in G$. By 5.3, $Hg \in \text{Fix}_X(H)$ if and only if $H \leqslant H^g$, that is (since $|H| = |H^g|$), if and only if $H = H^g$, or, equivalently, if and only if $g \in N_G(H)$. Hence $|\text{Fix}_X(H)| = |N_G(H)/H|$. Now the result follows.

We state the most important special case of 5.5 as a separate result. (We shall see that this is actually equivalent to 5.5, once we have proved Sylow's theorem.)

5.6 Corollary. *In a finite p-group G, every proper subgroup is a proper subgroup of its normalizer in G.*

This property of groups of prime power orders is not in general shared by finite groups whose orders involve two or more prime numbers: for instance, the group Σ_3, of order 6, has 'self-normalizing' subgroups of order 2.

246 Suppose that G acts on the set X and $\text{Fix}_X(G) = \varnothing$. If $|G| = 35$ and $|X| = 19$, find the number of orbits of the action and the length of each orbit.

247 Let $H, J \leqslant G$ (where possibly $H = J$). A subset of G of the form $HgJ = \{hgj : h \in H, j \in J\}$, where $g \in G$, is called a *double coset* with respect to H and J. Let X be the set of right cosets of H in G and let J act on X as in 5.3. For each $g \in G$, show that HgJ is the union of the elements of X which form the orbit of Hg under this action, and hence (using 4.11) show that if H and J are finite then

$$|HgJ| = \frac{|H||J|}{|H^g \cap J|}.$$

Show also that if $g_1, g_2 \in G$ and $Hg_1J \neq Hg_2J$ then $Hg_1J \cap Hg_2J = \varnothing$. What is the equivalence relation on G for which the double cosets with respect to H and J are the equivalence classes?

248 Let G be a finite group. We make the following definitions:

(a) Suppose that G acts on the set X. The action is said to be a *Frobenius action* if it is transitive but not regular, $|X| > 1$, and whenever x_1, x_2 are distinct elements of X, $\text{Stab}_G(x_1) \cap \text{Stab}_G(x_2) = 1$.

(b) G is said to be a *Frobenius group* if it has a non-trivial proper subgroup H such that $N_G(H) = H$ and whenever H^{g_1}, H^{g_2} are distinct conjugates of H in G (with $g_1, g_2 \in G$), $H^{g_1} \cap H^{g_2} = 1$. Any such subgroup H is called a *Frobenius complement* in G.

Prove the following statements:

(i) G has a Frobenius action on some set if and only if G is a Frobenius group.

(ii) If G is a Frobenius group and H is a Frobenius complement in G then $|G : H| \equiv 1 \bmod |H|$.

(iii) If G is a Frobenius group then $Z(G) = 1$.

(iv) Let n be a positive integer, and consider the natural action of Σ_n on the set $\{1, 2, \ldots, n\}$. This is a Frobenius action if and only if $n = 3$.

(v) Let n be an integer, $n \geqslant 3$. Then the dihedral group D_{2n} is a Frobenius group if and only if n is odd.

(Hints. See **124, 187, 203**. For (ii), consider the restriction to H of a suitable Frobenius action of G, and count orbits. Remarks. Let H be a Frobenius complement in the Frobenius group G. F. G. Frobenius (1849–1917) proved in [a 31] the important theorem that then $K = \{1\} \cup (G \setminus \bigcup_{g \in G} H^g)$ is a normal subgroup of G. Moreover, $G = HK, H \cap K = 1$ and any Frobenius complement in G is conjugate to H. Hence any two Frobenius actions of G are equivalent.)

249 Suppose that $|G| = p^m r$, where m and r are positive integers and p does not divide r. Then G has a subgroup of order p^m. (Hint. Consider a p-subgroup H of G of greatest possible order and use 5.5 and Cauchy's theorem **220** to prove that $|H| = p^m$. This is part of Sylow's theorem, which will be proved in a different way in 5.9.)

Next we apply 5.2 to prove

5.7 Theorem. *Let $H \trianglelefteq G$, a finite group, and let J be a p-subgroup of G. If $|H| \not\equiv 1 \bmod p$ then $H \cap C_G(J) \neq 1$.*
Proof. Since $H \trianglelefteq G, G$ acts on H by conjugation. Then, by restriction of this action, J also acts on H. By definition,

$$\text{Fix}_H(J) = \{h \in H : h^j = h \text{ for all } j \in J\}$$
$$= \{h \in H : hj = jh \text{ for all } j \in J\}$$
$$= H \cap C_G(J).$$

Since J is a finite p-group and H is finite, 5.2 shows that

$$|H \cap C_G(J)| \equiv |H| \bmod p.$$

Therefore, since, by hypothesis, $|H| \not\equiv 1 \bmod p$, it follows that $H \cap C_G(J) \neq 1$.

As the most important special case we note

5.8 Corollary. *Let G be a finite p-group and let $1 < H \trianglelefteq G$. Then $H \cap Z(G) \neq 1$.* (This includes 4.28 as a special case.)
Proof. In 5.7, choose $J = G$.

Recall that by Lagrange's theorem, the order of every subgroup of a finite group G is a divisor of $|G|$. The converse is false in the sense that there may be missing divisors: G need not have a subgroup of order n for every divisor n of $|G|$ (see **185**). Sylow's theorem, which we prove next, establishes the existence of subgroups of particular orders and provides valuable information about such subgroups. This theorem established in 1872, is of fundamental importance in finite group theory; its discovery has had a decisive effect in determining the character of the subsequent development of the theory. Several different proofs are known: one method for part of the result has been indicated in **249**. The proof which we give here is a very successful application of group action methods, due to H. Wielandt in 1959.

5.9 Theorem (L. Sylow [a95]). *Let G be a finite group with $|G| = p^m r$, where m is a non-negative integer and r is a positive integer such that p does not divide r. Then*

(a) *G has a subgroup of order p^m. Such a subgroup is called a Sylow p-subgroup of G.*

(b) *If H is a Sylow p-subgroup of G and J is any p-subgroup of G then $J \leqslant H^g$ for some $g \in G$. In particular, the Sylow p-subgroups of G form a single conjugacy class of subgroups of G.*

(c) *Let n be the number of distinct Sylow p-subgroups of G. Then $n = |G : N_G(H)|$, where H is any particular Sylow p-subgroup of G; n divides r; and $n \equiv 1 \bmod p$.*
Proof. (a) (H. Wielandt [a 102]). We consider the set $\mathscr{X}$ of all subsets U of G with $|U| = p^m$. The number of such subsets is

$$|\mathscr{X}| = \binom{p^m r}{p^m} = \frac{p^m r}{p^m} \cdot \frac{p^m r - 1}{p^m - 1} \cdot \frac{p^m r - 2}{p^m - 2} \cdots \frac{p^m r - p^m + 1}{1}.$$

If in each term $(p^m r - j)/(p^m - j)$ of this product we make all possible cancellations of common divisors of numerator and denominator, p does not remain as a divisor of the numerator. This is clear for $j = 0$; and for $j > 0$, with, say, $j = p^l q$, where l is a non-negative integer and q a positive integer not divisible by p, then $l < m$,

$$\frac{p^m r - j}{p^m - j} = \frac{p^{m-l} r - q}{p^{m-l} - q},$$

and p does not divide $p^{m-l} r - q$. Since p is prime, it follows that p does not divide the product of cancelled numerators, and therefore

$$p \text{ does not divide } |\mathscr{X}|. \tag{i}$$

For $U \in \mathscr{X}$ and $g \in G$, $Ug = \{ug : u \in U\}$ is a subset of G with $|Ug| = p^m$: thus $Ug \in \mathscr{X}$. Now it is clear that G acts on the set $\mathscr{X}$ by right multiplication. By this action, $\mathscr{X}$ is partitioned into orbits, and it follows from (i) that

$$\text{there is an orbit } \mathscr{X}_1 \text{ such that } p \text{ does not divide } |\mathscr{X}_1|. \tag{ii}$$

Let $V \in \mathscr{X}_1$, so that $\mathscr{X}_1$ is the orbit of V, and let $H = \mathrm{Stab}_G(V) \leqslant G$. By 4.11,

$$|\mathscr{X}_1| = |G : H|. \tag{iii}$$

Since $|G : H| \, |H| = |G| = p^m r$, it follows from (ii) and (iii) and the fact that p is prime that

$$p^m \text{ divides } |H|. \tag{iv}$$

Now let
$$V = \{x_1, x_2, \ldots, x_{p^m}\}.$$

Then for any $h \in H$, $\quad Vh = V$,

that is, $\{x_1 h, x_2 h, \ldots, x_{p^m} h\} = \{x_1, x_2, \ldots, x_{p^m}\}$.

Hence $x_1 h = x_i$ for some i, where $1 \leqslant i \leqslant p^m$, and then

$$h = x_1^{-1} x_i.$$

Thus
$$|H| \leqslant p^m. \tag{v}$$

By (iv) and (v), $|H| = p^m$: thus H is a Sylow p-subgroup of G.

(b) Now let H be any Sylow p-subgroup of G and let J be any p-subgroup of G. Since $|G : H| = |G|/|H| = r$ and p does not divide r, 5.4 applies to show that $J \leqslant H^g$ for some $g \in G$, as asserted. In particular, if J is a Sylow p-subgroup of G then, since $|J| = |H| = |H^g|$, by 2.20, it follows that $J = H^g$, a subgroup of G in the same conjugacy class as H. Since every subgroup of G conjugate to H certainly has the same order as H and is therefore a Sylow p-subgroup of G, it follows that the Sylow p-subgroups of G form a single conjugacy class.

Before proving (c), we note a consequence of (b):

5.10 Lemma. *Suppose that H is a Sylow p-subgroup of a finite group G. Then H is the unique Sylow p-subgroup of $N_G(H)$.*

Proof. It is easy to see that H is a Sylow p-subgroup of every subgroup of G which contains H (**252**); in particular, H is a Sylow p-subgroup of $N_G(H)$. Let K be any Sylow p-subgroup of $N_G(H)$. By 5.9(b), there is an element $g \in N_G(H)$ such that $K = H^g$. But then, since $g \in N_G(H)$, $H^g = H$. Hence H is the unique Sylow p-subgroup of $N_G(H)$.

Remark. 5.10 can also be proved easily by means of 3.38 and 3.40, without invoking 5.9(b).

Proof of 5.9(c). Let $\mathscr{S}$ denote the set of all Sylow p-subgroups of G and let $H \in \mathscr{S}$. By (b), $\mathscr{S}$ is the conjugacy class of H in G, and so, by 4.33,

$$n = |\mathscr{S}| = |G : N_G(H)|.$$

Since $\qquad r = |G : H| = |G : N_G(H)|\,|N_G(H) : H|,$

it follows that n divides r.

Now G acts transitively on $\mathscr{S}$ *by conjugation.* Then, by restriction of this action, H acts on $\mathscr{S}$ – though not necessarily transitively. By 5.2,

$$|\mathrm{Fix}_{\mathscr{S}}(H)| \equiv |\mathscr{S}| \bmod p.$$

Let $K \in \mathscr{S}$. Then $K \in \mathrm{Fix}_{\mathscr{S}}(H)$ if and only if $K^h = K$ for every $h \in H$, that is, if and only if $H \leqslant N_G(K)$. But, by 5.10, $H \leqslant N_G(K)$ if and only if $H = K$. Hence $\mathrm{Fix}_{\mathscr{S}}(H) = \{H\}$, so that $|\mathrm{Fix}_{\mathscr{S}}(H)| = 1$ and

$$n \equiv 1 \bmod p, \quad \text{as claimed.}$$

*250 (i) Let G be a finite group with a p-subgroup J such that $C_G(J)$ is also a p-group. Show that for every normal subgroup H of G of order not divisible by p, $|H| \equiv 1 \bmod p$.

(ii) Use (i) to show that the only possible order for a non-trivial normal subgroup of Σ_4 of order not divisible by 3 is 4. (Hint. See **185**.)

251 Let A be an abelian normal subgroup of G. We say that A is a *maximal abelian normal subgroup of G* if there is no abelian normal subgroup of G which contains A properly (cf. **235**(ii)).

Prove that if G is a finite p-group and A is a maximal abelian normal subgroup of G, then A is a maximal abelian subgroup of G. (Hints. Suppose that $A < C_G(A)$, and consider $\bar{G} = G/A$. Derive a contradiction by means of 3.30, 4.36, 5.8 and **125**. Remark. If H is a finite non-abelian simple group then 1 is a maximal abelian normal subgroup of H, but 1 is certainly not a maximal abelian subgroup of H. See also **392, 400, 644, 645**.)

*252 Let G be a finite group and H a Sylow p-subgroup of G.

(i) If $H \leqslant L \leqslant G$, then H is a Sylow p-subgroup of L and every Sylow p-subgroup of L is a Sylow p-subgroup of G.

(ii) If $K \trianglelefteq G$, then $H \cap K$ is a Sylow p-subgroup of K and HK/K is a Sylow p-subgroup of G/K. Moreover, every Sylow p-subgroup of K is of the form $H^* \cap K$, where H^* is a Sylow p-subgroup of G; and every Sylow p-subgroup of G/K is of the form H^*K/K, where again H^* is a Sylow p-subgroup of G.

(iii) Show by an example that if $K \ntrianglelefteq G, H \cap K$ need not be a Sylow p-subgroup of K.

(iv) $O_p(G) \leqslant H \leqslant O^{\varpi}(G)$, where ϖ is any set of primes which does not contain p. Moreover, $G = HO^p(G)$.

253 Let G be a finite group. If G has a normal Sylow p-subgroup then so has every subgroup and every quotient group of G.

254 Let P be a Sylow p-subgroup of the finite group G, and let $P \leqslant J \leqslant H \leqslant G$. Then p does not divide $|H : J|$.

255 Let K be a finite normal subgroup of G. If K has a normal Sylow p-subgroup P then $P \trianglelefteq G$.

256 Let $H \leqslant G$, a finite group. Let P_0 be a Sylow p-subgroup of H and let P be a Sylow p-subgroup of G with $P_0 \leqslant P$. (Such a subgroup P exists, by 5.9(b).) Then $P_0 = P \cap H$.

257 Let G be a finite group, and let H and K be normal subgroups of G, and P a Sylow p-subgroup of G. Then $(PH) \cap (PK) = P(H \cap K)$.

258 Let G be a finite group and suppose that H and K are subgroups of G such that $G = HK$.

(i) If H and K are normal in G then, for every Sylow p-subgroup P of $G, P = (P \cap H)(P \cap K)$.

(ii) Show by an example that if H and K are not both normal in G, the conclusion of (i) need not hold.

(iii) However, there is always *some* Sylow p-subgroup P of G for which $P = (P \cap H)(P \cap K)$.

(Hints. For (iii), use 5.9(b) and **256** to show that there is a Sylow p-subgroup P of G such that $P \cap H$ is a Sylow p-subgroup of H and $P \cap K$ is a Sylow p-subgroup of K. Then use **98**.)

*__259__ Let n be an odd integer, $n \geqslant 3$. Then every Sylow subgroup of the dihedral group D_{2n} of order $2n$ is cyclic.

260 Let U be the subgroup of $GL_3(F)$ defined in **120**. Show that when $F = \mathbf{Z}_p$, U is a Sylow p-subgroup of $GL_3(\mathbf{Z}_p)$.

*__261__ Find a Sylow 2-subgroup T of Σ_4, and show that $T \cong D_8$. How many Sylow 2-subgroups does Σ_4 have? (Hint. Σ_4 has a cyclic subgroup U of order 4, and, by Sylow's theorem, U lies in a Sylow 2-subgroup T of Σ_4.)

*__262__ Suppose that $|G| = 2^m r$, where m and r are positive integers with r odd. Suppose further that the Sylow 2-subgroups of G are cyclic. By generalizing the argument of **205**, show that G has a subgroup of index 2. Hence, by induction on m, prove that G has a normal subgroup of order r. (Hint. Note that if $H \trianglelefteq G$ with $|H| = r$ then in fact H is characteristic in G.)

263 Let G be a finite group of even order and let $x \in G$ with $o(x) = 2$. If $C_G(x)$ has a cyclic Sylow 2-subgroup then G has a subgroup of index 2 (and so, if $|G| > 2, G$ is not simple). (Hint. Use 3.32, 4.28 and **262**.)

264 (i) Suppose that $G = HK$, with $H < G$ and $K < G$. If $H \cap K$ contains a non-trivial normal subgroup L of H then $L^G \leqslant K$ (where L^G denotes the normal closure of L in G: see **180**); hence G is not simple.

(ii) A simple group G cannot be expressed in the form $G = HK$ with $H < G$, $K < G, H$ abelian and $H \cap K \neq 1$. (Remark. This would fail if we were to allow $H \cap K = 1$. For instance, with $G = A_5$, a simple group of order 60 – see 5.24 – there is a cyclic subgroup H of order 5 and a subgroup $K \cong A_4$, of order 12, and $G = HK$; but of course $H \cap K = 1$.)

(iii) Let G be a finite non-abelian simple group with an abelian Sylow p-subgroup H and a proper subgroup K of index a power of p. Then $|G : K| = |H|$, p does not divide $|K|$, and H is a maximal abelian subgroup of G. (See **235**. Hint. Use **98** and **100**.)

We have mentioned Cauchy's theorem on orders of elements in **220**. This can be used in the proof of existence of Sylow subgroups: see **249**. However, Cauchy's theorem was not needed in the proof of Sylow's theorem given in 5.9 and so we now deduce Cauchy's theorem from Sylow's theorem.

5.11 Theorem (A. Cauchy, 1844). *If G is a finite group such that p divides $|G|$ then G has an element of order p.*
Proof. Let H be a Sylow p-subgroup of G. Since p divides $|G|, H \neq 1$. Choose $x \in H$ with $x \neq 1$. Then $o(x) > 1$ and $o(x)$ divides $|H|$. Hence $o(x) = p^s$ for some positive integer s. Then $x^{p^{s-1}}$ is an element of G of order p.

5.12 Corollary. *Let G be a finite group. Then G is a ϖ-group if and only if the order of every element of G is a ϖ-number* (see 3.41).
Proof. If $|G|$ is a ϖ-number then, by Lagrange's theorem, the order of every element of G is a ϖ-number. On the other hand, if $|G|$ is not a ϖ-number then there is a prime $p \notin \varpi$ such that p divides $|G|$. Then, by Cauchy's theorem, it follows that G has an element whose order is not a ϖ-number.

The following simple result is very useful: it is often referred to as 'the Frattini argument'.

5.13 Lemma (G. Frattini [a29], 1885). *If K is a finite normal subgroup of G and P is a Sylow p-subgroup of K then $G = N_G(P)K$.*
Proof. Let $g \in G$. Then

$$P^g \leqslant K^g = K,$$

since $K \trianglelefteq G$. Therefore, since $|P^g| = |P|, P^g$ is also a Sylow p-subgroup of K. Hence, by Sylow's theorem, P and P^g are conjugate subgroups of K; thus

$$P^g = P^k \quad \text{for some } k \in K.$$

Then $\qquad\qquad P^{gk^{-1}} = P,$

so that $$gk^{-1} \in N_G(P).$$

Hence $$g \in N_G(P)K.$$

This is true for every $g \in G$, and so the result is proved.

5.14 Corollary. *Let G be a finite group and P a Sylow p-subgroup of G. Then, for every subgroup H of G which contains $N_G(P)$, $N_G(H) = H$.*
Proof. Let $N_G(P) \leqslant H \leqslant G$. Then, since $P \leqslant H \leqslant G$, P is certainly a Sylow p-subgroup of H (**252**). Let $L = N_G(H)$. Now apply 5.13 with H in place of K and L in place of G: this gives $L = N_L(P)H$. Since $N_L(P) \leqslant N_G(P) \leqslant H$, it follows that $L = H$, as asserted.

265 A group G, finite or infinite, is said to be a ϖ-*group* if the order of every element of G is finite and a ϖ-number. (If G is finite, 5.12 shows that this definition is consistent with the definition given in 3.41.)

(i) The set of all complex numbers z which satisfy an equation $z^n = 1$, where n ranges over all ϖ-numbers, forms an infinite ϖ-subgroup of $\mathbf{C}^\times$, provided that $\varpi \neq \oslash$.

(ii) All subgroups and all quotient groups of a ϖ-group are ϖ-groups.

(iii) If H is a subgroup of finite index in a ϖ-group G then $|G : H|$ in a ϖ-number.

*266 Let G be a finite group, p a prime divisor of $|G|$, and n the number of distinct Sylow p-subgroups of G. Then the normalizers in G of the Sylow p-subgroups of G form a single conjugacy class of n subgroups of G.

267 Let $H \leqslant G$. Then H is said to be *intravariant* in G if, for every $\alpha \in \text{Aut } G$, α maps H to a conjugate of H in G.

(i) If G is finite, every Sylow subgroup is intravariant in G.

(ii) If $H \leqslant K \trianglelefteq G$ and H is intravariant in K then $G = N_G(H)K$. (This generalizes 5.13.)

268 Let $J \leqslant G$. Then J is said to be *pronormal* in G if, for every $g \in G$, $J^g = J^x$ for some $x \in \langle J, J^g \rangle$; and to be *abnormal* in G if, for every $g \in G$, $g \in \langle J, J^g \rangle$.

(i) If K is a finite normal subgroup of G then every Sylow subgroup of K is pronormal in G.

(ii) If J is pronormal in G then $N_G(J)$ is abnormal in G.

(iii) If $J \leqslant H \leqslant G$ and J is abnormal in G then $N_G(H) = H$.
(This generalizes 5.14. See also **270**.)

269 Let J be a pronormal subgroup of G (**268**). Let n be the number of distinct subgroups in the conjugacy class of J in G and suppose that $n < \infty$. Then $n \neq 2$.
(Remark. In the particular case when G is finite and J is a Sylow subgroup of G, the assertion follows immediately from 5.9(c).)

270 Let $J \leqslant G$. The following statements are equivalent:
(a) J is abnormal in G (**268**).
(b) Whenever $H \leqslant G$, $g \in G$ and $J \leqslant H \cap H^g$ then $g \in H$.

We now show how Sylow's theorem can be applied to prove that certain numbers are ineligible as orders of finite simple groups. First we note

5.15. *Let G be a finite non-abelian simple group and let p be a prime divisor*

of $|G|$. *Then the number n of Sylow p-subgroups of G is greater than* 1.

Proof. Let P be a Sylow p-subgroup of G. By 4.29, $|G|$ is divisible by at least two distinct primes, and so $1 < P < G$. If P were the only subgroup of G of order $|P|$ then P would be normal in G, in contradiction to the simplicity of G. Hence $n > 1$.

5.16 Theorem. *If* $|G| = pq$, *where* p, q *are distinct primes such that* $q \not\equiv 1$ *mod* p, *then G has a normal Sylow p-subgroup.*

Proof. By Sylow's theorem, the number n of distinct Sylow p-subgroups of G is a divisor of q, and $n \equiv 1$ mod p. Since q is prime, n is either 1 or q, and since, by hypothesis, $q \not\equiv 1$ mod p, it follows that $n = 1$. Thus G has a unique Sylow p-subgroup, P say, and so $P \trianglelefteq G$.

5.17 Corollary. *If* $|G| = pq$, *where* p, q *are distinct primes, then G is not simple.*

Proof. We may assume without loss of generality that $p > q$. Then $q - 1$ cannot be divisible by p, and so, by 5.16, G has a normal Sylow p-subgroup P. Since $1 < P < G$, G is not simple.

We know that $v(p) = 1$ for every p (1.3). From 5.16 we can show also (cf. 1.4, **215**, **575**)

5.18 Corollary. *If p and q are distinct primes such that* $p \not\equiv 1$ *mod q and* $q \not\equiv 1$ *mod p then* $v(pq) = 1$; *that is, every group of order pq is cyclic.*

Proof. Suppose that $|G| = pq$. By 5.16, G has a normal Sylow p-subgroup P and a normal Sylow q-subgroup Q. Since P and Q have prime orders, they are cyclic: say

$$P = \langle x \rangle \quad \text{and} \quad Q = \langle y \rangle.$$

By Lagrange's theorem, $P \cap Q = 1$. Hence, by 3.53,

$$xy = yx.$$

Now it follows that the element xy of G has order pq (**6**), and so

$$\langle xy \rangle = G.$$

Thus G is cyclic.

5.19 Theorem. *If* $|G| = p^2 q$, *where* p, q *are distinct primes, then G has either a normal Sylow p-subgroup or a normal Sylow q-subgroup; and so G is not simple.*

Proof. Let n_p and n_q be, respectively, the number of Sylow p-subgroups and the number of Sylow q-subgroups of G. Suppose, contrary to what we wish to show, that $n_p > 1$ and $n_q > 1$. By Sylow's theorem, n_p divides q, which is prime: hence $n_p = q$. Also $n_p \equiv 1$ mod p, so it follows that $q > p$. Again by Sylow's theorem, n_q divides p^2, so that n_q is either p or p^2.

Now any element of order q in G generates a subgroup of order q, which is a Sylow q-subgroup of G. Any two distinct subgroups of G of order q intersect in 1, and so there are in G $n_q(q-1)$ distinct elements of order q. Hence, if $n_q = p^2$, there are in G just $p^2 q - p^2(q-1) = p^2$ elements which are *not* of order q. But then, since no element of a Sylow p-subgroup P of G has order q and since $|P| = p^2$, P must be the unique Sylow p-subgroup of G, in contradiction to the supposition that $n_p > 1$. Therefore $n_q = p$. But since also $n_q \equiv 1 \bmod q$, this implies that $p > q$, a final contradiction.

5.20 Theorem. *If $|G| = pqr$, where p, q, r are distinct primes, then G is not simple.*

Proof. We may assume that $p > q > r$. Suppose, contrary to what we want to show, that there is a simple group G of order pqr. Let n_p, n_q, n_r be, respectively, the numbers of Sylow p-, Sylow q-, Sylow r-subgroups of G. By 5.15, these numbers are all greater than 1. Since they have order p, any two distinct Sylow p-subgroups of G intersect in 1. Hence the n_p Sylow p-subgroups of G contain $n_p(p-1)$ distinct elements of order p. Similarly, the n_q Sylow q-subgroups of G contain $n_q(q-1)$ distinct elements of order q and the n_r Sylow r-subgroups of G contain $n_r(r-1)$ distinct elements of order r. Therefore

$$|G| = pqr \geqslant 1 + n_p(p-1) + n_q(q-1) + n_r(r-1).$$

By Sylow's theorem, n_p divides qr and $n_p \equiv 1 \bmod p$. Since $n_p > 1$ and $p > q, p > r$, it follows that $n_p = qr$. Also, n_q divides pr and $n_q \equiv 1 \bmod q$. Since $n_q > 1$ and $q > r$, $n_q \geqslant p$. Finally, $n_r > 1$ and n_r divides pq, so that $n_r \geqslant q$. Now we have

$$pqr \geqslant 1 + qr(p-1) + p(q-1) + q(r-1),$$

and hence

$$0 \geqslant (p-1)(q-1),$$

which is plainly false.

271 There is no simple group of order 1000.

272 There is no simple group of order 300. (Hint. Use 4.14 to show that if there were such a group, it could be embedded in Σ_6; but this is impossible.)

273 There is no simple group of order 132.

274 Suppose that G has normal subgroups H, J, L such that $L < J < H$ and $|H/J| = p, |J/L| = q$, where p, q are distinct primes. Show that if $p > q$ then there is a normal subgroup K of G such that $L < K < H$ and $|H/K| = q, |K/L| = p$.

275 Suppose that G is a simple group of order 60.

(i) Find the number of subgroups of G of order 5 and show that G has exactly 24 elements of order 5.

(ii) Show that G has no subgroup of order 15.

(iii) Show that G has exactly 20 elements of order 3.

276 If $|G| = p^2q$, where p and q are distinct primes such that $p^2 \not\equiv 1 \bmod q$ and $q \not\equiv 1 \bmod p$, then G is abelian. (Hint. Use 3.54.)

277 Suppose that $|G| = p^mq$, where p and q are distinct primes and m is a positive integer. Let Q be a Sylow q-subgroup of G, and suppose also that $N_G(Q) = Q$. Then G has a normal Sylow p-subgroup.

278 Every group of order 255 is cyclic. (Hints. Let $|G| = 255$. Show that G has a normal subgroup of order 17 and a subgroup K of order 85. Groups of order 85 are cyclic. Use 4.18 to show that $K \trianglelefteq G$. Then, by 4.36, **46**, **78** and **94**, $K \leqslant Z(G)$.)

*279 (i) Let $n = p^mr$, where m is a positive integer and r is an integer greater than 1 such that p does not divide r. If there is a simple group of order n then p^m divides $(r - 1)!$ (Hint. Use 4.14.)

(ii) There is no simple group of order $2^m \times 5$ for any integer $m \geqslant 4$.

We are going to show that the only non-abelian simple group of order at most 100 is A_5. We begin by noting

5.21 Lemma. *When $n \leqslant 4$, Σ_n has no non-abelian simple subgroup.*

Proof. This is clear for $n \leqslant 3$, so we consider Σ_4. We know that Σ_4 is not itself simple, for the alternating group A_4 is a non-trivial proper normal subgroup of Σ_4 (3.58). If H were a non-abelian simple subgroup of Σ_4, then, since $|\Sigma_4| = 2^3 \times 3$ and by 4.29, $|H|$ would be divisible by both 2 and 3. Hence $|H|$ would be either 2×3 or $2^2 \times 3$. But these possibilities are ruled out by 5.17 and 5.19 (or by **279**).

5.22 Corollary. *If G is a finite non-abelian simple group and $H < G$ then $|G : H| \geqslant 5$ (cf. **207**).*

Proof. Let $|G : H| = n$. Then, by 4.14, G/H_G can be embedded in Σ_n. Since $H_G \leqslant H < G$ and G is simple, $H_G = 1$. Thus G can be embedded in Σ_n. Hence, by 5.21, $n \geqslant 5$.

5.23 Lemma. *Let n be a positive integer such that $n \leqslant 100$ and $n \neq 60$. Then there is no non-abelian simple group of order n.*

Proof. Suppose that there is a non-abelian simple group G of order n. Then $n > 1$ and we can express n in the form

$$n = \prod_{i=1}^{s} p_i^{m_i},$$

where $s, m_1, \ldots, m_s$ are positive integers and $p_1, \ldots, p_s$ distinct primes. We may assume that $p_1 < p_2 < \ldots < p_s$. By 4.29, $s \geqslant 2$. If $s \geqslant 4$ then $n \geqslant 2 \times 3 \times 5 \times 7 > 100$, a contradiction. Hence s is either 2 or 3. By 5.17, 5.19 and 5.20, $\sum_{i=1}^{s} m_i > 3$. If $\sum_{i=1}^{s} m_i \geqslant 7$ then $n > 2^7 > 100$, a contradiction. Hence

$$4 \leqslant \sum_{i=1}^{s} m_i \leqslant 6.$$

Suppose first that $s = 2$. If p_1 and p_2 were both odd then $n \geqslant 3^3 \times 5 > 100$, a contradiction. Hence $p_1 = 2$ and p_2 is some odd prime, say p. Write $m_1 = l$ and $m_2 = m$. Then

$$n = 2^l p^m,$$

where l and m are positive integers such that $4 \leqslant l + m \leqslant 6$. If $l \leqslant 2$ then a Sylow p-subgroup of G would be a proper subgroup of index at most 4 in G, and this is ruled out by 5.22. Hence

$$3 \leqslant l \leqslant 5 \quad \text{and} \quad 1 \leqslant m \leqslant 3.$$

Let the number of Sylow p-subgroups of G be n_p. By 5.15, $n_p > 1$; and, by Sylow's theorem, n_p is the index in G of a subgroup of G (namely, of the normalizer in G of a Sylow p-subgroup of G), n_p divides 2^l and $n_p \equiv 1 \bmod p$. Hence n_p divides 2^5 and, by 5.22, $n_p > 4$. We cannot have $n_p = 32$, for this would imply that $p = 31$ and therefore that $n \geqslant 32 \times 31 > 100$. Hence we must have

$$n_p = 8 \quad \text{and} \quad p = 7, \quad \text{or} \quad n_p = 16 \quad \text{and} \quad p = 3 \text{ or } 5.$$

If $n_p = 8$ and $p = 7$ then, since $|G| \leqslant 100, |G| = 56$. Then, since the Sylow 7-subgroups of G have order 7 and $n_7 = 8$, there are in G $8 \times 6 = 48$ elements of order 7. But then there are in G just $56 - 48 = 8$ elements which are not of order 7, and these 8 elements must form the unique Sylow 2-subgroup of G: this is in contradiction to the simplicity of G. If $n_p = 16$, then $l \geqslant 4$: hence $m = 1$, since otherwise $n \geqslant 2^4 \times 3^2 > 100$, a contradiction. Thus, if $n_p = 16$ and $p = 3$, a Sylow 2-subgroup of G has index 3 in G, in contradiction to 5.22. Finally, if $n_p = 16$ and $p = 5$ then, since $|G| \leqslant 100, |G| = 80$. But this possibility is ruled out by **279**.

Now suppose that $s = 3$. Since $n \leqslant 100$ and $n \neq 60$, this implies that either $n = 2^2 \times 3 \times 7 = 84$ or $n = 2 \times 3^2 \times 5 = 90$. If $|G| = 84$, then the number, n_7 say, of Sylow 7-subgroups of G is, by Sylow's theorem, a divisor of 12 and $n_7 \equiv 1 \bmod 7$. Moreover, $n_7 > 1$, by 5.15. These conditions on n_7 are incompatible. Finally, since $90 = 2 \times 45$ and 45 is an odd number, we know by **205** that there is no simple group of order 90. This establishes the lemma.

280 There is no simple group of order $6 \times p^m$ for any prime p and positive integer m. (Hint. Use **279**, 5.20 and 5.22.)

281 Suppose that there is a simple group G of order 144. Then
   (i) G has 16 Sylow 3-subgroups.
   (ii) Let H_1 and H_2 be distinct Sylow 3-subgroups of G. Then $\langle H_1, H_2 \rangle = G$.

Hence if $H_1 \cap H_2 \neq 1$ then $Z(G) \neq 1$: a contradiction.

(iii) By (ii), any two distinct Sylow 3-subgroups of G intersect trivially. Deduce that G has only one Sylow 2-subgroup: a contradiction. Conclude that there is no simple group of order 144. (Hints. Use **99**, 4.30 and 5.22.)

282 Suppose that there is a simple group G of order 112. Let T_1 and T_2 be distinct Sylow 2-subgroups of G, chosen so that $|T_1 \cap T_2|$ is as large as possible. Then

(i) $|T_1 \cap T_2| \geqslant 4$.

(ii) $N_G(T_1 \cap T_2)$ is a 2-subgroup of G.

(iii) By Sylow's theorem, there is a Sylow 2-subgroup S of G containing $N_G(T_1 \cap T_2)$. Then $T_1 = S = T_2$: a contradiction.

Conclude that there is no simple group of order 112. (Hints. Use **99**, 5.6 and 5.22.) (Remark. A variant of this argument proves that there is no simple group of order $p^m q$, where p and q are distinct primes and m is a positive integer. See Huppert [b21] p. 41 or Zassenhaus [b41] p. 138.)

5.24 Lemma. *A_5 is simple.*

Proof. Suppose to the contrary that A_5 is not simple. Let $G = A_5$ and choose a proper normal subgroup K of G of largest possible order. Then $K \neq 1$. The quotient group G/K is simple: for otherwise G/K would have a non-trivial proper normal subgroup H/K and then, by 3.30, H would be a proper normal subgroup of G with $|H| > |K|$, contrary to the choice of K. Since G/K is simple and $|G/K| < |G| = 60$, it follows from 5.23 that G/K is abelian. Hence, by 3.52,

$$[x, y] \in K \quad \text{for all } x, y \in G.$$

Let
$$\{1, 2, 3, 4, 5\} = \{a, b, c, d, e\}.$$

We see (using 3.59) that the non-trivial elements of A_5 are of three kinds:

$(abcde)$: there are $\dfrac{5 \times 4 \times 3 \times 2 \times 1}{5} = 24$ such elements;

(abc) : there are $\dfrac{5 \times 4 \times 3}{3} = 20$ such elements;

$(ab)(cd)$: there are $\dfrac{5 \times 4 \times 3 \times 2}{2 \times 2 \times 2} = 15$ such elements.

We may choose for x and y above any elements of these kinds. Now
$$[(aeb), (aecbd)] = (bea)(dbcea)(aeb)(aecbd)$$
$$= (abcde),$$

$$[(adb), (bce)] = (bda)(ecb)(adb)(bce)$$
$$= (abc),$$

and
$$[(abc), (abd)] = (cba)(dba)(abc)(abd)$$
$$= (ab)(cd).$$

Hence (by choosing a, b, c, d, e appropriately) we see that every non-trivial element of A_5 belongs to K. Thus $K = G$, a contradiction. Therefore we must conclude that A_5 is simple.

It is fair to comment that there are more direct and elementary proofs of the simplicity of A_5. One such proof is given in outline in **287**. The proof given above has been placed in the present context in this book because the applications of Sylow's theorem used in the proof of 5.23 are central to the point of view adopted here, and are indeed an essential part of finite group theory. The simplicity of A_5 is then an easy deduction, as we have seen in 5.24.

We shall now illustrate by an example how Sylow's theorem may be used to obtain information about subgroups of a finite group other than p-subgroups.

5.25 Example. *Find the types of all the proper subgroups of A_5 whose orders are divisible by at least two distinct primes, and find the numbers of subgroups in each of the conjugacy classes of subgroups in A_5 into which they fall.*

(i) Let $G = A_5$. Then, since G is simple, 5.22 shows that for every $H < G, |G : H| \geqslant 5$ and therefore $|H| \leqslant 12$. Hence, since $|G| = 60$, the possible orders of subgroups to be considered are 6, 10, 12.

(ii) Now G has subgroups of order 12: for if we consider the natural action of G on the set $\{1, 2, 3, 4, 5\}$ then clearly $\text{Stab}_G(5) \cong A_4$. Let $H < G$ with $H \cong A_4$. Then H has a normal subgroup T of order 4 (**185**), and since $60 = 2^2 \times 3 \times 5$, T is a Sylow 2-subgroup of G. Now $H \leqslant N_G(T) < G$, since G is simple, and so, by (i), $12 = |H| \leqslant |N_G(T)| \leqslant 12$. Hence $N_G(T) = H$, and the number of Sylow 2-subgroups of G is equal to $|G : H| = 5$. Thus the normalizer in G of a Sylow 2-subgroup of G is isomorphic to A_4, and every subgroup of G isomorphic to A_4 is the normalizer in G of some Sylow 2-subgroup of G. It follows by **266** that the subgroups of G isomorphic to A_4 form a single conjugacy class of 5 subgroups.

(iii) Let the numbers of Sylow 3-subgroups and Sylow 5-subgroups of G be, respectively, n_3 and n_5. Then n_3 divides $2^2 \times 5$ and $n_3 \equiv 1 \bmod 3$, n_5 divides $2^2 \times 3$ and $n_5 \equiv 1 \bmod 5$. Since also n_3 and n_5 are both at least 5, we have $n_3 = 10$ and $n_5 = 6$. Let U be any subgroup of G of order 3, V any subgroup of G of order 5, and $J = N_G(U), K = N_G(V)$. Then $|G : J| = n_3 = 10$ and $|G : K| = n_5 = 6$. Hence $|J| = 6$ and $|K| = 10$, so that G has subgroups of orders 6 and 10. Moreover, **266** shows that the normalizers in G of the Sylow 3-subgroups of G form a single conjugacy class of 10 subgroups of G of order 6, and the normalizers in G of the Sylow 5-subgroups of G form a single conjugacy class of 6 subgroups of G of order 10.

(iv) Now let J be any subgroup of G of order 6 and let K be any subgroup of G of order 10. Then, by 5.16, J has a normal subgroup U of order 3 and K has a normal subgroup V of order 5. Then, by (iii), $|N_G(U)| =$

6 and $|N_G(V)| = 10$, so that $J = N_G(U)$ and $K = N_G(V)$. Hence the only subgroups of G of order 6 are the normalizers in G of the Sylow 3-subgroups of G, and the only subgroups of G of order 10 are the normalizers in G of the Sylow 5-subgroups of G.

(v) By considering the expressions of elements of A_5 as products of disjoint cycles, we see that G has no element of order greater than 5. Hence the subgroups of G of orders 6 and 10 are not cyclic. Therefore, **242** shows that if $J, K \leqslant G$ with $|J| = 6, |K| = 10$ then $J \cong D_6$ and $K \cong D_{10}$.

(vi) Finally, let H be any subgroup of G of order 12. Let T be a Sylow 2-subgroup of H and let U be a Sylow 3-subgroup of H. By 5.19, either $T \trianglelefteq H$ or $U \trianglelefteq H$. If $U \trianglelefteq H$ then $|N_G(U)| \geqslant |H| = 12$; but, since $|U| = 3$, we know by (iii) that $|N_G(U)| = 6$, a contradiction. Hence $T \trianglelefteq H$. Since $|T| = 4, T$ is a Sylow 2-subgroup of G. Thus, by (ii), $H \leqslant N_G(T) \cong A_4$. Since $|H| = 12$, it follows that $H = N_G(T)$.

The required list is therefore as follows:

    (a) a conjugacy class of 10 subgroups isomorphic to D_6,

    (b) a conjugacy class of 6 subgroups isomorphic to D_{10},

    (c) a conjugacy class of 5 subgroups isomorphic to A_4.

We now use 5.24 as the basis for an inductive proof that A_n is simple whenever $n \geqslant 5$. For this purpose we need to describe how Σ_n is partitioned into conjugacy classes.

5.26 Lemma. *Let n be a positive integer and let $\sigma, \tau \in \Sigma_n$. Let the expression of σ as a product of disjoint cycles be*

$$\sigma = (a_{11}a_{12}\ldots a_{1n_1})(a_{21}\ldots a_{2n_2})\ldots(a_{s1}\ldots a_{sn_s}),$$

where $s, n_1, \ldots, n_s$ are positive integers such that $n_1 + \ldots + n_s = n$, and let

$$\tau = \begin{pmatrix} a_{11} & a_{12}\ldots a_{sn_s} \\ b_{11} & b_{12}\ldots b_{sn_s} \end{pmatrix}.$$

Then

$$\sigma^\tau = (b_{11}b_{12}\ldots b_{1n_1})(b_{21}\ldots b_{2n_2})\ldots(b_{s1}\ldots b_{sn_s})$$

is the expression of σ^τ as a product of disjoint cycles.
Proof. For each $i = 1, \ldots, s$ and each $j = 1, \ldots, n_i$,

$$b_{ij}\sigma^\tau = b_{ij}\tau^{-1}\sigma\tau = a_{ij}\sigma\tau = a_{i,j+1}\tau = b_{i,j+1}$$

(where, if $j = n_i$, we replace the subscript $i, j+1$ by $i1$).

5.27 Corollary. *Let n be a positive integer and let $\sigma, \sigma' \in \Sigma_n$. Then σ and σ' are conjugate in Σ_n if and only if σ and σ' have the same cycle type; that is, if and only if the expressions of σ and σ' as products of disjoint cycles contain the same number of cycles of length m, for each integer m such that $1 \leqslant m \leqslant n$.*

Proof. Let the expression of σ as a product of disjoint cycles be

$$\sigma = (a_{11}a_{12}\ldots a_{1n_1})(a_{21}\ldots a_{2n_2})\ldots(a_{s1}\ldots a_{sn_s}),$$

where $s, n_1, \ldots, n_s$ are positive integers such that $n_1 + \ldots + n_s = n$. If $\sigma' = \sigma^\tau$ for some $\tau \in \Sigma_n$ then 5.26 shows that σ' has the same cycle type as σ. If, conversely, σ' has the same cycle type as σ then the expression of σ' as a product of disjoint cycles is of the form

$$\sigma' = (b_{11}b_{12}\ldots b_{1n_1})(b_{21}\ldots b_{2n_2})\ldots(b_{s1}\ldots b_{sn_s}).$$

Then, if we set

$$\tau = \begin{pmatrix} a_{11} & a_{12}\ldots a_{sn_s} \\ b_{11} & b_{12}\ldots b_{sn_s} \end{pmatrix} \in \Sigma_n,$$

5.26 shows that $\sigma' = \sigma^\tau$, so that σ and σ' are conjugate in Σ_n.

Remark. It follows that the class number of Σ_n is equal to the number of partitions of n (cf. 1.5).

For instance, in Σ_4 the possible cycle types are

$$(\times)(\times)(\times)(\times), \quad (\times\ \times), \quad (\times\ \times)(\times\ \times), \quad (\times\ \times\ \times), \quad (\times\ \times\ \times\ \times),$$

so that the class number of Σ_4 is 5. The numbers of elements of these types are, respectively,

$$1, \quad \frac{4 \times 3}{2} = 6, \quad \frac{4 \times 3 \times 2 \times 1}{2 \times 2 \times 2} = 3, \quad \frac{4 \times 3 \times 2}{3} = 8, \quad \frac{4 \times 3 \times 2 \times 1}{4} = 6.$$

Note that these numbers are all divisors of $|\Sigma_4| = 24$, as they ought to be, by 4.26, and

$$1 + 6 + 3 + 8 + 6 = 24.$$

283 Find the class number of Σ_5. Find the numbers of elements in the conjugacy classes of elements in Σ_5, and verify that these numbers are divisors of 120 whose sum is 120.

284 Find two elements of A_4 which are conjugate in Σ_4 but are not conjugate in A_4.

*285 Show that for every integer $n \geqslant 3$, $Z(\Sigma_n) = 1$.

286 Let n be a positive integer. The number of distinct conjugacy classes of elements of order 2 in Σ_n is equal to $n/2$ if n is even, and to $(n-1)/2$ if n is odd. (cf. **213**. Hint. See **22**.)

287 Let $G = A_5$. Let $\{1, 2, 3, 4, 5\} = \{a, b, c, d, e\}$.
  (i) Verify that

$$(ab)(cd) = (acd)(acb)$$

and

$$(abcde) = (abc)(ade).$$

Hence show that $G = \langle X \rangle$, where $X = \{x \in G : x^3 = 1\}$.

(ii) Let $x \in G$ with $o(x) = 3$. Prove that $C_G(x) = \langle x \rangle$. Deduce that the elements of order 3 form a single conjugacy class of elements of G. (Hint. Use 4.26 and 5.26.)

(iii) Let $u = (ab)(cd), v = (abcde)$ and $g = (ab)(de)$. Verify that

$$uu^g = (cde) \quad \text{and} \quad vv^g = (bec).$$

Hence show that any non-trivial normal subgroup of G must contain an element of order 3.

(iv) Conclude that G is simple.

*288 Verify that the following is a complete list of the conjugacy classes of non-trivial proper subgroups of A_4: (i) a normal subgroup of order 4, isomorphic to $C_2 \times C_2$, (ii) a class of 4 subgroups of order 3, and (iii) a class of 3 subgroups of order 2. (Hint. To eliminate the possibility that A_4 has a subgroup of order 6, see **185**. Alternatively, apply Sylow's theorem.)

*289 Verify that the following is a complete list of the conjugacy classes of non-trivial proper subgroups of Σ_4: (i) A_4, a normal subgroup of order 12, (ii) a class of 3 subgroups of order 8, isomorphic to D_8, (iii) a class of 4 subgroups of order 3, (iv) a class of 4 subgroups of order 6, isomorphic to Σ_3, (v) a normal subgroup of order 4, isomorphic to $C_2 \times C_2$, (vi) a class of 3 cyclic subgroups of order 4, (vii) a class of 6 subgroups of order 2, (viii) a class of 3 subgroups of order 2, and (ix) a class of 3 non-cyclic subgroups of order 4, isomorphic to $C_2 \times C_2$. (Hints. To show that A_4 is the only subgroup of Σ_4 of order 12, use **185**. See also **229, 250, 261, 266**.)

5.28 Theorem (3.60). *A_n is simple for every integer $n \geqslant 5$.*

Proof. We argue by induction on n. The assertion is true when $n = 5$, by 5.24. Assume that $n > 5$ and, inductively, that A_{n-1} is simple. Let $G = A_n$, and consider the natural action of G on the set $X = \{1, 2, \ldots, n\}$. For each $i = 1, \ldots, n$, let $H_i = \mathrm{Stab}_G(i)$. Note that G acts transitively on X: see the remarks following 4.12. Hence, by **187** (i), $H_1, \ldots, H_n$ all belong to the same conjugacy class of subgroups of G. Hence (2.20) for every $i = 1, \ldots, n$, $H_i \cong H_n \cong A_{n-1}$, and so H_i is simple.

Suppose, contrary to what we wish to show, that G has a non-trivial proper normal subgroup K. Then, for every $i = 1, \ldots n, H_i \cap K \trianglelefteq H_i$ so that, since H_i is simple, $H_i \cap K$ is either 1 or H_i. In fact, $H_i \cap K = 1$ for every i. For suppose that there were a j such that $H_j \cap K = H_j$; that is, such that $H_j \leqslant K$. Then, by the remarks above, for any i there is an element $\gamma \in G$ such that $H_i = H_j^\gamma$; and so, since $K \trianglelefteq G, H_i \leqslant K^\gamma = K$. Thus K would contain every H_i. But this would imply that $K = G$, a contradiction. For if $\sigma \in G$ then either $1\sigma = 1$, in which case $\sigma \in H_1 \leqslant K$, or $1\sigma = j$ for some $j \neq 1$. In the latter case, we can choose $i \in X$ with $i \neq 1$, $i \neq j$. Then $(j1i) \in G$ and $\sigma = \sigma(j1i)(j1i)^{-1}$. But then $\sigma(j1i)$ is an element of G which fixes the point 1, so that $\sigma(j1i) \in H_1 \leqslant K$; and, since $|X| > 3$, there is a point $l \in X$ such that $(j1i)^{-1} = (i1j) \in H_1 \leqslant K$. Hence also $\sigma \in K$.

Thus $H_i \cap K = 1$ for every $i = 1, \ldots, n$. Now let $1 \neq \sigma \in K$. Thus, for every $i, \sigma \notin H_i$: that is, σ fixes no point of X. Let $a \in X$, and let $a\sigma = b \neq a$. Since $|X| > 3$, there is a point $c \in X$ such that $c \neq a, c \neq b$ and $c \neq a\sigma^{-1}$. Let $c\sigma = d$: then, since σ is a permutation of X which does not fix c,

d is distinct from a, b and c. Since in fact $|X| \geqslant 6$, we can choose two more distinct points $e, f \in X$, both distinct from a, b, c, d. Now let

$$\tau = (ab)(cdef) \in G \quad \text{(by 3.59)}.$$

Then, since σ moves a to b and c to d, 5.26 shows that σ^{τ} moves b to a and d to e. Moreover, $\sigma^{\tau} \in K$, since $K \trianglelefteq G$. Hence $\sigma\sigma^{\tau} \in K$ and $\sigma\sigma^{\tau}$ fixes a and moves c to $e (\neq c)$. Thus $1 \neq \sigma\sigma^{\tau} \in H_a \cap K$: a contradiction. We conclude that G is simple, and so the induction argument goes through.

It is clear that for each integer $n > 2$, A_n has subgroups isomorphic to A_{n-1}; and these subgroups have index n in A_n, since $|A_n| = n|A_{n-1}|$. Now we note

5.29 Lemma. *Let n be any integer with $n > 2$. Every subgroup of index n in A_n is isomorphic to A_{n-1}.*
Proof. This is clear if $n < 5$, so assume that $n \geqslant 5$. Let $H < G = A_n$ with $|G : H| = n$. Consider the action of G by right multiplication on the set of right cosets of H in G. Since G is simple, by 5.28, this action is faithful. Hence, by **198**, the action is equivalent to the natural action of a suitable subgroup, J say, of Σ_n on the set $\{1, 2, \dots, n\}$. Since $|J| = |A_n|$ and (see **290** (ii)) the only subgroup of index 2 in Σ_n is A_n, we see that $J = A_n$. The actions in question are transitive and so, by **199**,

$$H = \operatorname{Stab}_G(H) \cong \operatorname{Stab}_{A_n}(n) \cong A_{n-1}.$$

Now we can complete the proof of

5.30 Theorem. *Let G be a finite non-abelian simple group of order at most 100. Then $G \cong A_5$.*
Proof. By 5.23, $|G| = 60$. Let n be the number of Sylow 5-subgroups of G. By 5.15, $n > 1$, and by Sylow's theorem, G has a subgroup of index n, n divides 12 and $n \equiv 1 \bmod 5$. Hence $n = 6$. Then, since G has a subgroup of index 6 and G is simple, it follows from 4.14 that G can be embedded in Σ_6. Let G^* be a subgroup of Σ_6 isomorphic to G. Then, by **184**, since G^* is simple, G^* does not contain an odd permutation: thus $G^* \leqslant A_6$. Now $|G^*| = 60$ and $|A_6| = 360$, so that $|A_6 : G^*| = 6$. Hence 5.29 shows that $G^* \cong A_5$.

*290 (i) For every integer $n \geqslant 5$, A_n is the only non-trivial proper normal subgroup of Σ_n. (Hint. Use **119** and **285**.)

(ii) For every integer $n \geqslant 2$, A_n is the only subgroup of index 2 in Σ_n. (Hint. For the case $n = 4$, see **289**.)

291 Let N denote the set of all positive integers. As in **148**, let $G = \Sigma_{(N)}$, the restricted symmetric group on N, and, for each $n \in N$, let

$$G_n = \{\sigma \in G : j\sigma = j \text{ for every } j \in N \text{ with } j > n\}.$$

Thus, by **148**, $G = \bigcup_{n=1}^{\infty} G_n$ and $G_n \cong \Sigma_n$ for every $n \in N$.

Now let $H_1 = 1$ and, for each integer $n > 1$, let H_n be the unique subgroup of index 2 in G_n: see **290** (ii). Then, for each integer $n > 1, H_n \cong A_n$; and $H_1, H_2, H_3, \ldots$ is an ascending sequence of subgroups of G.

Let $H = \bigcup_{n=1}^{\infty} H_n \leqslant G$ (3.34). Show that $|G : H| = 2$ and that H is simple. Show also that H has an infinite abelian subgroup.

(The infinite simple group H is denoted by A_N. Hints. See **155**. To demonstrate the existence of an infinite abelian subgroup of H, note that, by 2.28, it is enough to show that there is an infinite commuting set of elements in H. Remark. The group A_N is not finitely generated (by 3.36). The existence of finitely generated infinite simple groups was first established by G. Higman [a57].)

292 Let n be an integer, $n \geqslant 2$. Every subgroup of index n in Σ_n is isomorphic to Σ_{n-1}. (Hint. For $n \geqslant 5$ use an argument similar to the one in 5.29. The appropriate action is faithful by **290**.)

293 There is no simple group of order 120. (Hint. Show that if there were such a group, it could be embedded in Σ_6, and then apply **292**.)

294 If G is a simple group of order $12 \times p^m$, for some prime p and positive integer m, then $G \cong A_5$. (Hint. Use **272, 273, 279**, 5.22 and 5.30.)

295 Suppose that there is a simple group G of order 180 and let n_3 and n_5 be, respectively, the numbers of Sylow 3-subgroups and Sylow 5-subgroups of G. Then
  (i) $n_3 = 10$ and n_5 is either 6 or 36.
  (ii) If $n_5 = 6$ then G can be embedded in A_6; but this would contradict the simplicity of A_6.
  (iii) Hence $n_5 = 36$ and each Sylow 5-subgroup of G coincides with its normalizer in G.
  (iv) Let H_1 and H_2 be distinct Sylow 3-subgroups of $G, J = \langle H_1, H_2 \rangle$ and $D = H_1 \cap H_2$. Then $D \leqslant Z(J)$ and $|J : H_1| \geqslant 4$.
  (v) If $D \neq 1$ then 5 does not divide $|J|$ and $|J : H_1| = 4$; but this would imply that G could be embedded in Σ_5, which is impossible.
  (vi) Hence any two distinct Sylow 3-subgroups of G intersect trivially. Then there are in G 144 distinct elements of order 5 and 81 distinct elements of orders dividing 3^2. This is too many elements! Conclude that there is no simple group of order 180. (Hints. Use **99, 184**, 4.14, 4.30 and 5.22.)

296 Let n be an integer such that $100 < n \leqslant 200$ and $n \neq 168$. Then there is no non-abelian simple group of order n. (Hint. Argue as in 5.23 and use **205, 279, 281, 282, 293, 294** and **295**. Remark. There is a simple group of order 168, namely the group $PSL_2(\mathbf{Z}_7)$: see 3.61. This group is in fact isomorphic to $GL_3(\mathbf{Z}_2)$. See also **385**.)

Sylow's theorem provides no information about the internal structure of a finite p-group G, for then G itself is the unique Sylow p-subgroup of G. However, the theorem points to the desirability of an investigation of finite p-groups, since we may expect that their properties will have an important bearing on the structure of finite groups in general. By means of group action arguments, we have already established some special properties of finite p-groups. If G is a non-trivial finite p-group, we know that
  (i) whenever $H < G, H < N_G(H)$ (5.6),

(ii) whenever $1 < H \trianglelefteq G, H \cap Z(G) \neq 1$ (5.8),
and in particular $Z(G) \neq 1$.

We shall have some more to say about finite p-groups in chapter 11. (For further information, see Gorenstein [b13] chapter 5, and Huppert [b21] chapter 3.) We end this chapter by proving the following result, which shows that a finite p-group has *normal* subgroups of all possible orders.

5.31 Theorem. *Let G be a finite p-group with, say, $|G| = p^m$. Then G has normal subgroups $G_0, G_1, \ldots, G_m$ such that*

$$1 = G_0 < G_1 < \ldots < G_{m-1} < G_m = G$$

and $|G_i| = p^i$ for every $i = 0, 1, \ldots, m$.

Proof. We argue by induction on m. The assertion is trivial if $m \leqslant 1$. Suppose that $m > 1$, and assume inductively that the result is true for any group of order p^{m-1}. By 5.8, $Z(G) \neq 1$. Let $1 \neq z \in Z(G)$. Then $o(z) = p^n$ for some integer $n > 0$. Let $G_1 = \langle z^{p^{n-1}} \rangle \leqslant Z(G)$. Then $|G_1| = p$ and (by 118) $G_1 \trianglelefteq G$. Let $\bar{G} = G/G_1$. Then $|\bar{G}| = p^{m-1}$ and, by the inductive assumption, $\bar{G}$ has normal subgroups $\bar{G}_i (i = 0, 1, \ldots, m-1)$ with

$$1 = \bar{G}_0 < \bar{G}_1 < \ldots < \bar{G}_{m-1} = \bar{G} \quad \text{and} \quad |\bar{G}_i| = p^i \text{ for every } i.$$

By 3.30, each $\bar{G}_i$ is of the form

$$\bar{G}_i = G_{i+1}/G_1,$$

where $G_1 \leqslant G_{i+1} \trianglelefteq G$. Moreover, by 3.29,

$$G_1 < G_2 < \ldots < G_m = G,$$

and for every $i, |G_{i+1}| = |\bar{G}_i| \, |G_1| = p^{i+1}$. Now (with $G_0 = 1$) the subgroups $G_0, G_1, \ldots, G_m$ of G satisfy the conditions stated, and so the induction argument goes through.

5.32 Corollary (1.6). *Let G be a finite group and let p^m be any prime power divisor of $|G|$. Then G has a subgroup of order p^m.*

Proof. Let H be a Sylow p-subgroup of G and say $|H| = p^l$. Then $m \leqslant l$ and therefore, by 5.31, H has a subgroup J of order p^m. Then J is a subgroup of G of order p^m.

Remark. If p^m is a divisor of $|G|$, but not the highest power of p dividing $|G|$, then the subgroups of G of order p^m need not form a single conjugacy class of subgroups of G: see **289** or consider the group $C_2 \times C_2$. Thus if we think of 5.32 as a generalization of 5.9(a), the analogous generalization of 5.9(b) fails. However, part of the analogous generalization of 5.9(c) holds true: the number of distinct subgroups of G of order p^m is congruent to 1 mod p. For a proof of this (due to Wielandt) see Ledermann [b29], theorem 27.

*297 Let G be a finite group.

(i) If G is a non-trivial p-group then G has a normal subgroup of index p.

(ii) If G is abelian and p divides $|G|$, then G has a subgroup of index p. (Hint. Apply **135**.)

(iii) In general, G has a normal subgroup of index p if and only if p divides $|G/G'|$. (Hint. See 3.52.)

298 Let l, m, n be positive integers such that $l \geqslant m \geqslant n$, and suppose that P is a group of order p^l. If R is a subgroup of P of order p^n then there is a subgroup Q of P such that $R \leqslant Q \leqslant P$ and $|Q| = p^m$.

299 (i) Let G be a finite group and p a prime divisor of $|G|$. If there is in G a conjugacy class of elements containing just $|G|/p$ elements then p^2 does not divide $|G|$.

(ii) The class number of A_4 is 4.

(iii) If G is a finite group of order 12 and class number 4 then $G \cong A_4$.

(Hints. For (i), use **298** and 4.30. For (iii), use (i) to show that G has a non-normal subgroup of order 3, and then use 4.14 and **290**.)

300 Let G be a finite group and let p^m be a divisor of $|G|$. Let P be a Sylow p-subgroup of G and assume that the number of *normal* subgroups of P of order p^m is congruent to 1 mod p. (This is in fact always true.) Deduce that the number of subgroups of G of order p^m is congruent to 1 mod p. (Hint. Let P act by conjugation on the set of all subgroups of G of order p^m.)

301 Let G be a finite group and let $d(G)$ denote the least positive integer n such that G has a set of n generators.

(i) If $H \trianglelefteq G$ then $d(G) \leqslant d(H) + d(G/H)$.

(ii) If $|G| = p^m$ for some positive integer m then $d(G) \leqslant m$.

(Note that this bound cannot be improved, as is shown by the additive group of a vector space of dimension m over the field $\mathbf{Z}_p$.)

(iii) If $|G| = \prod_{i=1}^{s} p_i^{m_i}$, where $s, m_1, \ldots, m_s$ are positive integers and $p_1, \ldots, p_s$ are distinct primes, then $d(G) \leqslant \sum_{i=1}^{s} m_i \leqslant \log_2 |G|$.

(iv) For every integer $n > 1$,

$$\nu(n) \leqslant (n!)^{\log_2 n} < n^{n \log_2 n}.$$

(Hints. For (ii), use 5.31; for (iii), note that if P_i is a Sylow p_i-subgroup of G for each $i = 1, \ldots, s$ then $G = \langle P_1, P_2, \ldots, P_m \rangle$; and for (iv), use Cayley's theorem 4.24.)

6

GROUPS OF EVEN ORDERS

Throughout this chapter, let G be a finite group of even *order.* Then we know by 1.13 (or by 5.11) that G contains at least one involution. Suppose that there are in total n involutions in G, and let them be denoted by

$$t_1, \ldots, t_n.$$

Let the class number of G be k and choose one element from each of the k conjugacy classes: let these elements be denoted by

$$1 = x_0, x_1, \ldots, x_{k-1}.$$

For $i = 0, 1, \ldots, k - 1$, let

c_i be the number of ordered pairs (u, v) of involutions in G such that
$$uv = x_i:$$

then c_i is a non-negative integer.

We shall prove the result of R. Brauer and K. A. Fowler stated in 1.14. In order to do this we need several preliminary results.

6.1.
$$n^2 = \sum_{i=0}^{k-1} c_i |G : C_G(x_i)|.$$

Proof. Consider the n^2 products $t_j t_k$ $(j, k = 1, \ldots, n)$. By definition, precisely c_i of these are equal to x_i. For any $g \in G$, t_j^g and t_k^g are involutions, and $t_j t_k = x_i$ if and only if $t_j^g t_k^g = x_i^g$. Hence precisely c_i of the n^2 products are equal to x_i^g. Since the number of elements in the conjugacy class of x_i is $|G : C_G(x_i)|$ (4.26), the formula stated is correct.

6.2. (i) *If $x_i^2 \neq 1$ then c_i is the number of involutions t_j such that $x_i^{t_j} = x_i^{-1}$.*
(ii) *If x_i is an involution then $c_i + 1$ is the number of involutions in $C_G(x_i)$.*
(iii) $c_0 = n$.
Proof. Suppose that (u, v) is an ordered pair of involutions in G such that $uv = x_i$. Then

$$x_i^u = (uv)^u = vu = v^{-1}u^{-1} = x_i^{-1}.$$

Thus we can define a map

$$\varphi : (u, v) \mapsto u$$

from the set of ordered pairs of involutions (u, v) with $uv = x_i$ to the set of involutions u with $x_i^u = x_i^{-1}$. This map φ is injective, for if (u, v) and (u', v') are pairs of involutions with $uv = x_i = u'v'$ and $u = u'$ then also $v = v'$.

Now suppose that u is an involution in G such that $x_i^u = x_i^{-1}$. Let $v = u x_i$. Then

$$uv = x_i \quad \text{and} \quad v^2 = x_i^u x_i = 1.$$

Thus, providing $v \neq 1$, (u, v) is an ordered pair of involutions such that $uv = x_i$ and $(u, v)\varphi = u$. If $v = 1$ then $x_i = u$, an involution. Hence, when x_i is *not* an involution, the map φ is bijective. This proves (i) and (iii). (In any case, (iii) is obvious.)

Now assume that x_i is an involution. Then $x_i^{-1} = x_i$, so that any involution u such that $x_i^u = x_i^{-1}$ must belong to $C_G(x_i)$. Every involution in $C_G(x_i)$, except x_i itself, appears as the image under φ of some pair: for if u is such an involution and we set $v = u x_i$, then, because u and x_i commute, $v^2 = 1$ and, since $u \neq x_i, v \neq 1$; then also $uv = x_i$ and $(u, v)\varphi = u$. (Note that if $u = x_i$ and $v \in G$ with $uv = x_i$ then $v = 1$, not an involution: this shows that x_i does not appear in the image of φ.) This proves (ii).

6.3 Definition. Let $x \in G$. Then x is said to be *real in G* if x and x^{-1} are conjugate in G. (The reason for the use of the word 'real' here lies in character theory: see for instance Huppert [b21] p. 537.) Note that if $x^2 = 1$ then, trivially, x is real. In general G may have real elements x with $x^2 \neq 1$.

6.4. *Let $x \in G$ and let*

$$C_G^*(x) = \{g \in G : x^g \text{ is either } x \text{ or } x^{-1}\},$$

the extended centralizer of x in G (**234**). *Then $C_G^*(x) \leqslant G$ and*
  (i) *if either $x^2 = 1$ or x is non-real in G then $C_G^*(x) = C_G(x)$,*
  (ii) *if $x^2 \neq 1$ and x is real in G then $|C_G^*(x) : C_G(x)| = 2$.*
Proof. It is straightforward to verify directly that $C_G^*(x) \leqslant G$. (Alternatively, apply **234**.) The statement (i) is obvious. Suppose that $x^2 \neq 1$ and x is real in G. Then $x^{-1} \neq x$ and there is an element $g \in G$ such that $x^g = x^{-1}$. Thus $g \in C_G^*(x) \backslash C_G(x)$. Now let g' be any element of $C_G^*(x) \backslash C_G(x)$. Then $x^{g'} = x^{-1} = x^g$ and so $g'g^{-1} \in C_G(x)$, hence $g' \in C_G(x)g$. Therefore $C_G^*(x) = C_G(x) \cup C_G(x)g$, and so $|C_G^*(x) : C_G(x)| = 2$. This proves (ii).

6.5. *Let $x \in G$. If x is real in G then the number of elements $g \in G$ such that $x^g = x^{-1}$ is equal to $|C_G(x)|$.*
Proof. The result is clear if $x^2 = 1$. Suppose that $x^2 \neq 1$. Then, as in the proof of 6.4, the set of elements $g \in G$ such that $x^g = x^{-1}$ is a coset of $C_G(x)$ in $C_G^*(x)$, and therefore contains $|C_G(x)|$ elements. (This is just a particular case of **210**.)

6.6. *For each* $i = 0, 1, \ldots, k - 1, c_i \leqslant |C_G(x_i)|$. *Moreover, if* x_i *is non-real in* G *then* $c_i = 0$, *while if* x_i *is an involution then* $c_i \leqslant |C_G(x_i)| - 2$.

Proof. The fact that if x_i is non-real in G then $c_i = 0$ is immediate from 6.2(i) and 6.3. Now suppose that x_i is real in G. By 6.5, the number of elements $g \in G$ such that $x_i^g = x_i^{-1}$ is equal to $|C_G(x_i)|$. Hence the number of *involutions* t_j such that $x_i^{t_j} = x_i^{-1}$ is at most $|C_G(x_i)|$. Then if $x_i^2 \neq 1$, $c_i \leqslant |C_G(x_i)|$, by 6.2(i). If x_i is an involution then $c_i \leqslant |C_G(x_i)| - 2$, by 6.2(ii) and since $1 \in C_G(x_i)$ and 1 is not an involution. Finally, if $x_i = 1$, that is if $i = 0$, then, by 6.2(iii), $c_0 = n < |G| = |C_G(x_0)|$. This covers all cases.

302 Let J be a finite group. If there is a non-trivial element $x \in J$ such that x and x^{-1} are conjugate in J then $|J|$ is even.

303 Suppose that H is a cyclic normal subgroup of G of order 4 such that $H \not\leqslant Z(G)$. Then every element of H is real in G.

304 Let n be a positive integer.
 (i) In Σ_n every element is real.
 (ii) Let $n \geqslant 3$. In the dihedral group D_{2n} every element is real.
 (iii) In the quaternion group Q_8 (**181**) every element is real. (Hint. Apply **303**.)

305 In the alternating group A_4 of degree 4, the only elements x which are real are those satisfying $x^2 = 1$. (Hint. See **185**.)

306 Verify the formula of 6.1 for each of the groups Σ_3, Q_8, A_4.

307 Let $x \in G$. Then x is said to be *strongly real in* G if there is an *involution* $t \in G$ such that $x^t = x^{-1}$.
 (i) If $x^2 = 1$ then x is strongly real in G.
 (ii) If $x \neq 1$, x is real in G and $|C_G(x)|$ is odd, then x is strongly real in G.
 (iii) Also, in the notation of this chapter, if x_i is strongly real in G then either $c_i > 0$ or x_i is an involution and the only involution in $C_G(x_i)$.
 (iv) If x_i is not strongly real in G then $c_i = 0$.

308 Let n be an integer, $n \geqslant 2$.
 (i) In Σ_n every element is strongly real.
 (ii) Let $n \geqslant 3$. In D_{2n} every element is strongly real.
 (iii) In Q_8 the only elements x which are strongly real are those satisfying $x^2 = 1$ (cf. **304**).

309 Suppose that G has n involutions. For any subgroup H such that $|H| > |G|/(n + 1)$, there is an element $h \in H$ such that $h \neq 1$ and h is strongly real in G. (Hint. If $|H|$ is odd, let I be the set of involutions of G and show that there are distinct elements $x, y \in H$ such that $(xI) \cap (yI) \neq \emptyset$.)

310 Let $T = \{x \in G : x^2 = 1\}$ and suppose that T is a commuting set of elements. Prove that $T \trianglelefteq G$ and that no element of $G \backslash T$ is strongly real in G. Show by an example that it can happen under these conditions that $T < G$ and all elements of G are real in G.
 Prove also that if y is an element of G which is real in G then $T \cap C_G(y) \neq 1$. (See **316** for a converse result. Hint. If y has odd order greater than 1, apply **307** (ii).)

6.7 Theorem. (R. Brauer and K. A. Fowler [a8], 1955). *Let G be a group of even order with precisely n involutions, and suppose that $|Z(G)|$ is odd.*

Let $a = |G|/n$ (*not an integer in general*). *Then G has a proper subgroup H such that either* $|G : H| = 2$ *or* $|G : H| < \frac{1}{2}a(a + 1)$.

Proof. We may suppose the elements $x_0, x_1, \ldots, x_{k-1}$ labelled so that $x_1, \ldots, x_s$ are involutions, $x_{s+1}, \ldots, x_{r-1}$ are real but not involutions, and $x_r, \ldots, x_{k-1}$ are non-real in G, where s and r are integers such that $0 < s \leqslant r - 1 \leqslant k - 1$. Since each of the n involutions in G is conjugate in G to just one of $x_1, \ldots, x_s$, and by 4.26,

$$n = \sum_{i=1}^{s} |G : C_G(x_i)|. \tag{i}$$

Also, by 6.1,

$$n^2 = \sum_{i=0}^{k-1} c_i |G : C_G(x_i)|.$$

Hence, by 6.2(iii) and 6.6,

$$n^2 \leqslant n + \sum_{i=1}^{s} (|C_G(x_i)| - 2)|G : C_G(x_i)| + \sum_{i=s+1}^{r-1} |C_G(x_i)| \, |G : C_G(x_i)|$$

$$= n + (r-1)|G| - 2 \sum_{i=1}^{s} |G : C_G(x_i)|$$

$$= n + (r-1)|G| - 2n, \quad \text{by (i)}.$$

Thus

$$n^2 \leqslant (r-1)|G| - n. \tag{ii}$$

Let $$j = \min\{|G : H| : H < G\}.$$

If $j = 2$, there is nothing more to prove; so suppose $j > 2$. Since $|Z(G)|$ is odd, no involution lies in $Z(G)$ and therefore

$$C_G(x_i) < G \quad \text{for } i = 1, \ldots, s.$$

Hence $$j \leqslant |G : C_G(x_i)| \quad \text{for } i = 1, \ldots, s.$$

Therefore, by (i),

$$sj \leqslant n. \tag{iii}$$

For $i = s + 1, \ldots, r - 1$, x_i is real in G and $x_i^2 \neq 1$. Hence, by 6.4,

$$|C_G^*(x_i) : C_G(x_i)| = 2.$$

Since $j > 2$, G has no subgroup of index 2 and therefore

$$C_G^*(x_i) < G \quad \text{for } i = s + 1, \ldots, r - 1.$$

Hence $$j \leqslant |G : C_G^*(x_i)|,$$

that is,

$$j \leqslant \tfrac{1}{2}|G : C_G(x_i)| \text{ for } i = s + 1, \ldots, r - 1. \tag{iv}$$

The total number of real elements in G is

$$1 + n + \sum_{i=s+1}^{r-1} |G : C_G(x_i)| \leqslant |G|.$$

Hence, by (iv),

$$1 + n + 2j(r - s - 1) \leqslant |G|. \tag{v}$$

From (ii),

$$n^2 \leqslant s|G| + (r - s - 1)|G| - n$$

$$\leqslant \frac{n|G|}{j} + \frac{(|G| - 1 - n)|G|}{2j} - n,$$

by (iii) and (v); thus

$$n^2 \leqslant \frac{n|G|}{2j} + \frac{|G|^2}{2j} - \frac{|G|}{2j} - n.$$

Multiply this last inequality by $|G|/n^2$. Then, with $|G|/n = a$,

$$|G| \leqslant \frac{|G|a}{2j} + \frac{|G|a^2}{2j} - \frac{a^2}{2j} - a$$

$$< \frac{|G|a}{2j} + \frac{|G|a^2}{2j}.$$

Hence

$$2j < a + a^2,$$

and so $\quad\quad\quad j < \tfrac{1}{2}a(a + 1)$, as asserted.

6.8 Corollary. *Let G and a be as in 6.7. Then G has a proper normal subgroup K such that either $|G/K| = 2$ or $|G/K| \leqslant [\tfrac{1}{2}a(a + 1)]$! (where, for any real number b, $[b]$ denotes the largest integer not greater than b).*
Proof. Let H be as in the statement of 6.7 and let $K = H_G$, the core of H in G. Then K is a proper normal subgroup of G and, by 4.14, G/K can be embedded in $\Sigma_{|G:H|}$. The result follows.

6.9 Corollary (1.14). *Let G be a simple group of even order greater than 2, let t be any involution in G, and let $m = |C_G(t)|$. Then $C_G(t) < G$ and $|G| \leqslant (\tfrac{1}{2}m(m + 1))$!*
Proof. Since G is simple and has even order greater than 2, G is non-abelian. Therefore $Z(G)$ is a proper normal subgroup of G and so $Z(G) = 1$. Hence $C_G(t) < G$. Now we use the notation of the proof of 6.7. The involution t is conjugate in G to some x_l with $1 \leqslant l \leqslant s$. Then

$$m = |C_G(t)| = |C_G(x_i)| \quad \textbf{(229)}.$$

Since

$$n = \sum_{i=1}^{s} |G : C_G(x_i)|$$

(equation (i) of the proof of 6.7),

$$\frac{1}{a} = \frac{n}{|G|} = \sum_{i=1}^{s} \frac{1}{|C_G(x_i)|} \geqslant \frac{1}{|C_G(x_i)|} = \frac{1}{m}.$$

Hence

$$a \leqslant m.$$

We now apply 6.8. The conditions are satisfied and, since K is a proper normal subgroup of G and G is simple, $K = 1$. Then, since $|G| > 2$,

$$|G| \leqslant [\tfrac{1}{2}a(a + 1)]! \leqslant (\tfrac{1}{2}m(m + 1))!$$

Recall the deduction of 1.15 from 1.14.

311 We have tacitly assumed that the proof of 6.7 remains valid when G has no real element of order greater than 2, that is, when $s + 1 = r$. Check through the proof of 6.7 and show that the argument yields the following sharper result in this case:

Let G be a group of even order with precisely n involutions and let $a = |G|/n$. Suppose that $|Z(G)|$ is odd and that there is in G no real element of order greater than 2. Then G has a proper subgroup H such that either $|G : H| = 2$ or $|G : H| < a$.

312 (i) In the alternating group A_5 of degree 5, there are just 15 involutions.

(ii) Let G be a simple group of even order greater than 2 and suppose that G has n involutions. Then $n < |G|/3$.

313 Let G be a simple group of even order greater than 2, and let t be an involution in G. Use 6.9 to show that $|C_G(t)| > 2$. (Remark. A stronger result than this has been obtained by other methods in **263**; cf. **287**(ii).)

One of the most striking special properties of groups of even orders is that the structure of subgroups generated by 2 involutions can be characterized precisely. Nothing comparable is available for subgroups generated by elements of orders greater than 2.

6.10 Definition. We shall say that a group D is *of dihedral type* if it is non-abelian and has a set of 2 generators $\{x, t\}$ such that t is an involution and $x^t = x^{-1}$.

Note that for each integer $n \geqslant 3$, the dihedral group D_{2n} of order $2n$ (defined in 2.24) is of dihedral type. We shall now show conversely that any finite group of dihedral type is isomorphic to D_{2n} for some integer $n \geqslant 3$. (Note also that the infinite dihedral group D_∞ (**57**) is of dihedral type. Conversely, any infinite group of dihedral type is isomorphic to D_∞ : see **314**.)

6.11. *Let D be a finite group of dihedral type. Then $D \cong D_{2n}$ for some integer $n \geqslant 3$.*

Proof. By definition, D is non-abelian and there are elements $x, t \in D$ such that $D = \langle x, t \rangle, o(t) = 2$ and $x^t = x^{-1}$. Let $o(x) = n$ and let $X = \langle x \rangle$. Since D is non-abelian, $X < D$. Since $D = \langle x, t \rangle$ and $x^t = x^{-1}$, it is clear that $X \lhd D$. Now (by **108**) $D/X = \langle xX, tX \rangle = \langle tX \rangle$, since $x \in X$. Since $t \notin X$ but $t^2 \in X$, it follows that $|D/X| = 2$. Hence $|D| = 2|X| = 2n$ and

$$D = \{1, x, x^2, \dots, x^{n-1}, t, xt, x^2t, \dots, x^{n-1}t\}.$$

Since D is non-abelian, $n \geqslant 3$, by 4.30 (or **77**).

Now let $G = D_{2n}$, the dihedral group of order $2n$. In the notation of 2.24,

$$G = \{1, \rho, \rho^2, \dots, \rho^{n-1}, \varepsilon, \rho\varepsilon, \rho^2\varepsilon, \dots, \rho^{n-1}\varepsilon\},$$

where $\rho^n = 1 = \varepsilon^2$ and $\rho^\varepsilon = \rho^{-1}$. It is now easy to check that the map

$$x^i t^j \mapsto \rho^i \varepsilon^j \quad (i = 0, 1, \dots, n - 1; \quad j = 0, 1)$$

is an isomorphism of D onto G.

6.12. *Every group D of dihedral type can be generated by 2 involutions.*

Proof. Suppose that $D = \langle x, t \rangle$, where t is an involution and $x^t = x^{-1}$. Then also $D = \langle xt, t \rangle$, since $x = (xt)t$. Certainly $xt \neq 1$, since D is non-abelian, and

$$(xt)^2 = xx^t = 1.$$

Thus xt is an involution.

The remarkable fact is that the converse is true. The proof is extremely easy.

6.13 Theorem. *Suppose that D is a non-abelian group which can be generated by 2 involutions. Then D is of dihedral type.*

Proof. Suppose that $D = \langle s, t \rangle$, where $o(s) = 2 = o(t)$. Then also $D = \langle st, t \rangle$, since $s = (st)t$. Moreover,

$$(st)^t = tst^2 = t^{-1}s^{-1} = (st)^{-1}.$$

Hence D is of dihedral type.

Remark. If A is an *abelian* group which can be generated by 2 involutions, say $A = \langle s, t \rangle$, where $o(s) = 2 = o(t)$, then either $s = t$ and $A = \langle s \rangle \cong C_2$ or $s \neq t$ and $A = \{1, s, t, st\} \cong C_2 \times C_2$.

6.14 Corollary. *Let s and t be involutions in G. Then either s and t are conjugate in G or there is an involution $u \in \langle s, t \rangle$ such that u commutes with both s and t.*

Proof. Let $D = \langle s,t \rangle$. If D is abelian, there is nothing to prove. Assume then that D is non-abelian, hence, by 6.13, that D is of dihedral type. Let $x = st$. Since G is finite, x has finite order, say n. As in the proof of 6.13, $D = \langle x,t \rangle$ and $x^t = x^{-1}$, and, as in the proof of 6.11, $|D| = 2n$. If n is odd, then $\langle s \rangle$ and $\langle t \rangle$ are Sylow 2-subgroups of D and are therefore conjugate in D. Because s, t are the only non-trivial elements of $\langle s \rangle, \langle t \rangle$ respectively, the conjugacy of the subgroups $\langle s \rangle, \langle t \rangle$ implies the conjugacy of the elements s, t.

Now suppose that n is even and let $u = x^{n/2}$. Then $u \in D$ and $o(u) = 2$. Since

$$x^s = ts = x^{-1} \quad \text{and} \quad x^t = x^{-1},$$

$$(x^{n/2})^s = x^{-n/2} = (x^{n/2})^t,$$

that is,
$$u^s = u^{-1} = u = u^t.$$

Thus u commutes with both s and t.

We use this to prove

6.15 Theorem (R. Brauer). *Suppose that G has at least 2 conjugacy classes of involutions. Let t be an involution in G such that $|C_G(t)|$ is as large as possible. Then*

$$|G| < |C_G(t)|^3.$$

Remark. This result is of course trivial if $|Z(G)|$ is even, for then $C_G(t) = G$; but if $|Z(G)|$ is odd then $C_G(t) < G$.

Proof. Let $|C_G(t)| = m$ and $|G : C_G(t)| = j$. Then $mj = |G|$ and we want to prove that $j < m^2$. By hypothesis, there is an involution s in G which is not conjugate to t. Let the distinct involutions in $C_G(s)$ be

$$s = s_1, s_2, \ldots, s_l.$$

By choice of t, for each $k = 1, \ldots, l$,

$$|C_G(s_k)| \leqslant m.$$

In particular,

$$l < |C_G(s)| \leqslant m.$$

Hence the number of distinct non-trivial elements in the set $\bigcup_{k=1}^{l} C_G(s_k)$ is at most equal to

$$\sum_{k=1}^{l} |C_G(s_k)| - 1 \leqslant lm - 1 < m^2.$$

By 4.26, t has exactly j distinct conjugates in G, say

$$t = t_1, t_2, \ldots, t_j.$$

Each of these elements is non-trivial, and so the required inequality $j < m^2$ will follow if we show that every t_i lies in $\bigcup_{k=1}^{l} C_G(s_k)$. Since each t_i is conjugate to t but t is not conjugate to s in G, t_i is not conjugate to s. Hence, by 6.14, there is in G an involution u_i which commutes with both s and t_i. Then $u_i \in C_G(s)$ and therefore

$$u_i = s_k \quad \text{for some } k \text{ with } 1 \leqslant k \leqslant l.$$

Hence $t_i \in C_G(u_i) = C_G(s_k)$.

Remarks. The conclusion of 6.15 does not hold in general for a group G with only 1 conjugacy class of involutions: see **318**. However, it is true that for any group G of even order greater than 2, there is a proper subgroup H of G such that $|G| < |H|^3$: see **319**. A corresponding result holds for groups of odd orders. If K is a non-trivial group of odd order, with $|K|$ not a prime, then there is a proper subgroup L of K such that $|K| < |L|^3$. In fact, in this case the assertion can be improved to $|K| \leqslant |L|^2$, and further to $|K| < |L|^2$ unless $|K| = p^2$ for some prime p: see **503**, **664**. But the only known proofs of these facts for groups of odd orders are made by invoking the very deep Feit–Thompson theorem (1.12). It would be interesting to find proofs independent of this. For further information about groups of even orders, we refer to R. Brauer and K. A. Fowler [a8] and Gorenstein [b13] chapter 9.

314 Let D be an infinite group of dihedral type. Then $D \cong D_\infty$. (See **57**. Hint. Argue as in 6.11.)

315 Let $G = A_5$, the alternating group of degree 5. There are in G elements t, x, y such that $o(t) = 2$, $o(x) = 3 = o(y)$ and $G = \langle x, t \rangle = \langle x, y \rangle$. (cf. 6.13, **179**. Hint. Use 5.25.)

316 Suppose that G has no strongly real element of order greater than 2. Then the set of all involutions in G is a commuting set of elements. (Hint. See 6.13. Remark. This is a converse to **310**.)

317 Let T be a Sylow 2-subgroup of G. Suppose that $T \ntrianglelefteq G$ and that $T \cap T^g = 1$ whenever $g \in G$ and $T^g \neq T$. Then G has just 1 conjugacy class of involutions. (Hint. Apply 6.14.)

318 Let $G = D_{2n}$, the dihedral group of order $2n$, where n is an odd integer, $n \geqslant 3$. Then
    (i) G has just 1 conjugacy class of involutions (cf. **214**).
    (ii) Let t be an involution in G. Then $C_G(t) = \langle t \rangle$. Hence $|G| < |C_G(t)|^3$ if and only if $n = 3$.
(Note also that $Z(G) = 1$: see **124**; cf. 6.15.)

319 (a) Suppose that G has just 1 conjugacy class of involutions and that $|Z(G)|$ is odd. Then there is a real element $x \in G$ such that $C_G(x) < G$ and $|G| < |C_G(x)|^3$. (Hints. Let the notation be as in the proof of 6.7, where now $s = 1$. Define $m_1 = |C_G(x_1)|$ and, if $r = 2$, $m = 0$, while if $r > 2$, $m = \max\{|C_G(x_i)| : i = 2, \ldots, r - 1\}$. Use 6.1, 6.2 and 6.6 to show that $n^2 \leqslant (m_1 - 1)n + (r - 2)|G|$. By counting the total number of real elements in G, show that $(r-2)|G| \leqslant m(|G|-1-n)$. From these two inequalities and the fact that $m_1 n = |G|$, deduce that $|G| < \max\{m_1^3, m^3\}$. Finally use 6.4.)

(b) For any group G of even order greater than 2, there is a proper subgroup H such that $|G| < |H|^3$. (Hints. Argue by induction on $|G|$. If $|Z(G)|$ is odd, apply (a) and 6.15. If $|Z(G)|$ is even, there is an involution $z \in Z(G)$. Then consider $\bar{G} = G/\langle z \rangle$. If $|\bar{G}|$ is even, apply the induction hypothesis to $\bar{G}$, while if $|\bar{G}|$ is odd, use **205** and **184**).

7

SERIES

In this chapter we shall develop further the theory introduced in chapter 3 of the *normal* structure of a group (as distinct from the *arithmetical* structure discussed in chapters 4, 5, 6). We shall prove the Jordan–Hölder theorem (1.9) and introduce two important classes of groups, the classes of *nilpotent* groups and *soluble* groups.

It is sometimes convenient to write $G \geqslant H$ to have the same meaning as $H \leqslant G$; and similarly $G > H$ for $H < G$ and $G \trianglerighteq H$ for $H \trianglelefteq G$. Recall also from chapter 1 that we use the notation $K \lhd G$ to mean 'K is a proper normal subgroup of G'.

7.1 Definitions. Let $H \leqslant G$. Suppose that there is a finite sequence $(H_i)_{0 \leqslant i \leqslant n}$ of subgroups of G, such that

$$H = H_0 \trianglelefteq H_1 \trianglelefteq \dots \trianglelefteq H_{n-1} \trianglelefteq H_n = G. \qquad \text{(a)}$$

Then we call (a) a *series of length n from H to G* (or *from G to H* if we wish to think of the series as 'descending' rather than 'ascending'). The subgroups $H_0, H_1, \dots, H_n$ are called the *terms* of the series and the quotient groups $H_i/H_{i-1}(i = 1, \dots, n)$ the *factors* of the series. When we refer to a *series of G* (without qualification), we mean a series from 1 to G (or from G to 1). (Warning. What is here called a series is called by some authors, e.g. Macdonald [b30], a 'normal series'. We shall reserve the latter name for a series in which each term is normal in the whole group: see 7.33.)

The series (a) is called *proper* if

$$H_{i-1} \lhd H_i \quad \text{for every } i = 1, \dots, n.$$

Another series

$$H = J_0 \trianglelefteq J_1 \trianglelefteq \dots \trianglelefteq J_m = G \qquad \text{(b)}$$

from H to G is said to be a *refinement* of (a) if $n \leqslant m$ and there are non-negative integers $j_0 < j_1 < \dots < j_n \leqslant m$ such that

$$H_i = J_{j_i} \quad \text{for } i = 0, 1, \dots, n;$$

that is, if (a) can be obtained from (b) by deleting terms of (b). Then (b) is said to be a *proper refinement* of (a) if there is a $j \in \{0, 1, \dots, m\}$ such that

$$H_i \neq J_j \quad \text{for } i = 0, 1, \dots, n.$$

A proper series of G which has no proper refinement is called a *composition series* of G. The factors of a composition series of G are called *composition factors* of G.

Note that every finite group has a composition series: see 1.8. An infinite group need not have a composition series. For instance, the infinite cyclic group $\mathbf{Z}^+$ is not simple and every non-trivial subgroup of $\mathbf{Z}^+$ is isomorphic to $\mathbf{Z}^+$ (3.25); therefore, any series of $\mathbf{Z}^+$ has a proper refinement. On the other hand, we know that there are infinite groups which have composition series, for we know that there are infinite simple groups: see 3.61 and **291**.

7.2 (cf. 1.10). *A series*

$$1 = G_0 \trianglelefteq G_1 \trianglelefteq \ldots \trianglelefteq G_n = G$$

of G is a composition series of G if and only if the factors G_i/G_{i-1} of the series are all simple ($i = 1, 2, \ldots, n$).

Proof. If the series is a composition series of G then it is by definition proper, and so the factors G_i/G_{i-1} are all non-trivial ($i = 1, \ldots, n$). If some factor G_i/G_{i-1} were not simple, it would have a proper non-trivial normal subgroup, say H/G_{i-1}. But then we should have $G_{i-1} \triangleleft H \triangleleft G_i$ (by 3.30) and hence obtain a proper refinement of the original series by inserting H as an extra term between G_{i-1} and G_i. This is contrary to the definition of composition series. Hence the factors of a composition series are all simple.

On the other hand, if the given series is not a composition series then either the series is not proper, in which case one of its factors is trivial and therefore not simple; or the series is proper and has a proper refinement, say

$$1 = H_0 \trianglelefteq H_1 \trianglelefteq \ldots \trianglelefteq H_m = G.$$

In this latter case, let l be the largest positive integer for which H_l is not equal to any term of the original series. Then $0 < l < m$ and $H_{l+1} = G_k$ for some integer k with $0 < k \leqslant m$. Since the series $(H_j)_{0 \leqslant j \leqslant m}$ is a refinement of $(G_i)_{0 \leqslant i \leqslant n}$ and by the choice of l,

$$G_{k-1} < H_l \triangleleft H_{l+1} = G_k.$$

Then, by 3.30, H_l/G_{k-1} is a proper non-trivial normal subgroup of G_k/G_{k-1}, and thus the factor G_k/G_{k-1} is not simple.

320 (i) An abelian group A has a composition series if and only if A is finite. (Hint. Use 7.2.)

(ii) A subgroup of a group with a composition series need not have a composition series (see **291**).

321 Let n be a positive integer.

(i) A cyclic group of order p^n has just 1 composition series.

(ii) The direct product of n groups of order p has $\psi(p^n)$ distinct composition series, where

$$\psi(p^n) = \frac{(p^n - 1)(p^{n-1} - 1)\ldots(p - 1)}{(p - 1)^n}.$$

322 How many composition series has each of the following groups: Σ_3, A_4, Σ_4?

The following elementary result is often useful.

7.3 (Dedekind's rule). *Let A, B, C be subgroups of G such that $B \leqslant A$. Then*

$$A \cap (BC) = B(A \cap C).$$

(Here we do not assume that BC and $B(A \cap C)$ are subgroups of G.)
Proof. Certainly $B(A \cap C) \subseteq A \cap (BC)$, since $B \leqslant A$. Let $a \in A \cap (BC)$. Then

$$a = bc \quad \text{for some } b \in B \text{ and } c \in C.$$

Then $b^{-1}a = c \in A \cap C$ (since $B \leqslant A$).

Hence $a \in B(A \cap C)$.

Thus $A \cap (BC) = B(A \cap C)$.

In proving some of the main results of this chapter, we shall need certain deductions from 3.40.

7.4 Lemma. *Suppose that $B \trianglelefteq A \leqslant G$ and $C \leqslant G$. Then*

(i) $(B \cap C) \trianglelefteq (A \cap C)$ and $(A \cap C)/(B \cap C) \cong B(A \cap C)/B$.

(ii) If also $C \trianglelefteq G$ then $BC \trianglelefteq AC$ and $AC/BC \cong A/B(A \cap C)$.
Proof. (i) Since $A \cap C \leqslant A$ and $B \trianglelefteq A$, we may apply 3.40 with G replaced by A, H by $A \cap C$ and K by B. This gives $B \cap C = ((A \cap C) \cap B) \trianglelefteq (A \cap C)$ and

$$(A \cap C)/(B \cap C) \cong (A \cap C)B/B = B(A \cap C)/B.$$

(ii) Now suppose that $C \trianglelefteq G$. By 3.38, AC and BC are subgroups of G and then clearly $BC \leqslant AC$. Since A normalizes B and C and since C normalizes every subgroup of G containing C, it follows that $BC \trianglelefteq AC$. Certainly $A \leqslant AC$. Now we apply 3.40 with G replaced by AC, H by A and K by BC. This gives

$$A \cap (BC) \trianglelefteq A \quad \text{and}$$

$$A/(A \cap (BC)) \cong A(BC)/BC.$$

With 7.3 and since $A(BC) = AC$, this gives the result.

7.5 Lemma (H. J. Zassenhaus [a 107], 1934). *Let $C_1 \trianglelefteq A_1 \leqslant G$ and*

$C_2 \unlhd A_2 \leqslant G.$ Then

$$(A_1 \cap C_2)C_1 \unlhd (A_1 \cap A_2)C_1, \qquad (A_2 \cap C_1)C_2 \unlhd (A_2 \cap A_1)C_2$$

and

$$(A_1 \cap A_2)C_1/(A_1 \cap C_2)C_1 \cong (A_2 \cap A_1)C_2/(A_2 \cap C_1)C_2.$$

Proof. We have

$$A_1 \cap C_2 = ((A_1 \cap A_2) \cap C_2) \unlhd (A_1 \cap A_2),$$

by 3.40, and similarly

$$(A_2 \cap C_1) \unlhd (A_2 \cap A_1) = A_1 \cap A_2.$$

Let $$B = (A_1 \cap C_2)(A_2 \cap C_1) \unlhd (A_1 \cap A_2),$$

by 3.39. Now, by 7.4(ii), applied to A_1 in place of G,

$$(A_1 \cap C_2)C_1 = BC_1 \unlhd (A_1 \cap A_2)C_1$$

and

$$(A_1 \cap A_2)C_1/BC_1 \cong (A_1 \cap A_2)/B(A_1 \cap A_2 \cap C_1) = (A_1 \cap A_2)/B. \qquad \text{(i)}$$

Similarly,

$$(A_2 \cap C_1)C_2 = BC_2 \unlhd (A_2 \cap A_1)C_2$$

and

$$(A_2 \cap A_1)C_2/BC_2 \cong (A_2 \cap A_1)/B. \qquad \text{(ii)}$$

Now the stated isomorphism follows from (i) and (ii).

323 Suppose that $G = HK$, where $H \leqslant G$ and $K \unlhd G$. Then
   (i) $\{V : K \leqslant V \leqslant G\} = \{JK : J \leqslant H\}$.
   (ii) Every subgroup of G/K is isomorphic to a quotient group of a subgroup of H.

*324 A *section* of G is a group A/B such that $B \unlhd A \leqslant G$. A subgroup C of G is said to *cover* a section A/B of G if $A \subseteq BC$, and C is said to *avoid* A/B if $A \cap C \leqslant B$.
   Let A/B be a section of G and let $C \leqslant G$. Then
   (i) C both covers and avoids A/B if and only if $A = B$.
   (ii) If C covers A/B then $(A \cap C)/(B \cap C) \cong A/B$, while if C avoids A/B then $(A \cap C)/(B \cap C)$ is trivial.
   (iii) Suppose that $C \unlhd G$. If C covers A/B then AC/BC is trivial, while if C avoids A/B then $AC/BC \cong A/B$.
   (iv) If $C \unlhd G$ and A/B is simple, then C either covers or avoids A/B.

325 Let $K \unlhd G$. Suppose that G has a composition series. Then (see **324**)
   (i) K has a composition series in which every factor is isomorphic to a composition factor of G covered by K (cf. **320** (ii)). Moreover, every composition factor of G covered by K is isomorphic to a composition factor of K.
   (ii) G/K has a composition series in which every factor is isomorphic to a composition factor of G avoided by K. Moreover, every composition factor of G avoided by K is isomorphic to a composition factor of G/K.

7.6 Definition. Two series of G, say

$$1 = G_0 \trianglelefteq G_1 \trianglelefteq \ldots \trianglelefteq G_n = G$$

and

$$1 = H_0 \trianglelefteq H_1 \trianglelefteq \ldots \trianglelefteq H_m = G,$$

are said to be *equivalent* if $m = n$ and there is a permutation π of the set $\{1, 2, \ldots, n\}$ such that

$$G_i/G_{i-1} \cong H_{i\pi}/H_{i\pi-1} \quad \text{for every } i = 1, \ldots, n.$$

This obviously defines an equivalence relation on the set of series of G.

In the form stated in 1.9, the Jordan–Hölder theorem asserts that any two composition series of a finite group are equivalent. Before proving this in a slightly more general form, we first establish a fundamental general result about series, proved by Schreier in 1928.

7.7 Theorem (O. Schreier [a88]). *Any two series of G have equivalent refinements.*

Proof. Consider two series of G, say

$$1 = G_0 \trianglelefteq G_1 \trianglelefteq \ldots \trianglelefteq G_n = G \tag{a}$$

and

$$1 = H_0 \trianglelefteq H_1 \trianglelefteq \ldots \trianglelefteq H_m = G. \tag{b}$$

We shall construct a refinement of (a) by inserting $m - 1$ subgroups $G_{ij}(j = 1, \ldots, m-1)$ between G_{i-1} and G_i, for each $i = 1, \ldots, n$; and a refinement of (b) by inserting $n - 1$ subgroups $H_{ij}(i = 1, \ldots, n-1)$ between H_{j-1} and H_j, for each $j = 1, \ldots, m$. Then these refinements will both have mn factors:

$$1 = G_0 \trianglelefteq G_{11} \trianglelefteq G_{12} \trianglelefteq \ldots \trianglelefteq G_{1,m-1} \trianglelefteq G_1 \trianglelefteq G_{21} \trianglelefteq \ldots \trianglelefteq G_{n,m-1} \trianglelefteq G_n = G \tag{c}$$

and

$$1 = H_0 \trianglelefteq H_{11} \trianglelefteq H_{21} \trianglelefteq \ldots \trianglelefteq H_{n-1,1} \trianglelefteq H_1 \trianglelefteq H_{12} \trianglelefteq \ldots \trianglelefteq H_{n-1,m} \trianglelefteq H_m = G. \tag{d}$$

We shall arrange that (c) and (d) are equivalent.

For this purpose we define, for every $i = 1, \ldots, n$ and every $j = 1, \ldots, m$,

$$G_{ij} = (G_i \cap H_j)G_{i-1}$$

and

$$H_{ij} = (H_j \cap G_i)H_{j-1}.$$

Note that, by 3.38, G_{ij} and H_{ij} are subgroups of G, since, for example, $G_i \cap H_j \leqslant G_i$ and $G_{i-1} \trianglelefteq G_i$. Note also that

$$G_{im} = G_i \quad \text{and} \quad H_{nj} = H_j.$$

For $i = 1, \ldots, n$ and $j = 1, \ldots, m$,

$$G_{i-1} \leqslant G_{i1} \leqslant G_{i2} \leqslant \ldots \leqslant G_{im} = G_i$$

and

$$H_{j-1} \leqslant H_{1j} \leqslant H_{2j} \leqslant \ldots \leqslant H_{nj} = H_j.$$

It is convenient also to set

$$G_{i0} = G_{i-1} = (G_i \cap H_0)G_{i-1} \quad \text{and} \quad H_{0j} = H_{j-1} = (H_j \cap G_0)H_{j-1}.$$

We now apply 7.5, choosing

$$C_1 = G_{i-1}, \qquad A_1 = G_i, \qquad C_2 = H_{j-1}, \qquad A_2 = H_j.$$

Then 7.5 gives

$$G_{i,j-1} \trianglelefteq G_{ij}, \qquad H_{i-1,j} \trianglelefteq H_{ij}$$

and

$$G_{ij}/G_{i,j-1} \cong H_{ij}/H_{i-1,j}. \tag{$\dagger$}$$

Thus our definitions of G_{ij}, H_{ij} do yield series (c) and (d) of G, and these are obviously refinements of (a) and (b), respectively. Moreover, the isomorphism ($\dagger$)(valid for every $i = 1, \ldots, n$ and every $j = 1, \ldots, m$) shows that (c) and (d) are equivalent series.

Next we note

7.8. *Any series of G which is equivalent to a composition series of G is also a composition series of G.*

Proof. This follows immediately from the definitions of 'equivalent' and 'composition series', together with 7.2.

The following theorem includes the fundamental result on composition series, established in part by Jordan in 1869 and completely for finite groups by Hölder in 1889.

7.9 Theorem. *Suppose that G has a composition series.*

(i) *Every proper series of G has a refinement which is a composition series of G.*

(ii) (C. Jordan [a66], O. Hölder [a59]; cf. 1.9.) *Any two composition series of G are equivalent.*

Proof. Consider a proper series (a) and a composition series (c) of G. By Schreier's theorem 7.7, these two series have equivalent refinements, say (a*) and (c*) respectively. If now from (a*) and (c*) we discard trivial factors, that is, delete repetitions of terms, we obtain two *equivalent proper* series, say (a') and (c') respectively. Since (a) and (c) are, by hypothesis, proper series, (a') and (c') are refinements of (a) and (c) respectively; and since (c') is a proper series while, by hypothesis, (c) has no proper refinement, (c') must coincide with (c). Thus (a') is equivalent to a composition series of G, hence (7.8) is itself a composition series of G. This proves (i).

If also (a) is a composition series of G then, by the same argument, (a') must coincide with (a). Then (a) and (c) are equivalent. This proves (ii).

7.10 Definition. Suppose that G has a composition series. It follows in particular from 7.9(ii) that any two composition series of G have the same number, n say, of factors. We call n the *composition length* of G.

326 Let p and q be distinct primes. Find equivalent refinements of the following two series of $\mathbf{Z}^+$:

$$0 \lhd p\mathbf{Z}^+ \lhd \mathbf{Z}^+ \text{ and } 0 \lhd q\mathbf{Z}^+ \lhd \mathbf{Z}^+.$$

(Note that Schreier's theorem 7.7 does not require the hypothesis that G has a composition series.)

327 (i) Every composition factor of a finite abelian group has prime order.
 (ii) Every composition factor of a finite p-group has order p.

328 (i) Every abelian group of order $\prod_{j=1}^{s} p_j^{m_j}$ has composition length $\sum_{j=1}^{s} m_j$ (where $s, m_1, \ldots, m_s$ are positive integers and $p_1, \ldots, p_s$ distinct primes).
 (ii) Every group of order p^n has composition length n (where n is a positive integer).
 (iii) Give an example of a finite group G whose composition length is not equal to the sum of the exponents of the distinct primes in the factorization of $|G|$ as a product of powers of prime numbers.

329 Suppose that G has a composition series and that the composition length of G is 2. Prove that either G has just 1 composition series or G is isomorphic to a direct product of simple groups.
 Show that there is no upper bound on the number of distinct composition series which G can have.

The following generalization of the concept of normal subgroup has proved to be of considerable importance.

7.11 Definition. Let $H \leqslant G$. We say that H is a *subnormal* subgroup of G if there is a series from H to G.

Certainly any normal subgroup of G is a subnormal subgroup; that the converse is not true in general is shown by 3.14. In fact, the definition of subnormality is made precisely in order to repair the deficiency of normality in failing to be a transitive relation. It is immediate from the definition that subnormality is transitive.

7.12. *Let* $K \leqslant H \leqslant G$. *If* K *is subnormal in* H *and* H *is subnormal in* G *then* K *is subnormal in* G.

Clearly any term of a series of G must be a term of some *proper* series of G. Then, in view of 7.9(i), we have

7.13. *Let* $H \leqslant G$ *and suppose that* G *has a composition series. Then* H *is subnormal in* G *if and only if* H *is a term of some composition series of* G.
 We note the following consequence of 5.6.

7.14. *Let* G *be a finite* p-group. *Then every subgroup of* G *is subnormal in* G.

Proof. Let $H \leqslant G$. We prove by induction on $|G:H|$ that H is subnormal in G. If $|G:H| = 1$ then $H = G$ and the assertion is trivial. Therefore we assume that $|G:H| > 1$ and, inductively, that K is subnormal in G whenever $K \leqslant G$ and $|G:K| < |G:H|$. Then $H < G$ and so, by 5.6, $H < N_G(H) \leqslant G$. Hence $|G:N_G(H)| < |G:H|$ and so, by the inductive assumption, $N_G(H)$ is subnormal in G. Since $H \lhd N_G(H)$, it follows that H is subnormal in G. This completes the induction argument.

The first systematic development of a theory of subnormal subgroups was made by H. Wielandt [a100]. We include several of his results here.

7.15 (cf. 3.40). *Let $H, K \leqslant G$. If K is subnormal in G then $H \cap K$ is subnormal in H.*
Proof. By hypothesis, there are subgroups K_i of G $(i = 0, 1, \ldots, n)$ such that

$$K = K_0 \lhd K_1 \lhd \ldots \lhd K_n = G.$$

Then certainly

$$H \cap K = H \cap K_0 \leqslant H \cap K_1 \leqslant \ldots \leqslant H \cap K_n = H.$$

Moreover, for $i = 1, \ldots, n$,

$$H \cap K_{i-1} = ((H \cap K_i) \cap K_{i-1}) \lhd (H \cap K_i),$$

since $K_{i-1} \lhd K_i$. Thus $H \cap K$ is subnormal in H.

7.16. *Let $K \leqslant H \leqslant G$. If K is subnormal in G then K is subnormal in H.*
Proof. This follows immediately from 7.15.

7.17. *Let $H, K \leqslant G$. If H and K are both subnormal in G then $H \cap K$ is subnormal in G.*
Proof. This follows immediately from 7.15 and 7.12.

Recall that if $H \leqslant G$ and $K \lhd G$ then $HK \leqslant G$ (3.38); and then of course $HK = \langle H, K \rangle$ (see **71**, **95**). We now show by an example that HK need not be a subgroup of G when H and K are *subnormal* subgroups of G.

7.18. Let n be an integer, $n \geqslant 3$, and let $G = D_{2^n}$, the dihedral group of order 2^n. By 6.12, there are involutions, say h and k, in G such that $G = \langle h, k \rangle$. Let $H = \langle h \rangle \leqslant G$ and $K = \langle k \rangle \leqslant G$. Then $|H| = |K| = 2$ and $HK = \{1, h, k, hk\}$: HK is not a subgroup of G, since the smallest subgroup of G containing both H and K is G itself and $|G| = 2^n \geqslant 8$. On the other hand, by 7.14, H and K are both subnormal in G.

We shall, however, prove an analogue of 3.39 by showing that if H and K are subnormal subgroups of a group G with a composition series then $\langle H, K \rangle$ is subnormal in G. We begin by proving a special case, for which we do not need the assumption that G has a composition series.

7.19 Lemma. *If H is a subnormal subgroup of G and $K \trianglelefteq G$ then HK is subnormal in G.*

Proof. By hypothesis, there is a series

$$H = H_0 \trianglelefteq H_1 \trianglelefteq \ldots \trianglelefteq H_n = G$$

from H to G. Then, by 7.4 (ii),

$$HK = H_0 K \trianglelefteq H_1 K \trianglelefteq \ldots \trianglelefteq H_n K = G.$$

This is a series from HK to G. Hence HK is subnormal in G.

*330 Let $H < G$. Suppose that G has a composition series and that H is subnormal in G. Then H has a composition series. Moreover, if G has composition length n and H has composition length m then $m < n$ (cf. 320(ii).)

331 Let $H \leqslant G$.

(i) Consider the following sequence of subgroups of G:

$$G = J_0 \geqslant J_1 \geqslant J_2 \geqslant \ldots,$$

where, for each integer $i > 0$,

$$J_i = H^{J_{i-1}},$$

the normal closure of H in J_{i-1} (see **180**). Then H is subnormal in G if and only if there is a non-negative integer n such that $J_n = H$.

If H is subnormal in G and n is the least integer such that $J_n = H$ then

$$G = J_0 \trianglerighteq J_1 \trianglerighteq \ldots \trianglerighteq J_n = H$$

is called the *standard series from G to H.*

(ii) Suppose that H is subnormal in G. The *defect* (or *index of subnormality*) of H in G is defined to be the least non-negative integer n for which there is a series of length n from H to G. Then the defect of H in G is equal to the length of the standard series from G to H.

(iii) Suppose that H is subnormal in G and that $H \leqslant L \leqslant G$. Then the defect of H in L is not greater than the defect of H in G.

332 Let n be an integer, $n \geqslant 2$, and let $G = D^{2^{n+1}}$, the dihedral group of order 2^{n+1}. Then there are elements $x, t \in G$ such that $G = \langle x, t \rangle$, $x^{2^n} = 1 = t^2$ and $x^t = x^{-1}$: see 2.24, 6.10, 6.11. Let $H = \langle t \rangle$. By 7.14, H is subnormal in G.

Let $J = H^G$, the normal closure of H in G. Prove that $J = \langle x^2, t \rangle$. Hence prove that the defect of H in G is n (see **331**).

333 Let $G = \Sigma_4$. Find a subnormal subgroup H of G such that $N_G(H)$ is a Sylow 2-subgroup of G.

(Remark. Note that $N_G(N_G(H)) = N_G(H)$. This example shows that for $H \leqslant G$, the ascending sequence of subgroups of G formed from H by taking successive normalizers in G need not reach G, even though H is subnormal in G: cf. **331**. Hint. See **289**.)

334 Let $H \leqslant G$. Then H is both pronormal and subnormal in G if and only if $H \trianglelefteq G$. In particular, if G is finite and H is a subnormal Sylow subgroup of G then $H \trianglelefteq G$. (See **268**. Hint. Argue by induction on the length of a series from H to G.)

*335 Let H be a subnormal subgroup of G. Then, for any homomorphism φ of G onto a group $\bar{G}$, $H\varphi$ is a subnormal subgroup of $\bar{G}$. In particular, for every $g \in G$, H^g is subnormal in G.

336 Let $\mathfrak{X}$ and $\mathfrak{Y}$ be classes of groups with the following two properties.
  (i) Every normal subgroup of every $\mathfrak{X}$-group is an $\mathfrak{X}$-group.
  (ii) Every $\mathfrak{X}$-group has a $\mathfrak{Y}$-radical. (See 3.45.)
  Then if G is an $\mathfrak{X}$-group, every subnormal $\mathfrak{Y}$-subgroup H of G is contained in the $\mathfrak{Y}$-radical of G.
  In particular, if G is any finite group then every subnormal ϖ-subgroup of G is contained in $O_\varpi(G)$.
(Hint. Argue by induction on the length of a series from H to G and use **160**.)

337 Let $\mathfrak{X}$ and $\mathfrak{Y}$ be classes of groups with the following three properties.
  (i) Every normal subgroup of every $\mathfrak{X}$-group is an $\mathfrak{X}$-group.
  (ii) Every $\mathfrak{X}$-group has a $\mathfrak{Y}$-residual. (See 3.45.)
  (iii) Whenever $J \trianglelefteq G$ and both J and G/J are $\mathfrak{Y}$-groups, G is a $\mathfrak{Y}$-group.
  Let G be an $\mathfrak{X}$-group and let G/K be the $\mathfrak{Y}$-residual of G. Suppose that H is a subnormal subgroup of G such that there is a series from H to G, the factors of which are all $\mathfrak{Y}$-groups. Then $K \leqslant H$.
  In particular, if H is a subnormal subgroup of a finite group G such that there is a series from H to G, all the factors of which are ϖ-groups, then $O_\varpi(G) \leqslant H$ (see **160**).

For the purpose of proving the general result, it will be convenient to associate to a subnormal subgroup H in a group G with a composition series a certain non-negative integer $j(G:H)$, which we now define.

7.20 Definition. Let $H \leqslant G$. Suppose that G has a composition series and that H is subnormal in G. Then H is a term of a composition series of G (7.13). Let

$$1 = G_0 \lhd G_1 \lhd \ldots \lhd G_n = G \qquad \text{(a)}$$

be a composition series of G with H as a term: say

$$G_k = H,$$

where $0 \leqslant k \leqslant n$.
  Now consider any other composition series of G of which H is a term. Let the part of the series from H to G be

$$H = H_0 \lhd H_1 \lhd \ldots \lhd H_m = G.$$

Then, by 7.2, the series

$$1 = G_0 \lhd G_1 \lhd \ldots \lhd G_k \lhd H_1 \lhd \ldots \lhd H_m = G \qquad \text{(b)}$$

is also a composition series of G. Therefore, by the Jordan–Hölder theorem (7.9), the series (a) and (b) are equivalent. In particular,

$$n = k + m.$$

We define $j(G:H) = n - k$. In words, $j(G:H)$ is the number of composition factors *above* H in any composition series of G of which H is a term.

The argument above shows that this is well defined. Note that $j(G:1)$ is the composition length of G.

The following properties are immediate from the definition, together with 7.16 and 7.9(i). (See also **330**.)

7.21. Let $K \leqslant H \leqslant G$. Suppose that G has a composition series and that H and K are subnormal in G. Then

(i) $j(G:K) = j(G:H) + j(H:K),$

(ii) $j(G:H) = 0$ if and only if $H = G$.

7.22 Theorem (H. Wielandt [a 100]). Suppose that G has a composition series. If H and K are subnormal subgroups of G then $\langle H, K \rangle$ is subnormal in G.

Proof. We argue by induction on $j(G:K) = n$, say.

If $K \trianglelefteq G$ then $\langle H, K \rangle = HK$ and the result has been proved in 7.19. In particular, this gives the result if $n \leqslant 1$.

Now suppose that $K \ntrianglelefteq G$. Then $n > 1$. There is a series

$$K = K_0 \lhd K_1 \lhd \dots \lhd K_{n-1} \lhd K_n = G$$

from K to G which is part of a composition series of G. Let $G_1 = \langle H, K_1 \rangle$. By 7.21, $j(G:K_1) = n-1$ and so, by the induction hypothesis, G_1 is subnormal in G. Hence G_1 has a composition series (**330**). By 7.16, H and K are subnormal subgroups of G_1.

If $G_1 < G$ then, by 7.21, $j(G_1 : K) < n$. Then, by the induction hypothesis, $\langle H, K \rangle$ is subnormal in G_1, hence also subnormal in G, by 7.12.

Now suppose that $G_1 = G$. If

$$K^h \leqslant K \quad \text{for every } h \in H,$$

then in fact

$$K^h = K \quad \text{for every } h \in H :$$

see **175** or 7.24 (ii).
Then

$$N_G(K) \geqslant \langle H, K_1 \rangle = G.$$

This contradicts the supposition that $K \ntrianglelefteq G$.
Therefore

$$K^h \nleqslant K \quad \text{for some } h \in H.$$

Let

$$K^* = \langle K, K^h \rangle > K.$$

Now K^h is subnormal in G (**335**). Moreover,

$$K^h \leqslant K_{n-1}^h = K_{n-1},$$

since $K_{n-1} \lhd G$, and so, by 7.16, K^h is subnormal in K_{n-1}. Certainly K_{n-1} has a composition series (330). Since $j(K_{n-1} : K) = n - 1$, it follows by the induction hypothesis that K^* is subnormal in K_{n-1}. Hence also K^* is subnormal in G. Since $K^* > K$, it follows from 7.21 that $j(G : K^*) < n$. Hence, by the induction hypothesis, $\langle H, K^* \rangle$ is subnormal in G. However, $K^h \leqslant \langle H, K \rangle$, and so

$$\langle H, K^* \rangle = \langle H, K, K^h \rangle = \langle H, K \rangle.$$

This completes the induction argument.

Remark. This result fails in the absence of the condition that G has a composition series. Zassenhaus ([b41] p. 235, ex. 23) has given an example of a group G with subnormal subgroups H and K such that $\langle H, K \rangle$ is not subnormal in G. See also 345.

338 Suppose that H and K are subnormal subgroups of the finite group G. If $(|H|, |K|) = 1$ then $\langle H, K \rangle = HK \cong H \times K$. (cf. 7.18. Hint. Use 336, 3.53, 3.54.)

339 (i) Let $H, K \leqslant G$, a finite group, with $\langle H, K \rangle = G$. Suppose that K is subnormal in G and that every composition factor of G above K has prime order. Then $j(H : H \cap K) \leqslant j(G : K)$. (cf. 7.15. Remark. This inequality no longer holds in general in the absence of the condition that every composition factor of G above K has prime order. For an example see 532.)

(ii) Show by an example that for any integer $n > 1$, it is possible to satisfy all the conditions of (i) and have $j(H : H \cap K) = 1$ and $j(G : K) = n$.

We end this selection of results by proving an analogue of the isomorphism part of the statement of 3.40. What we shall show is that if H and K are subnormal subgroups of G and if G has a composition series then the set of types of composition factors of G between $H \cap K$ and H coincides with the set of types of composition factors of G between K and $\langle H, K \rangle$.

7.23 Definition. Suppose that G has a composition series and let H be a subnormal subgroup of G. Then H is a term of a composition series of G (7.13). We denote by $\mathcal{K}(G, H)$ the set of composition factors of G above H, where factors of the same type are identified. Thus $\mathcal{K}(G, H)$ is a set of pairwise non-isomorphic simple groups; it takes no account of the multiplicity of occurrences of any particular simple group in a composition series of G.

A similar argument to the one used in 7.20 to justify the definition of $j(G : H)$ shows that $\mathcal{K}(G, H)$ is determined by the factors in any particular composition series of G through H.

If also K is a subnormal subgroup of G with $K \leqslant H$ then, by 7.16 and 7.9(i), there is a composition series of G through both H and K, and then clearly $\mathcal{K}(G, K) = \mathcal{K}(G, H) \cup \mathcal{K}(H, K)$.

We have $\mathcal{K}(G, H) = \emptyset$ if and only if $H = G$.

We shall use the following lemma.

7.24 Lemma. *Suppose that G has a composition series and let H be a subnormal subgroup of G and $g \in G$. Then*

    (i) *H^g is subnormal in G and $j(G:H) = j(G:H^g)$.*

    (ii) *If $H \neq H^g$ then $H^g \not\leqslant H$. (cf. **174, 175**.)*

    (iii) *$\mathscr{K}(H, H \cap H^g) = \mathscr{K}(H^g, H \cap H^g)$.*

Proof. We may assume that $H \neq H^g$. Let $j(G:H) = n$. There is a series

$$H = H_0 \lhd H_1 \lhd \dots \lhd H_n = G$$

from H to G which is part of a composition series of G. It follows from 3.29 that

$$H^g = H_0^g \lhd H_1^g \lhd \dots \lhd H_n^g = G$$

with $\qquad\qquad H_i^g / H_{i-1}^g \cong H_i / H_{i-1}$ (†)

for each $i = 1, \dots, n$. Now (i) follows, by 7.2.

If $H^g \leqslant H$ then it follows from (i) and 7.21 that

$$j(G:H) = j(G:H^g) = j(G:H) + j(H:H^g),$$

hence that $\qquad\qquad\qquad j(H:H^g) = 0$.

Thus $\qquad\qquad\qquad\qquad H = H^g$.

This is contrary to hypothesis, and so we conclude that $H^g \not\leqslant H$. This proves (ii).

Let $L = H \cap H^g$. By (i) and 7.15, L is subnormal in both H and H^g. By 7.9(i), there is a composition series of G of the form

$$1 = G_0 \lhd G_1 \lhd \dots \lhd G_l \lhd G_{l+1} \lhd \dots \lhd G_m \lhd H_1 \lhd \dots \lhd H_n = G, \quad \text{(a)}$$

where $0 \leqslant l < m$, $G_l = L$ and $G_m = H$. Then there is also a composition series of G of the form

$$1 = G_0 \lhd G_1 \lhd \dots \lhd G_l \lhd G_{l+1}^* \lhd \dots \lhd G_{m*}^* \lhd H_1^g \lhd \dots \lhd H_n^g = G, \quad \text{(b)}$$

where $l < m^*$ and $G_{m*}^* = H^g$. By the Jordan–Hölder theorem, the series (a) and (b) are equivalent. Hence $m = m^*$, and, in view of (†), we must have

$$\mathscr{K}(H, L) = \mathscr{K}(H^g, L).$$

340 Let φ be an isomorphism of G_1 onto G_2. Suppose that G_1 has a composition series, let H_1 be a subnormal subgroup of G_1 and let $H_2 = H_1 \varphi \leqslant G_2$. Then G_2 has a composition series, H_2 is subnormal in G_2, $\mathscr{K}(G_1, H_1) = \mathscr{K}(G_2, H_2)$ and $\mathscr{K}(H_1, 1) = \mathscr{K}(H_2, 1)$.

341 (i) Let $H, K \leqslant G$, a finite group. Suppose that K is subnormal in G and that every composition factor of G above K has prime order. Then $\mathscr{K}(H, H \cap K) \subseteq \mathscr{K}(G, K)$.

    (ii) Let $G = A_4$, the alternating group of degree 4. Every composition factor of G has prime order and there are subgroups H, K of G such that $\langle H, K \rangle = G$, K is subnormal in G, and $\mathscr{K}(H, H \cap K) \neq \mathscr{K}(G, K)$. (cf. **339**.)

342 Suppose that G has a composition series, no two distinct factors of which are isomorphic. Then

(i) No two distinct normal subgroups of G are isomorphic.

(Hints. Note that if H and K are isomorphic normal subgroups of G then $\mathcal{K}(H, H \cap K) = \mathcal{K}(K, H \cap K)$. If $H \neq K$, consider a composition series of G through $H \cap K, K$ and HK, and derive a contradiction to the hypothesis on G.)

(ii) Every normal subgroup of G is characteristic in G.

(iii) Every subnormal subgroup of G is normal in G (and hence characteristic in G).

(Hint. Argue by induction on the length of a series from H to G.)

7.25 Theorem (H. Wielandt [a100]). *Suppose that G has a composition series and let H and K be subnormal subgroups of G. Then $\mathcal{K}(H, H \cap K) = \mathcal{K}(\langle H, K \rangle, K)$.*

Proof. By 7.15 and 7.16, $H \cap K$ is subnormal in H and K is subnormal in $\langle H, K \rangle$. Since H and $\langle H, K \rangle$ are both subnormal in G (by 7.22), H and $\langle H, K \rangle$ both have composition series (**330**). Therefore the assertion of the theorem makes sense. Furthermore, we may assume without loss of generality that $\langle H, K \rangle = G$.

We show first that

$$\mathcal{K}(H, H \cap K) \subseteq \mathcal{K}(G, K). \tag{i}$$

Let

$$1 = G_0 \lhd G_1 \lhd \ldots \lhd G_n = G \tag{a}$$

be a composition series of G with K as a term: say

$$G_m = K,$$

where $0 \leqslant m \leqslant n$. By 7.4(i),

$$1 = (G_0 \cap H) \unlhd (G_1 \cap H) \unlhd \ldots \unlhd (G_n \cap H) = H, \tag{b}$$

and for each $i = 1, \ldots, n$,

$$(G_i \cap H)/(G_{i-1} \cap H) \cong G_{i-1}(G_i \cap H)/G_{i-1}.$$

By 7.15, $G_i \cap H$ is subnormal in G_i. Hence, since $G_i \cap H$ is mapped to $G_{i-1}(G_i \cap H)/G_{i-1}$ by the natural homomorphism of G_i onto G_i/G_{i-1}, $G_{i-1}(G_i \cap H)/G_{i-1}$ is subnormal in G_i/G_{i-1} (**335**). But G_i/G_{i-1} is simple (7.2) and therefore, for each $i = 1, \ldots, n$,

either $(G_i \cap H)/(G_{i-1} \cap H)$ is trivial

or $(G_i \cap H)/(G_{i-1} \cap H) \cong G_i/G_{i-1}.$

Therefore, by 7.2, if from the series (b) we delete repetitions of terms, we obtain a composition series (c) of H. Moreover, $H \cap K$ is a term of (c) and each factor of (c) above $H \cap K$ is isomorphic to a factor of (a) above K. This establishes (i).

Now we show that

$$\mathcal{K}(G, K) \subseteq \mathcal{K}(H, H \cap K). \tag{ii}$$

Suppose first that $H \trianglelefteq G$. Then consider any composition factor A/B of G above K: thus

$$K \leqslant B \triangleleft A \leqslant G,$$

A and B are subnormal in G, and A/B is simple. By 7.4(i), $(B \cap H) \trianglelefteq (A \cap H)$ and

$$(A \cap H)/(B \cap H) \cong B(A \cap H)/B.$$

Since $H \trianglelefteq G, BH \leqslant G$. Then, since $BH \geqslant \langle H, K \rangle = G$,

$$BH = G.$$

Therefore, by 7.3,

$$B(A \cap H) = A \cap (BH) = A.$$

Hence

$$(A \cap H)/(B \cap H) \cong A/B.$$

Thus $(A \cap H)/(B \cap H)$ is simple and is therefore a composition factor of G, since, by 7.17, $A \cap H$ is subnormal in G. Hence every composition factor of G above K is isomorphic to a composition factor of G between $H \cap K$ and H. This establishes (ii) in this special case.

For the general case, we argue by induction on $j(G : H) = m$, say. If $m \leqslant 1$ then $H \trianglelefteq G$ and there is nothing more to prove. Therefore we assume that $m > 1$ and also that $H \ntrianglelefteq G$. There is a series

$$H = H_0 \triangleleft H_1 \triangleleft \ldots \triangleleft H_{m-1} \triangleleft H_m = G$$

from H to G which is part of a composition series of G. Let $L = H_{m-1} \cap K$. Then

$$H \cap K \leqslant L \trianglelefteq K.$$

Suppose that $L \nleqslant H$. Then let

$$J = \langle H, L \rangle > H.$$

By 7.17 and 7.22, J is subnormal in G and, by 7.21,

$$j(G : J) < j(G : H) = m.$$

Certainly $\langle J, K \rangle = \langle H, K \rangle = G$. Therefore, by the induction hypothesis,

$$\mathscr{K}(G, K) \subseteq \mathscr{K}(J, J \cap K). \tag{iii}$$

Since

$$L \leqslant J \cap K \leqslant J,$$

and all these subgroups are subnormal in G,

$$\mathscr{K}(J, L) = \mathscr{K}(J, J \cap K) \cup \mathscr{K}(J \cap K, L). \tag{iv}$$

Now $J \leqslant H_{m-1} \vartriangleleft G$ and therefore, by 7.21,

$$j(J, H) < j(G, H) = m.$$

Certainly $H \cap L = H \cap K$, and of course H and L are subnormal in $J = \langle H, L \rangle$. Therefore, by the induction hypothesis,

$$\mathscr{K}(J, L) \subseteq \mathscr{K}(H, H \cap K). \tag{v}$$

Now (ii) follows from (iii), (iv) and (v).

Suppose on the other hand that $L \leqslant H$. Then

$$L = H \cap K,$$

and so

$$H \cap K \vartriangleleft K.$$

Since $H \ntrianglelefteq G = \langle H, K \rangle$, $K \nleqslant N_G(H)$. Therefore

$$H^k \neq H \text{ for some } k \in K.$$

Hence, by 7.24 (ii),

$$H^k \nleqslant H.$$

Then let $H^* = \langle H, H^k \rangle > H.$

Since H^k is subnormal in G (by **335** or 7.24(i)), H^* is subnormal in G, by 7.22; and by 7.21,

$$j(G : H^*) < j(G : H) = m.$$

Certainly $\langle H^*, K \rangle = \langle H, K \rangle = G$. Therefore, by the induction hypothesis,

$$\mathscr{K}(G, K) \subseteq \mathscr{K}(H^*, H^* \cap K). \tag{vi}$$

Now

$$H \cap K \leqslant H^* \cap K \leqslant H^* \quad \text{and} \quad H \cap K \leqslant H \leqslant H^*,$$

and all these subgroups are subnormal in G. Therefore

$$\mathscr{K}(H^*, H \cap K) = \mathscr{K}(H^*, H^* \cap K) \cup \mathscr{K}(H^* \cap K, H \cap K) \tag{vii}$$

$$= \mathscr{K}(H^*, H) \cup \mathscr{K}(H, H \cap K). \tag{viii}$$

Now H^k and H are subnormal in H^*. Moreover, since $H \leqslant H_{m-1} \vartriangleleft G$, $H^* \leqslant H_{m-1}$ and therefore, by 7.21 and 7.24(i),

$$j(H^* : H^k) < j(G : H^k) = m.$$

Hence, by the induction hypothesis (applied to H^* and H^k in place of G and H) and 7.24(ii),

$$\mathscr{K}(H^*, H) \subseteq \mathscr{K}(H^k, H \cap H^k) = \mathscr{K}(H, H \cap H^k). \tag{ix}$$

Since in the present case $H \cap K \trianglelefteq K$,

$$H \cap K = (H \cap K)^k \leqslant H \cap H^k \leqslant H.$$

These subgroups are subnormal in G, and so

$$\mathscr{K}(H, H \cap K) = \mathscr{K}(H, H \cap H^k) \cup \mathscr{K}(H \cap H^k, H \cap K). \qquad (\text{x})$$

Now (ii) follows from (vi), (vii), (viii), (ix) and (x).

This completes the proof of (ii) in all cases. Together with (i), this gives the result.

7.26 Corollary. *Suppose that G has a composition series and let H and K be subnormal subgroups of G. Then*

$$\mathscr{K}(G, H \cap K) = \mathscr{K}(G, H) \cup \mathscr{K}(G, K).$$

Proof. By 7.17 and 7.22, $H \cap K$ and $\langle H, K \rangle$ are subnormal in G. Since $H \cap K \leqslant H \leqslant G$,

$$\mathscr{K}(G, H \cap K) = \mathscr{K}(G, H) \cup \mathscr{K}(H, H \cap K), \qquad (\text{i})$$

and since $K \leqslant \langle H, K \rangle \leqslant G$,

$$\mathscr{K}(G, K) = \mathscr{K}(G, \langle H, K \rangle) \cup \mathscr{K}(\langle H, K \rangle, K). \qquad (\text{ii})$$

Moreover, since $H \leqslant \langle H, K \rangle$,

$$\mathscr{K}(G, \langle H, K \rangle) \subseteq \mathscr{K}(G, H). \qquad (\text{iii})$$

By (i) and 7.25,

$$\begin{aligned}
\mathscr{K}(G, H \cap K) &= \mathscr{K}(G, H) \cup \mathscr{K}(\langle H, K \rangle, K) \\
&= \mathscr{K}(G, H) \cup \mathscr{K}(G, \langle H, K \rangle) \cup \mathscr{K}(\langle H, K \rangle, K) \quad \text{(by (iii))} \\
&= \mathscr{K}(G, H) \cup \mathscr{K}(G, K) \quad \text{(by (ii))}.
\end{aligned}$$

343 Suppose that G has a composition series and let H and K be subnormal subgroups of G.

(i) Then $\mathscr{K}(\langle H, K \rangle, 1) = \mathscr{K}(H, 1) \cup \mathscr{K}(K, 1)$ (cf. 7.26).

(ii) If H and K are finite then $\langle H, K \rangle$ is finite (cf. **71** (ii). Remark. The statement in (ii) is in fact true without the condition that G has a composition series. See **344, 345**.)

344 Suppose that $H, K \leqslant G$ with $G = \langle H, K \rangle$.

(i) Let $J = H^G$, the normal closure of H in G (**180**). Show that

$$J = \langle H^k : k \in K \rangle.$$

(Hint. Show that $K \leqslant N_G(J)$.)

(ii) Prove that if H and K are both finite and H is subnormal in G then G is finite. (cf. **71, 343**. See also **345**. Hints. Argue by induction on the defect of H in G: see **331**. Use (i) and induction to show that H^G is finite.)

345 Let $H, K \leqslant G$.

(i) Prove that if H is subnormal in G and $K \leqslant N_G(H)$, then K normalizes every

SERIES

term of the standard series from G to H. (See **331**. Hint. Use an induction argument.)
(ii) Prove that if H and K are both finite subnormal subgroups of G then $\langle H, K \rangle$ is a finite subnormal subgroup of G. (cf. 7.22, **343**, **344**. Hints. Let $L = \langle H, K \rangle$. Argue by induction on the defect, n say, of H in L. If $n \leqslant 1$, use (i). For $n > 1$, use **344** and induction to show that H^L is a finite subnormal subgroup of G. Then note that $L = H^L K$.)

346 Suppose that G has a composition series and let H, H^*, K, K^* be subnormal subgroups of G with $H^* \leqslant H$ and $K^* \leqslant K$. Then

$$\mathscr{K}(\langle H, K \rangle, \langle H^*, K^* \rangle) \subseteq \mathscr{K}(H, H^*) \cup \mathscr{K}(K, K^*).$$

(Hint. Note that $\langle H^*, K^* \rangle \leqslant \langle H, K^* \rangle \leqslant \langle H, K \rangle$ and apply 7.25 twice.)

347 Let H and K be subnormal subgroups of the finite group G. Then

$$O^\varpi(\langle H, K \rangle) = \langle O^\varpi(H), O^\varpi(K) \rangle.$$

(This generalizes **159** (ii). Hint. Apply **157**, **337**, and **346**.)

348 Let K be a subnormal subgroup of the finite group G, and let $J = K^G$, the normal closure of K in G (**180**), and $L = K_G$, the core of K in G (**90**).
   (i) Prove that $\mathscr{K}(J, 1) = \mathscr{K}(K, 1)$ and $\mathscr{K}(G, K) = \mathscr{K}(G, L)$.
(Hint. Apply **340**, **343**, and 7.26.)
   (ii) Show by an example that it can happen that

$$\mathscr{K}(G, J) \neq \mathscr{K}(G, K) \quad \text{and} \quad \mathscr{K}(K, 1) \neq \mathscr{K}(L, 1).$$

349 Let H and K be subnormal subgroups of the finite group G and suppose that

$$\mathscr{K}(H, 1) \cap \mathscr{K}(K, 1) = \emptyset.$$

Then $\langle H, K \rangle = HK \cong H \times K$.
(This generalizes **338**. Hint. Use **348**, 3.53, 3.54.)

Our next step in the development of the theory of normal structure is to observe that many of the results obtained are valid more generally for *groups with operators*.

7.27 Definitions. A *group with operators* consists of a group G and a set Ω (the *operator domain*) such that, to each $g \in G$ and each $\omega \in \Omega$, there corresponds a unique element $g^\omega \in G$, and such that

$$(g_1 g_2)^\omega = g_1^\omega g_2^\omega \quad \text{for all } g_1, g_2 \in G \text{ and } \omega \in \Omega.$$

We say then that G is an Ω-*group*.
   Let $H \leqslant G$. We say H is a *stable* (or *admissible*) subgroup, or explicitly that H is an Ω-*subgroup*, if

$$h^\omega \in H \quad \text{for every } h \in H \text{ and } \omega \in \Omega.$$

Note that if H is a stable subgroup of the Ω-group G then Ω is also an operator domain for H.

7.28 Remarks and examples. (1) If G is an Ω-group then, for each $\omega \in \Omega$, the map

$$g \mapsto g^{\omega},$$

defined for all $g \in G$, is an *endomorphism* of G (2.18).

If follows that the trivial subgroup 1 is stable; and of course G itself is certainly stable.

Clearly, any set of endomorphisms of G is a suitable operator domain for G. However, in 7.27 we do not restrict Ω to be a set of endomorphisms of G since we wish to allow the possibility that there are distinct elements ω_1, ω_2 of Ω which operate in the same way on G; that is, for which $g^{\omega_1} = g^{\omega_2}$ for every $g \in G$. (This is analogous to the situation in chapter 4, where, in considering the action of a group G on a set X, we do not restrict G to be a subgroup of Σ_X.)

If G is an Ω-group and $\emptyset \subset \Omega \subseteq \operatorname{Aut} G$ then the stable subgroups of G are just the Ω-invariant subgroups in the sense of 3.1.

(2) Trivially, we can consider any group G as an Ω-group with $\Omega = \emptyset$. Then every subgroup of G is stable and the theory of the Ω-group G is simply the familiar theory of G as a group without operators.

(3) G is an operator domain for itself when G acts on itself by conjugation, as in 4.25. For each $x \in G$ and each $g \in G$ the element of G which corresponds by the operation of g on x is x^g, the conjugate of x by g. In this case, the stable subgroups of G are just the *normal* subgroups of G.

(4) Let R be any ring. Then R is an operator domain for R^+, the additive group of R (2.11), by right multiplication: for, by ring axioms,

$$(x_1 + x_2)a = x_1 a + x_2 a$$

for all $x_1, x_2 \in R^+$ and $a \in R$. Then the stable subgroups of R^+ are precisely the right ideals of the ring R.

Similarly, R is an operator domain for R^+ by left multiplication, and in this case the stable subgroups are the left ideals of R.

(5) Let V be a vector space over a field F. Then F is an operator domain for V^+, the additive group of V (2.15), by scalar multiplication: for, by vector space axioms,

$$a(v_1 + v_2) = a v_1 + a v_2$$

for all $v_1, v_2 \in V^+$ and $a \in F$. Then the stable subgroups are just the subspaces of V.

7.29 Definition. Let G and H be groups with the same operator domain Ω. Then a homomorphism $\varphi : G \to H$ is said to be an Ω-*homomorphism* if

$$(g^{\omega})\varphi = (g\varphi)^{\omega}$$

for all $g \in G$ and $\omega \in \Omega$. It is immediate from the definition that then $\operatorname{Ker} \varphi$ is a *stable* normal subgroup of G and $\operatorname{Im} \varphi$ is a *stable* subgroup of H.

7.30. Let G be an Ω-group and suppose that K is a stable normal subgroup of G. Then Ω is in a natural way an operator domain for the quotient group G/K: we define

$$(gK)^\omega = g^\omega K$$

(for all $g \in G$ and $\omega \in \Omega$).

This is well defined, for if $g_1, g_2 \in G$ and $g_1 K = g_2 K$ then

$$g_1 = g_2 k$$

for some $k \in K$, hence

$$g_1^\omega K = g_2^\omega k^\omega K = g_2^\omega K,$$

since $k^\omega \in K$.

Note that this definition of G/K as an Ω-group makes the natural homomorphism $\nu : G \to G/K$ an Ω-homomorphism.

We shall now verify that the fundamental theorem on homomorphisms remains valid for groups with operators.

7.31 (cf. 3.24). *Let the set Ω be an operator domain for the groups G and H, and let $\varphi : G \to H$ be an Ω-homomorphism, $K = \mathrm{Ker}\,\varphi$ and $\nu : G \to G/K$ the natural homomorphism. Then K is stable, G/K is in a natural way an Ω-group, ν is an Ω-homomorphism, and there is an injective Ω-homomorphism $\psi : G/K \to H$ such that $\varphi = \nu\psi$. In particular, $\mathrm{Im}\,\varphi$ and $G/\mathrm{Ker}\,\varphi$ are Ω-isomorphic Ω-groups.*

Proof. In view of the remarks above, it is enough to verify that the injective homomorphism $\psi : G/K \to H$ defined in 3.24 is an Ω-homomorphism. Let $g \in G$ and $\omega \in \Omega$. By definition,

$$(gK)\psi = g\varphi,$$

and so $\qquad (gK)^\omega \psi = (g^\omega K)\psi = g^\omega \varphi = (g\varphi)^\omega,$

since φ is an Ω-homomorphism; that is,

$$(gK)^\omega \psi = ((gK)\psi)^\omega.$$

Thus ψ is an Ω-homomorphism.

350 Let G be an Ω-group. If H and K are stable subgroups of G then so are $H \cap K$, $\langle H, K \rangle$ and $[H, K]$.

351 If H is a stable subgroup of an Ω-group G, $N_G(H)$ need not be stable (cf. **176**). Show this by means of the following example.

Let $G = \Sigma_4$ and consider G as an Ω-group, where $\Omega = \{\omega\}$ and ω is an endomorphism of G such that $\mathrm{Ker}\,\omega = A_4$. Let H be a suitable Sylow 3-subgroup of G.

352 Let G be an Ω-group. If H is a stable subgroup and K a stable normal subgroup of G then HK is a stable subgroup of G. Moreover, $H \cap K$ is a stable normal subgroup of H and the Ω-groups $H/H \cap K$ and HK/K are Ω-isomorphic (see 3.40).

*353 (i) Let G be an Ω-group. Then the set of all Ω-endomorphisms of G forms a subsemigroup S of the semigroup of all endomorphisms of G (see 2.18). S has an identity element and the group of units of S consists of all Ω-automorphisms of G. Denote this group by $\mathrm{Aut}_\Omega (G)$.

(ii) Consider G as a G-group, as in 7.28 (3). Then

$$\mathrm{Aut}_G(G) = C_{\mathrm{Aut}G}(\mathrm{Inn}\ G) \quad \text{(cf. 245)}.$$

7.32 Remarks and definitions. (1) It is equally straightforward to verify that all the main results on normal structure of groups remain valid for groups with operators: thus 3.29, 3.30, 3.38, 3.39, 3.40, 7.7 and 7.9 are true when, in their statements and proofs, we replace groups by Ω-groups, subgroups by stable subgroups, homomorphisms by Ω-homomorphisms and isomorphisms by Ω-isomorphisms (see for instance **352**). We shall not write down explicit proofs but from now on use these results in their operator versions whenever we need them.

(2) We ought perhaps to say something more about the correct interpretation of 7.7 and 7.9 for groups with operators. Let G be an Ω-group. Then an Ω-*series* of G is a series of G the terms of which are stable subgroups of G. The definitions of 'proper', 'refinement' and 'proper refinement' given in 7.1 apply without change. An Ω-*composition series* of G is a proper Ω-series of G which has no proper refinement (as an Ω-series). The definition of 'equivalence' of two Ω-series of G is exactly as in 7.6, with the additional requirement that corresponding factors are not merely isomorphic but Ω-isomorphic. Now the versions of 7.7 and 7.9 for Ω-groups are clear.

(3) We say that a non-trivial Ω-group G is Ω-*simple* if the only *stable* normal subgroups of G are 1 and G.

Let G be any Ω-group. By previous remarks, we know that the factors of an Ω-series of G are in a natural way Ω-groups. Then a repetition of the proof of 7.2 shows that an Ω-series of G is an Ω-composition series of G if and only if the factors of the series are all Ω-simple.

(4) A simple Ω-group is certainly Ω-simple. However, an Ω-simple Ω-group need not be a simple group. For instance, let V be a vector space $\neq 0$ over a field F and view V^+ as an F-group, as in 7.28 (5). Since V^+ is abelian, all its subgroups are normal. The *stable* subgroups are the subspaces of V. Thus the F-group V^+ is F-simple if and only if the only subspaces of V are 0 and V; that is, if and only if V has dimension 1. Since V^+ is abelian, V^+ is a simple group if and only if V^+ is finite and of prime order (3.6). Hence, if V has dimension 1 then V^+ is an F-simple F-group which is not a simple group unless $F = \mathbf{Z}_p$ for some prime p.

Note in passing that for any finite-dimensional vector space V over F, the F-group V^+ has an F-composition series and the length of any F-composition series of V^+ is equal to the dimension of V. Indeed, the Jordan–Hölder theorem for groups with operators provides one method of proving that any two bases of a finite-dimensional vector space contain the same number of elements.

(5) As another example, consider again a vector space $V \neq 0$ over a field F. Linear maps of V into itself are certainly endomorphisms of V^+. Therefore the ring $\mathscr{L}(V)$ of all linear maps of V into itself is an operator domain for V^+. As an $\mathscr{L}(V)$-group, V^+ is $\mathscr{L}(V)$-simple; or, equivalently, since V^+ is abelian, the only stable subgroups of the $\mathscr{L}(V)$-group V^+ are 0 and V^+. To see this, consider any non-trivial stable subgroup W of V^+. Let $0 \neq w \in W$. Then for any $v \in V$, there is a linear map $\theta : V \to V$ such that $v = w\theta$. Since W is an $\mathscr{L}(V)$-subgroup of V^+, it follows that $v \in W$. Hence $W = V^+$.

Note that V^+ is $\mathscr{L}(V)$-simple without any condition on the dimension of V, while V^+ is a simple group only if V has dimension 1 and $F = \mathbf{Z}_p$ for some prime p.

7.33 Definitions. The most important special case for group theory of the idea of a group with operators occurs when we view G as a G-group, as in 7.28 (3). There is a special terminology for this case.

A G-series of G is called a *normal series* of G. Thus a normal series of G is simply a series of G in which each term is normal in G.

A G-composition series of G is called a *chief series* (or *principal series*) of G. The factors of a chief series of G are called *chief factors* of G.

By Schreier's theorem for groups with operators, any two normal series of G have equivalent refinements: these are normal series of G the factors of which are G-isomorphic in pairs. Also, if G has a chief series then every proper normal series of G has a refinement which is a chief series of G; and, by the Jordan–Hölder theorem for groups with operators, any two chief series of G are equivalent.

Note that every finite group G has a chief series. This is clear, for if G is finite, the process of refining a proper normal series of G must lead in a finite number of steps to a chief series of G.

7.34. Unlike composition factors, chief factors of a group need not be simple groups. For example, let $G = A_4$, the alternating group of degree 4. The only non-trivial proper normal subgroup of G is the unique subgroup V of order 4: see **185, 288**. Thus

$$1 < V < G$$

is a chief series of G. In this example, the chief factor G/V of G is simple but the chief factor $V/1$ is not.

354 Suppose that G has a chief series

$$1 = G_0 < G_1 < \ldots < G_n = G,$$

where n is a positive integer. Then G_n/G_{n-1} is simple (although the factors G_i/G_{i-1} with $0 < i < n$ need not be simple: see 7.34).

355 If G has a composition series then G has a chief series.
(Hint. Any proper normal series of G has length at most equal to the composition length of G. Remark. It is not true that if G has a chief series then G has a composition series: see **534**.)

356 Let H/J and K/L be G-isomorphic factors of normal series of G. If $H/J \leqslant Z(G/J)$ then $K/L \leqslant Z(G/L)$.

We shall need information about the structure of chief factors of finite groups. We note some preliminary results.

7.35 Definition. Let $1 < K \trianglelefteq G$. Then K is said to be a *minimal normal subgroup of G* if there is no normal subgroup L of G such that $1 < L < K$. (Compare this with the definition of 'maximal subgroup' in **140**. Just as a maximal subgroup is a subgroup maximal among *proper* subgroups, so a minimal normal subgroup is a normal subgroup minimal among all *non-trivial* normal subgroups.)

7.36. *Suppose that G has a chief series and let H, K be normal subgroups of G with $K < H$. Then H/K is a chief factor of G if and only if H/K is a minimal normal subgroup of G/K.*
Proof. If H/K is a chief factor of G, it follows immediately from 3.30 that H/K is a minimal normal subgroup of G/K.
Suppose conversely that H/K is a minimal normal subgroup of G/K. By the analogue of 7.9(i) for groups with operators, there is a chief series of G of which both K and H are terms. By 3.30, no term of this chief series lies strictly between K and H. Therefore H/K is a chief factor of G.

7.37 Definition. A non-trivial group G is said to be *characteristically simple* if the only characteristic subgroups of G are 1 and G.
Simple groups are certainly characteristically simple, but a characteristically simple group is not necessarily simple, as we shall see presently.

7.38. *Suppose that K is a minimal normal subgroup of G. Then K is a characteristically simple group. In particular (by 7.36), if G has a chief series then every chief factor of G is characteristically simple.*
Proof. Let L be a characteristic subgroup of K. Then, by 3.15, $L \trianglelefteq G$. Therefore, since $L \leqslant K$ and K is minimal normal in G, either $L = 1$ or $L = K$. Thus K is characteristically simple.
We defer the main result on the structure of characteristically simple groups to the next chapter (see 8.10). Here we shall describe the characteristically simple finite abelian groups.

7.39 Definition. An abelian group A is said to be *elementary* if there is a prime p such that $a^p = 1$ for every $a \in A$.

7.40. *Let A be an abelian group. Then the following two statements are equivalent:*

(i) *A is elementary.*

(ii) *There is a prime p and a vector space V over $\mathbf{Z}_p$ such that $A \cong V^+$.*

Proof. Suppose that A is elementary: thus there is a prime p such that

$$a^p = 1$$

for every $a \in A$. We define a vector space V over $\mathbf{Z}_p$ as follows (cf. **42**). The elements of V are the elements of A. The vector sum of two elements of V is defined to be the product of the elements in A. The elements of $\mathbf{Z}_p$ are residue classes of integers mod p. Let $\bar{n} \in \mathbf{Z}_p$ and let n be an integer in the residue class $\bar{n}$. Then, for each $a \in V$, the scalar product $\bar{n}a$ is defined to be the element $a^n \in A$. This does not depend on the choice of n in the residue class $\bar{n}$ because $a^p = 1$. Now it is easy to check that V is a vector space over $\mathbf{Z}_p$; and clearly, as groups,

$$A \cong V^+.$$

Conversely, let V be a vector space over $\mathbf{Z}_p$. Then V^+ is an abelian group. Moreover, since $pv = 0$ for every $v \in V$, V^+ is elementary. Hence if $A \cong V^+$ then A is elementary.

7.41. *Let A be a finite abelian group, $A \neq 1$. Then A is characteristically simple if and only if A is elementary.*

Proof. Suppose that A is elementary. By 7.40, we may suppose that $A = V^+$, where V is a vector space over $\mathbf{Z}_p$ for some p; and, since A is finite, V is finite-dimensional. Suppose that W is a non-trivial characteristic subgroup of V^+ and let $0 \neq w \in W$. Let $0 \neq v \in V^+$. Then v and w are elements of bases of V. Hence there is an invertible linear map $\theta : V \to V$ such that $v = w\theta$. Then since $\theta \in \operatorname{Aut} V^+$ and W is characteristic in V^+, $v \in W$. Hence $W = V^+$. Thus V^+ is characteristically simple.

Suppose conversely that A is characteristically simple. Let p be a prime divisor of $|A|$ and let

$$B = \{a \in A : a^p = 1\}.$$

Since A is abelian, it is clear that $B \leqslant A$. Let $b \in B$ and $\alpha \in \operatorname{Aut} A$. Then

$$(b^\alpha)^p = (b^p)^\alpha = 1,$$

so that

$$b^\alpha \in B.$$

Hence B is characteristic in A. By **107** or 5.11, there is in A an element a of order p. Then

$$1 \neq a \in B.$$

Therefore $B \neq 1$. Since A is characteristically simple, it follows that

$$B = A.$$

Thus A is elementary.

357 Suppose that K is a minimal normal subgroup of G. Then either K is abelian or $Z(K) = 1$.

358 Suppose that K and L are distinct minimal normal subgroups of G. Then $KL \cong K \times L$. (Hint. See 3.54.)

359 Suppose that $G = HK$, where $H < G$ and K is an abelian minimal normal subgroup of G. Then H is a maximal subgroup of G (see **140**) and $H \cap K = 1$. Moreover, if $K \leqslant Z(G)$ then $H \lhd G$ and $G \cong H \times K$.
(Hint. If $H \leqslant J \leqslant G$, apply Dedekind's rule and show that $J \cap K \lhd G$.)

360 $\mathbf{Z}^+$ has no minimal normal subgroup.

*361 (i) Let F be any field. Then the abelian group F^+ is characteristically simple. (Hint. See 2.11.)
(ii) $\mathbf{Z}^+$ is not characteristically simple.

362 Suppose that G is characteristically simple. Then $G \times G$ is also characteristically simple. (Hints. Let K be a characteristic subgroup of $G \times G$ and let $G_1 = G \times 1$, $G_2 = 1 \times G$. Show that $K \cap G_1$ is characteristic in G_1; see **94**. If $G_1 \leqslant K$ then $K = G \times G$; see 3.12. Then it may be assumed that $K \cap G_1 = 1 = K \cap G_2$. Let π_1 be the projection $G \times G \to G$ with Ker $\pi_1 = G_2$: see 3.11. Let $K_1 = K\pi_1$. Show that K_1 is characteristic in G. If $K_1 = 1$ then $K = 1$. Show that if $K_1 = G$ then Aut $G = 1$. Then use the result stated in **52**.)

363 Let $L \lhd G$. Then L is said to be a *maximal normal subgroup of G* if there is no normal subgroup K of G such that $L < K < G$ (cf. **140**, 7.35).
(a) Prove that a normal subgroup L of G is a maximal normal subgroup of G if and only if G/L is simple.
(b) Suppose that L is a maximal normal subgroup of G. Suppose also that there is a subnormal subgroup H of G such that
(i) $H \not\leqslant L$, and
(ii) whenever J is a subnormal subgroup of G such that $J < H$, $J \leqslant L$.
(Remark. If G is finite, there is such a subgroup H.) Prove that $H \cap L$ is the *unique* maximal normal subgroup of H.

We now introduce the classes of nilpotent and soluble groups.

7.42 Definitions. (i) A factor H/K of a series of G is said to be a *central factor* of G if K is normal in G and $H/K \leqslant Z(G/K)$.

(ii) G is said to be *nilpotent* if it has a series all of whose factors are central factors of G. Such a series is called a *central series*.

(iii) G is said to be *soluble* (or, by American authors, *solvable*) if it has a series all of whose factors are abelian. We shall call such a series an *abelian series*. Groups which are not soluble are said to be *insoluble*.

Note that a central series is necessarily a normal series. An abelian series need not be a normal series.

The notion of solubility of groups was formulated by Galois in the earliest stages of the development of group theory. Indeed, the name 'soluble' reflects the intimate connexion discovered by Galois between the possibility of solving polynomial equations by radicals and the solubility (in the sense defined above) of the groups associated by Galois with these equations. (See the references to Galois theory mentioned in 2.25.)

7.43. Not all groups are soluble, for it is clear that non-abelian simple groups are insoluble.

A central series is certainly an abelian series and therefore all nilpotent groups are soluble. However, soluble groups are not necessarily nilpotent. For example, let $G = \Sigma_3$ and let K be the unique subgroup of G of order 3 (see 3.7). Then

$$1 \lhd K \lhd G$$

is an abelian series of G, and therefore Σ_3 is soluble. On the other hand, Σ_3 is not nilpotent, for $Z(\Sigma_3) = 1$ and therefore Σ_3 cannot have a central series.

Clearly all abelian groups are nilpotent. However, there are also non-abelian nilpotent groups, for we show now that all finite p-groups are nilpotent.

7.44 Theorem. *Let G be a finite p-group. Then G is nilpotent.*
Proof. We argue by induction on $|G|$. The assertion is trivial if $|G| = 1$. Therefore we assume that $G \neq 1$ and, inductively, that all finite p-groups which have smaller orders than G are nilpotent. By 4.28, $Z(G) \neq 1$. Thus $|G/Z(G)| < |G|$. Since $G/Z(G)$ is a finite p-group, it follows by the inductive assumption that $G/Z(G)$ is nilpotent. Now it is clear by 3.30 that G is nilpotent. This completes the induction argument.

7.45 Lemma. *A series of G, say*

$$1 = G_0 \unlhd G_1 \unlhd \dots \unlhd G_n = G,$$

is a central series if and only if, for each $i = 1, \dots, n$,

$$[G_i, G] \leqslant G_{i-1}$$

(cf. **162**).
Proof. If the given series is a central series then, for each $i = 1, \dots, n$,

$$G_{i-1} \unlhd G \quad \text{and} \quad G_i/G_{i-1} \leqslant Z(G/G_{i-1}).$$

Then, for any $x \in G_i$ and any $y \in G$,

$$(xG_{i-1})(yG_{i-1}) = (yG_{i-1})(xG_{i-1}),$$

that is $\qquad\qquad xyG_{i-1} = yxG_{i-1}.$

Hence $\qquad\qquad [x, y] \in G_{i-1}.$

It follows that $[G_i, G] \leqslant G_{i-1}$.

Suppose conversely that for each $i = 1, \dots, n$,

$$[G_i, G] \leqslant G_{i-1}.$$

Let $x \in G_i$ and $y \in G$. Then

$$x^{-1}x^y = [x, y] \in G_{i-1}.$$

In particular, since $G_{i-1} \leqslant G_i$, if $x \in G_{i-1}$ then $x^y \in G_{i-1}$. Thus $G_{i-1} \trianglelefteq G$. Moreover,

$$xyG_{i-1} = yxG_{i-1}$$

for every $x \in G_i$ and $y \in G$, and so

$$G_i/G_{i-1} \leqslant Z(G/G_{i-1}).$$

Thus the series is a central series.

7.46 Theorem. (i) *If G is nilpotent then all subgroups and all quotient groups of G are nilpotent.*

(ii) *If G is soluble then all subgroups and all quotient groups of G are soluble.*

Proof. Consider a series of G, say

$$1 = G_0 \trianglelefteq G_1 \trianglelefteq \dots \trianglelefteq G_n = G. \tag{a}$$

Let $H \leqslant G$ and $K \trianglelefteq G$. Then, by 7.4(i),

$$1 = (G_0 \cap H) \trianglelefteq (G_1 \cap H) \trianglelefteq \dots \trianglelefteq (G_n \cap H) = H, \tag{b}$$

and by 7.4(ii) and 3.30,

$$K/K = G_0K/K \trianglelefteq G_1K/K \trianglelefteq \dots \trianglelefteq G_nK/K = G/K. \tag{c}$$

Suppose first that G is nilpotent and that (a) is a central series. Then, by 7.45, for each $i = 1, \dots, n$,

$$[G_i, G] \leqslant G_{i-1}.$$

Hence

$$[G_i \cap H, H] \leqslant H \cap [G_i, G] \leqslant H \cap G_{i-1},$$

and by **164**,

$$[G_iK/K, G/K] = [G_i, G]K/K \leqslant G_{i-1}K/K.$$

Therefore, by 7.45, (b) and (c) are central series, so that H and G/K are nilpotent.

Now suppose that G is soluble and that (a) is an abelian series. Then, by 7.4(i), for each $i = 1, \dots, n$,

$$(G_i \cap H)/(G_{i-1} \cap H) \cong G_{i-1}(G_i \cap H)/G_{i-1} \leqslant G_i/G_{i-1},$$

so that $(G_i \cap H)/(G_{i-1} \cap H)$ is abelian. Also, by 3.30 and 7.4(ii),

$$G_i K/K \,\Big/\, G_{i-1} K/K \cong G_i K/G_{i-1} K \cong G_i/G_{i-1}(G_i \cap K) \cong \text{a quotient group}$$

of G_i/G_{i-1}, so that $G_i K/K \,\Big/\, G_{i-1} K/K$ is abelian. Thus (b) and (c) are abelian series, so that H and G/K are soluble.

The class of soluble groups has the important property that every 'extension' of a soluble group by a soluble group is also soluble. This property is not shared by the class of nilpotent groups.

7.47 Theorem. *Let $K \trianglelefteq G$. If K and G/K are soluble then G is soluble.*
Proof. Suppose that K and G/K are soluble. Then K has an abelian series

$$1 = K_0 \trianglelefteq K_1 \trianglelefteq \ldots \trianglelefteq K_m = K,$$

and G/K has an abelian series

$$K/K = G_0/K \trianglelefteq G_1/K \trianglelefteq \ldots \trianglelefteq G_n/K = G/K.$$

Then, by 3.30,

$$1 = K_0 \trianglelefteq K_1 \trianglelefteq \ldots \trianglelefteq K_m = G_0 \trianglelefteq G_1 \trianglelefteq \ldots \trianglelefteq G_n = G,$$

and since, for each $i = 1, \ldots, n$, $G_i/G_{i-1} \cong G_i/K \,\Big/\, G_{i-1}/K$, this series of G is abelian. Hence G is soluble.
Remark. A non-nilpotent group G can have a normal subgroup K such that K and G/K are both nilpotent. For example, let $G = \Sigma_3$ and let K be the unique subgroup of G of order 3. Then K and G/K are both abelian, hence nilpotent; but, as we have noted in 7.43, G is not nilpotent.

7.48 Corollary. (i) *Suppose that H and K are soluble normal subgroups of G. Then HK is a soluble normal subgroup of G.*
  (ii) *Every finite group has a soluble radical* (see 3.45).
Proof. (i) By 3.39, $HK \trianglelefteq G$, and, by 3.40,

$$HK/K \cong H/H \cap K.$$

Since H is soluble, $H/H \cap K$ is soluble, by 7.46. Therefore, since K and HK/K are soluble, 7.47 shows that HK is soluble.
  (ii) Let G be a finite group and let K be a soluble normal subgroup of G of largest possible order. Then, if H is any soluble normal subgroup of G,

$$K \leqslant HK,$$

and, by (i), HK is a soluble normal subgroup of G. Hence, by choice of K, $K = HK$ and therefore $H \leqslant K$. Thus K has the right property to be the soluble radical of G.

The argument above fails if we try to replace 'soluble' by 'nilpotent'. Nevertheless, finite groups do have nilpotent radicals. This result lies rather deeper than 7.48 and we defer the proof to 7.63.

*364 Let n be a positive integer. Then Σ_n is soluble if $n \leqslant 4$, and insoluble if $n \geqslant 5$.

365 The groups $GL_2(\mathbf{Z}_2)$ and $GL_2(\mathbf{Z}_3)$ are soluble.
(Hint. See **44** and **193**. Remark. Let n be an integer, $n > 1$, and F a field. Since the group $GL_n(F)$ has a section isomorphic to $PSL_n(F)$, it follows from 3.61 and 7.46 that $GL_n(F)$ is insoluble whenever $n > 2$ and also when $n = 2$ and $|F| > 3$.)

366 (a) Every non-trivial nilpotent group has non-trivial centre (cf. 4.28).
(b) Let G be a finite group. Then the following two statements are equivalent:
(i) G is nilpotent.
(ii) Every non-trivial quotient group of G has non-trivial centre.

367 Let n be an even positive integer. Then every group of order n is nilpotent if and only if n is a power of 2. (Hint. See **124**.)

368 (i) Every quotient group of a finitely generated soluble group is finitely generated and soluble.
(ii) Let $K \trianglelefteq G$. If K and G/K are both finitely generated and soluble then G is finitely generated and soluble.
(iii) A normal subgroup of a finitely generated soluble group need not be finitely generated. (Hint. See 3.37, **145**, **146**.)

369 Let $\mathfrak{X}$ be a class of groups with the following two properties.
(i) Every quotient group of an $\mathfrak{X}$-group is an $\mathfrak{X}$-group.
(ii) Whenever $J \trianglelefteq H$ and both J and H/J are $\mathfrak{X}$-groups, H is an $\mathfrak{X}$-group.
Then the product of two normal $\mathfrak{X}$-subgroups of G is a normal $\mathfrak{X}$-subgroup of G; and every finite group has an $\mathfrak{X}$-radical.

370 Let $\mathfrak{X}$ denote the class of all groups with trivial centre.
(i) Prove that if $J \trianglelefteq H$ and both J and H/J are $\mathfrak{X}$-groups then H is an $\mathfrak{X}$-group.
(ii) Let $G = \Sigma_3 \times C_2$. Show that $|O^2(G)| = 3$ and that G has just three subgroups of index 2, of which one is isomorphic to C_6 while the other two are isomorphic to Σ_3. Hence show that G has normal $\mathfrak{X}$-subgroups H and K such that $G = HK$, but G is not an $\mathfrak{X}$-group.
(Remark. This shows that in **369**, property (i) is needed. Hint. See **60**.)

371 Let $H \leqslant G$. Suppose that H is a *maximal soluble subgroup* of G; that is, H is soluble and there is no soluble subgroup of G which contains H properly (cf. **235** (ii)). Then $N_G(H) = H$.

7.49 Theorem. *Suppose that $G = H \times K$.*
(i) *If H and K are both nilpotent then G is nilpotent.*
(ii) *If H and K are both soluble then G is soluble.*
Proof. (i) If H and K are both nilpotent then there are central series

$$1 = H_0 \trianglelefteq H_1 \trianglelefteq \ldots \trianglelefteq H_m = H$$
and
$$1 = K_0 \trianglelefteq K_1 \trianglelefteq \ldots \trianglelefteq K_n = K.$$

By inserting repetitions of terms if necessary, we may assume without loss of generality that $m = n$. Then, by **111**,

$$1 = (H_0 \times K_0) \trianglelefteq (H_1 \times K_1) \trianglelefteq \ldots \trianglelefteq (H_n \times K_n) = G,$$

and, by **165** and 7.45, for each $i = 1, \ldots, n$,

$$[H_i \times K_i, G] = ([H_i, H] \times [K_i, K]) \leqslant (H_{i-1} \times K_{i-1}).$$

Hence, by 7.45, G is nilpotent.

(ii) Suppose that H and K are both soluble. We can follow a similar argument to (i). Alternatively, we can apply 7.48: for, by 2.33 and 3.12,

$$H \cong (H \times 1) \trianglelefteq G \quad \text{and} \quad K \cong (1 \times K) \trianglelefteq G,$$

and clearly

$$G = (H \times 1)(1 \times K).$$

7.50 Theorem. *Suppose that H and K are normal subgroups of G.*

(i) *If G/H and G/K are both nilpotent then $G/(H \cap K)$ is nilpotent.*

(ii) *If G/H and G/K are both soluble then $G/(H \cap K)$ is soluble.*

(iii) *Every finite group has a nilpotent residual and a soluble residual* (see 3.45).

Proof. (i) By **109**, $G/(H \cap K)$ can be embedded in $(G/H) \times (G/K)$. Hence if G/H and G/K are both nilpotent, then it follows from 7.49 and 7.46 that $G/(H \cap K)$ is nilpotent.

(ii) A similar argument to (i) is applicable. Alternatively, we may argue as follows. By 3.40,

$$H/(H \cap K) \cong HK/K \leqslant G/K.$$

Therefore, if G/K is soluble, it follows from 7.46 that $H/(H \cap K)$ is soluble. If G/H is also soluble, 3.30 and 7.47 show that $G/(H \cap K)$ is soluble.

(iii) Let G be a finite group and let K be a normal subgroup of G of smallest possible order such that G/K is nilpotent. Then, if $H \trianglelefteq G$ and G/H is nilpotent,

$$H \cap K \leqslant K$$

and, by (i), $G/(H \cap K)$ is nilpotent. Hence, by choice of K, $H \cap K = K$ and so $K \leqslant H$. Thus G/K is the nilpotent residual of G. An exactly similar argument, using (ii), shows that G has a soluble residual.

We shall show that among all the abelian series of a soluble group there is one which descends most rapidly; and that among all the central series of a nilpotent group there is one which descends most rapidly and one which ascends most rapidly.

7.51 Definition. We define subgroups $G^{(n)}$ of G, one for each non-negative integer n, recursively as follows:

$$G^{(0)} = G,$$

and for each integer $n > 0$,

$$G^{(n)} = [G^{(n-1)}, G^{(n-1)}] = (G^{(n-1)})'.$$

Thus $G^{(1)} = G'$. It is customary to write $G^{(2)} = G''$ and $G^{(3)} = G'''$.

Every $G^{(n)}$ is characteristic in G: this follows from 3.51, by induction on n. By definition,

$$G = G^{(0)} \geqslant G^{(1)} \geqslant G^{(2)} \geqslant \dots.$$

This descending sequence of characteristic subgroups of G is called the *derived series* of G.

If, for some $n, G^{(n)} = G^{(n+1)}$ then clearly $G^{(n)} = G^{(r)}$ for every integer $r \geqslant n$. In this case we say that the derived series *terminates*. The derived series of a finite group must terminate; but if G is infinite, the derived series of G need not terminate and then is not strictly a series in the sense of 7.1. However, we show that if G is soluble then the derived series of G terminates in 1.

7.52 Theorem. (i) *G is soluble if and only if $G^{(n)} = 1$ for some integer n.*

(ii) *Suppose that G is soluble and let n be the least integer such that $G^{(n)} = 1$. Then n is called the* derived length *of G. For any abelian series of G, say*

$$G = G_0 \trianglerighteq G_1 \trianglerighteq \dots \trianglerighteq G_r = 1,$$

$$G_i \geqslant G^{(i)}$$

for each $i = 0, 1, \dots, r$. In particular, $r \geqslant n$.

Proof. If $G^{(n)} = 1$ for some n, then

$$G = G^{(0)} \geqslant G^{(1)} \geqslant \dots \geqslant G^{(n)} = 1$$

is a series of G, indeed a *normal* series of G; and, by 3.52, each factor $G^{(i-1)}/G^{(i)}$ is abelian $(i = 1, \dots, n)$. Therefore G is soluble.

Suppose conversely that G is soluble, and let

$$G = G_0 \trianglerighteq G_1 \trianglerighteq \dots \trianglerighteq G_r = 1$$

be an abelian series of G. We prove, by induction on i, that

$$G_i \geqslant G^{(i)} \quad \text{for each } i = 0, 1, \dots, r.$$

This is trivial for $i = 0$. Assume that $i > 0$ and, inductively, that

$$G_{i-1} \geqslant G^{(i-1)}.$$

Since $G_{i-1} \trianglerighteq G_i$ and G_{i-1}/G_i is abelian, 3.52 shows that

$$G_i \geqslant G'_{i-1} \geqslant (G^{(i-1)})' = G^{(i)}.$$

Thus the induction argument goes through. In particular, since $1 = G_r \geqslant G^{(r)}, G^{(r)} = 1$.

Remarks. The statement in (ii) is expressed briefly by the remark that the derived series of a soluble group is its most rapidly descending abelian series. Note that this result shows that a soluble group, which is defined to be a group with an abelian series, actually has an abelian *normal* series. Note also that the soluble groups of derived length 1 are just the non-trivial abelian groups. We shall show in 9.23 that there are soluble groups of derived length n for every positive integer n.

372 Find the derived lengths of Σ_3, A_4, Σ_4 (see **364** and 7.46).

*373 Let n and m be positive integers.

(i) Suppose that G is soluble, of derived length n. Every subgroup and every quotient group of G has derived length at most n.

(ii) Let $K \trianglelefteq G$. Suppose that K is soluble, of derived length n, and that G/K is soluble, of derived length m. Then G has derived length at most $n + m$.

(iii) Let $G = H \times K$. Suppose that H is soluble, of derived length m, and that K is soluble, of derived length n. Then the derived length of G is equal to $\max\{m, n\}$.

(See 7.46, 7.47, 7.49.)

*374 (a) Every non-trivial soluble group has a non-trivial abelian normal subgroup and a non-trivial abelian quotient group.

(b) Let G be a finite group. Then the following three statements are equivalent:

(i) G is soluble.

(ii) Every non-trivial normal subgroup of G has a non-trivial abelian quotient group.

(iii) Every non-trivial quotient group of G has a non-trivial abelian normal subgroup.

(cf. **366**.)

375 Let $H \leqslant G$.

(i) Then $H^G G' = HG'$, where H^G denotes the normal closure of H in G (see **180**).

(ii) Suppose that G is soluble. Then $H^G = G$ if and only if $HG' = G$.

7.53 Definitions. We define subgroups $\Gamma_n(G)$ and $Z_n(G)$ of G recursively as follows. Let $\Gamma_1(G) = G$ and $Z_0(G) = 1$. Then, for each integer $n > 1$, $\Gamma_n(G) = [\Gamma_{n-1}(G), G]$, and for each integer $n > 0, Z_n(G)/Z_{n-1}(G) = Z(G/Z_{n-1}(G))$. Then

$$G = \Gamma_1(G) \geqslant \Gamma_2(G) \geqslant \Gamma_3(G) \geqslant \dots \tag{a}$$

and $$1 = Z_0(G) \leqslant Z_1(G) \leqslant Z_2(G) \leqslant \dots \tag{b}$$

The descending sequence (a) is called the *lower central series* of G and the ascending sequence (b) is called the *upper central series* of G.

The terms of (a) and (b) are characteristic subgroups of G: this follows by induction on n, using 3.51 for (a) and using **118** and **136** for (b). The factors of (a) and (b) are all central, by **162** for (a) and immediately by the definition of (b). But (a) and (b) are not central series of G in the sense of 7.42 if G is non-nilpotent, since they do not terminate in 1 and G, respectively (and, indeed, they need not terminate if G is infinite). However, they do if G is nilpotent, as we now show.

Note that the numbering of the terms of (a) starts from 1 while the numbering of the terms of (b) starts from 0: this is conventional. Note also that by definition $\Gamma_2(G) = G'$. However, in general, $\Gamma_3(G) \neq G''$: see **376**.

7.54 Theorem. (a) *The following three statements are equivalent:*
  (i) *G is nilpotent.*
  (ii) *$\Gamma_n(G) = 1$ for some integer n.*
  (iii) *$Z_n(G) = G$ for some integer n.*
  (b) *Suppose that G is nilpotent. Then for any central series of G, say*

$$1 = G_0 \trianglelefteq G_1 \trianglelefteq \ldots \trianglelefteq G_r = G,$$

$$\Gamma_{r-i+1}(G) \leqslant G_i \leqslant Z_i(G) \quad \text{for each } i = 0, 1, \ldots, r.$$

Furthermore, the least integer c such that $\Gamma_{c+1}(G) = 1$ is equal to the least integer c such that $Z_c(G) = G$: this integer c is called the class *of the nilpotent group G.*

Proof. If $\Gamma_n(G) = 1$ for some n then

$$G = \Gamma_1(G) \geqslant \Gamma_2(G) \geqslant \ldots \geqslant \Gamma_n(G) = 1$$

is a central series of G, and so G is nilpotent.

Similarly, if $Z_n(G) = G$ for some n, then

$$1 = Z_0(G) \leqslant Z_1(G) \leqslant \ldots \leqslant Z_n(G) = G$$

is a central series of G, so that G is nilpotent.

Now suppose conversely that G is nilpotent and let

$$1 = G_0 \trianglelefteq G_1 \trianglelefteq \ldots \trianglelefteq G_r = G$$

be a central series of G. We prove first by induction on i that

$$G_i \leqslant Z_i(G)$$

for each $i = 0, 1, \ldots, r$. This is trivial for $i = 0$. Assume that $i > 0$ and, inductively, that

$$G_{i-1} \leqslant Z_{i-1}(G).$$

Then

$$G_{i-1} Z_{i-1}(G) = Z_{i-1}(G).$$

By hypothesis, $G_i/G_{i-1} \leqslant Z(G/G_{i-1})$. Therefore, by **151**,

$$G_i Z_{i-1}(G)/Z_{i-1}(G) \leqslant Z(G/Z_{i-1}(G)) = Z_i(G)/Z_{i-1}(G).$$

Hence

$$G_i \leqslant Z_i(G).$$

Thus the induction argument goes through. In particular, since $G = G_r \leqslant Z_r(G), Z_r(G) = G$.

Now we prove by induction on j that

$$\Gamma_{j+1}(G) \leqslant G_{r-j}$$

for each $j = 0, 1, \ldots, r$. This is trivial for $j = 0$. Assume that $j > 0$ and, inductively, that

$$\Gamma_j(G) \leqslant G_{r-j+1}.$$

By 7.45, $$[G_{r-j+1}, G] \leqslant G_{r-j}.$$

Hence $$\Gamma_{j+1}(G) = [\Gamma_j(G), G] \leqslant [G_{r-j+1}, G] \leqslant G_{r-j}.$$

Again the induction argument goes through. In particular, since $\Gamma_{r+1}(G) \leqslant G_0 = 1, \Gamma_{r+1}(G) = 1$.

Let c be the least integer such that $Z_c(G) = G$. Then we may choose $G_i = Z_i(G)$ for $i = 0, 1, \ldots, c$ and $r = c$. By what we have proved, it follows that $\Gamma_{c+1}(G) = 1$. We assert that if $c > 0, \Gamma_c(G) \neq 1$. Suppose to the contrary that $\Gamma_c(G) = 1$. Then we may choose $G_i = \Gamma_{c-i}(G)$ for $i = 0, 1, \ldots, c - 1$ and $r = c - 1$. But then, by what we have proved, it follows that $Z_{c-1}(G) = G$; this is contrary to the definition of c. Thus $\Gamma_c(G) \neq 1$ and c is also the least integer such that $\Gamma_{c+1}(G) = 1$.

Remarks. The theorem shows that if G is a nilpotent group then the lower central series of G is its most rapidly descending central series and the upper central series of G is its most rapidly ascending central series. Note that the nilpotent groups of class 1 are the non-trivial abelian groups.

*376 (i) For every positive integer $n, G^{(n-1)} \leqslant \Gamma_n(G)$.
  (ii) Give an example of a group G such that $G'' < \Gamma_3(G)$.

377 (i) Suppose that $|G| = p^n$, where n is an integer, $n \geqslant 2$. Then G is nilpotent of class at most $n - 1$.
  (ii) Let n be an integer, $n \geqslant 3$. Then the dihedral group D_{2^n} of order 2^n has class $n - 1$. (Hint. See **124**.)

378 Let n and m be positive integers.
  (i) Suppose that G is nilpotent of class n. Every subgroup and every quotient group of G has class at most n.
  (ii) Let $G = H \times K$. Suppose that H is nilpotent of class m and K is nilpotent of class n. Then the class of G is equal to $\max\{m, n\}$.
(See 7.46, 7.49: cf. **373**.)

We consider next the composition and chief factors of finite nilpotent and soluble groups.

7.55 (cf. 3.6). *The only soluble simple groups are the groups of prime orders.*

Proof. Let G be a soluble simple group. Since $G \neq 1$, 7.52 shows that $G' < G$. Then, since $G' \lhd G$ and G is simple, $G' = 1$. Thus G is abelian. It now follows from 3.6 that G is finite and of prime order.

7.56 Theorem. *Let G be a finite group. Then the following three statements are equivalent:*

(i) *G is soluble.*

(ii) *Every composition factor of G has prime order.*

(iii) *Every chief factor of G is elementary abelian.*

Proof. Suppose that G is soluble. A composition factor H/J of G is a quotient group of a subgroup of G and is therefore soluble, by 7.46. But also, by 7.2, H/J is simple. Therefore, by 7.55, H/J has prime order.

Let n be the derived length of G. Then the derived series of G,

$$1 = G^{(n)} \lhd G^{(n-1)} \lhd \ldots \lhd G' \lhd G$$

is a proper normal series of G, which can therefore be refined to a chief series of G (by the version of 7.9 for groups with operators). Since the factors of the derived series are abelian, so are the factors of this chief series. Moreover, by the Jordan–Hölder theorem, every chief factor of G is isomorphic to one of the factors of this particular chief series. Thus all chief factors of G are abelian. By 7.38, they are also characteristically simple. Hence, by 7.41, they are elementary.

Suppose conversely that either every composition factor of G has prime order or every chief factor of G is elementary abelian. Then either a composition series or a chief series of G is an abelian series of G. Hence G is soluble.

Remarks. The theorem shows in particular that a finite soluble group has a series all of whose factors are *cyclic*. This is not true in general for infinite soluble groups: see **387**.

A finite soluble group does not in general have a *normal* series all of whose factors are cyclic: 7.34 shows this. See also **389**.

7.57. *Suppose that G has a chief series. Any central chief factor of G is finite and has prime order.*

Proof. In view of 7.36, it is enough to consider a minimal normal subgroup L of G such that $L \leqslant Z(G)$ and to show that $|L| = p$ for some prime p. Since $L \leqslant Z(G)$, every subgroup of L is normal in G (**118**). Therefore, since L is minimal normal in G, the only subgroups of L are 1 and L. It follows (**29**) that $|L| = p$ for some prime p.

7.58 Theorem. *Let G be a finite group. Then the following two statements are equivalent:*

(i) *G is nilpotent.*

(ii) *Every chief factor of G is central.*

Proof. Suppose that G is nilpotent. Since quotient groups of nilpotent groups are nilpotent (7.46), it is enough, by 7.36, to prove that every minimal normal subgroup of G lies in the centre of G. Let L be a minimal normal subgroup of G. Then, by 3.51 and 3.53,

$$[L, G] \unlhd G \quad \text{and} \quad [L, G] \leqslant L.$$

Since L is minimal normal in G, it follows that either

$$[L, G] = 1 \text{ or } [L, G] = L.$$

Suppose that $[L, G] = L$. Then we show by induction on n that for every positive integer n,

$$L \leqslant \Gamma_n(G).$$

This is trivial for $n = 1$. Assume that $n > 1$ and, inductively, that

$$L \leqslant \Gamma_{n-1}(G).$$

Then

$$L = [L, G] \leqslant [\Gamma_{n-1}(G), G] = \Gamma_n(G), \text{ by definition.}$$

This completes the induction argument. Since G is nilpotent, 7.54 shows that, for some n,

$$\Gamma_n(G) = 1.$$

Thus it follows that $L = 1$. This is in contradiction to the definition of L. Therefore $[L, G] = 1$; that is, $L \leqslant Z(G)$, as required.

If, conversely, every chief factor of G is central then a chief series of G is a central series of G, and so G is nilpotent.

Remarks. By 7.57 and 7.58, every chief factor of a finite nilpotent group has prime order. However, it is not true that if every chief factor of a finite group G has prime order then G is nilpotent: for example, let $G = \Sigma_3$ and see 7.43. The implication (i) $\Rightarrow$ (ii) in 7.58 can also be proved by applying the Jordan–Hölder theorem for chief series, together with **356**.

379 Let G be a finite group and suppose that $|G| = p_1^{m_1} p_2^{m_2} \dots p_s^{m_s}$, where $s, m_1, \dots, m_s$ are positive integers and $p_1, \dots, p_s$ distinct primes. Then G is soluble if and only if the composition length of G is $\sum\limits_{j=1}^{s} m_j$.

380 Let G be a soluble group. Then G has a composition series if and only if G is finite (cf. **320**).

*381 Let G be a finite group.

(a) If G is soluble and non-trivial then there are prime divisors p, q of $|G|$ such that $1 < O_p(G)$ and $O^q(G) < G$. (Possibly $p = q$.)

(b) The following three statements are equivalent:

(i) G is soluble.

(ii) For every proper normal subgroup K of G, there is a prime divisor p of $|G/K|$ such that $K/K < O_p(G/K)$.

(iii) For every non-trivial characteristic subgroup K of G, there is a prime divisor q of $|K|$ such that $O^q(K) < K$.

(Hints. To show that (ii) $\Rightarrow$ (i) and (iii) $\Rightarrow$ (i) argue by induction on $|G|$. Use **93** and **156**.)

382 Let p and q be distinct primes and let $\varpi = \{p, q\}$. Then the following two statements are equivalent:

(i) The only finite simple ϖ-groups are the groups of orders p and q.

(ii) Every finite ϖ-group is soluble.

(Remark. Both these statements are true. They are equivalent versions of the theorem of Burnside mentioned after 4.29.)

*383 The following two statements are equivalent:

(i) Every non-abelian finite simple group has even order.

(ii) Every group of odd order is soluble.

(Remark. Both these statements are true. They are equivalent versions of the theorem of Feit and Thompson quoted in 1.12.)

384 If G is insoluble and $|G| \leqslant 100$ then $G \cong A_5$. (Hint. Apply 5.30.)

385 Let $G = \mathrm{GL}_3(\mathbf{Z}_2)$.

(i) Then $|G| = 168 = 2^3 \times 3 \times 7$ (see 2.16, 2.17).

(ii) A group of order 168 which is not simple must be soluble. (Hint. Use **384**.)

(iii) Find elements in G of orders 3 and 7. Show that G does not have a normal subgroup of order either 3 or 7. Deduce that if G is soluble then G has a non-trivial abelian normal 2-subgroup A. (Hint. Note that if G had a normal subgroup of order 3 it would be the unique Sylow 3-subgroup of G; similarly with 7 in place of 3.)

(iv) Let
$$g = \begin{pmatrix} 1 & 0 & 1 \\ 0 & 1 & 0 \\ 0 & 0 & 1 \end{pmatrix} \in G.$$

Deduce from (iii) that if G is soluble then $g \in A$. Show that g does not commute with every conjugate of g in G. (Hint. See **120**, **252**(iv), **260** and 5.8.)

(v) Conclude that G is simple.

386 (i) If G is insoluble and $|G| \leqslant 200$ then $|G| = 60$ or 120 or 168 or 180. (Hint. See **296**, **384**.)

(ii) There are insoluble groups of orders 60, 120, 168 and 180.

387 A group is said to be *polycyclic* if it has a series all of whose factors are cyclic. Thus every polycyclic group is soluble and, by 7.56, every finite soluble group is polycyclic.

(i) Let G be a polycyclic group. Suppose that G has a series of length n, all of whose factors are cyclic, where n is a positive integer. Then G is an n-generator group. Moreover, every subgroup and every quotient group of G has a series of length n, all of whose factors are cyclic; thus all subgroups and all quotient groups of G are n-generator polycyclic groups.

(ii) Let $K \trianglelefteq G$. If K and G/K are both polycyclic then G is polycyclic.

(iii) Not every finitely generated soluble group is polycyclic. (Hint. See **368**.)

388 (a) Let G be an n-generator abelian group, where n is a positive integer. Then G is polycyclic (**387**) and all subgroups of G are n-generator groups.

(b) The following three statements are equivalent:

(i) Every normal subgroup of G is finitely generated and soluble.

(ii) G is polycyclic.

(iii) Every subgroup of G is finitely generated and soluble.

(Hint. To prove that (i) $\Rightarrow$ (ii), consider the derived series of G and apply (a) and **387** (ii).)

*389 A group is said to be *supersoluble* if it has a *normal* series all of whose factors are cyclic. Thus a supersoluble group is in particular polycyclic (**387**); though not conversely, as 7.34 shows.

(i) If G is supersoluble then all subgroups and all quotient groups of G are supersoluble.

(ii) Let $G = H \times K$. If H and K are both supersoluble then G is supersoluble.

(iii) Show by an example that a non-supersoluble group G can have a normal subgroup K such that K and G/K are both supersoluble.

(iv) Every finite nilpotent group is supersoluble. Show by an example that not every finite supersoluble group is nilpotent.

(v) Suppose that G is finite. Then G is supersoluble if and only if every chief factor of G has prime order.

*390 (i) Let $K \trianglelefteq G$ and $K \leqslant H \leqslant G$. The *centralizer of H/K in G* is defined to be the subgroup J of G such that $K \leqslant J$ and $J/K = C_{G/K}(H/K)$. We write $J = C_G(H/K)$. Then also $C_G(H/K) = \{g \in G : [g, h] \in K \text{ for all } h \in H\}$ (cf. **162**(ii)). Moreover, if $H \trianglelefteq G$ then $C_G(H/K) \trianglelefteq G$ and $G/C_G(H/K)$ can be embedded in $\operatorname{Aut}(H/K)$. (Hint. Apply 3.30 and 4.36.)

(ii) Suppose that there is a normal series of G,

$$G = G_0 \geqslant G_1 \geqslant \dots \geqslant G_r = 1,$$

such that $G_1 = G'$ and, for each $i \geqslant 1$, G_i/G_{i+1} is cyclic. Then G' is nilpotent. (Hint. Apply (i), 4.38, 3.52 and 7.45.)

(iii) If G is supersoluble (**389**) then G' is nilpotent. (Hint. Apply Schreier's theorem for groups with operators, and (ii).)

We shall deal with the arithmetical structure of finite nilpotent and soluble groups in chapter 11.

7.59 Theorem. *Let G be a nilpotent group and let the class of G be c. Then, for every subgroup H of G, there is a series of length c from H to G. In particular, every subgroup of G is subnormal in G.* (Note that this result, together with 7.44, gives another proof of 7.14.)

Proof. For each integer $i \geqslant 0$, let $Z_i = Z_i(G)$. Then (see 7.54)

$$1 = Z_0 < Z_1 < \dots < Z_c = G.$$

This is a normal series of G, so that, by 3.38,

$$H = HZ_0 \leqslant HZ_1 \leqslant \dots \leqslant HZ_c = G. \tag{a}$$

Since the centre of a group normalizes every subgroup of the group and since, by definition, $Z_i/Z_{i-1} = Z(G/Z_{i-1})$, for each $i = 1, \dots, c$,

$$Z_i/Z_{i-1} \leqslant N_{G/Z_{i-1}}(HZ_{i-1}/Z_{i-1}).$$

Hence

$$HZ_{i-1}/Z_{i-1} \trianglelefteq (HZ_{i-1}/Z_{i-1})(Z_i/Z_{i-1}) = HZ_i/Z_{i-1},$$

and so, by 3.30, for each $i = 1, \dots, c$,

$$HZ_{i-1} \trianglelefteq HZ_i.$$

Thus (a) is a series of length c from H to G.

7.60. One might ask conversely whether a group in which every subgroup is subnormal is necessarily nilpotent. We shall prove in 11.3 that this is true for finite groups. The general question remained for long unresolved until in 1968 it was settled negatively by H. Heineken and I. J. Mohamed [a 54]. They proved that there are infinite soluble groups G, of derived length 2, with $Z(G) = 1$, hence which are not nilpotent, but such that all proper subgroups are nilpotent and subnormal in G.

All subgroups of an abelian group A are normal in A. We know also that there is a non-abelian group in which all subgroups are normal, namely the quaternion group Q_8: see **181**. There is a classical result, due to Dedekind, giving a complete description of the non-abelian groups in which all groups are normal: see Huppert [b21] p. 308, theorem 3.7.12, or Schenkman [b35] p. 195, theorem 6.4.g, or Scott [b36] p. 253, theorem 9.7.4, or Zassenhaus [b41] p. 159, §4.6. Such groups are nilpotent of class 2. Building on this result, J. E. Roseblade [a81] proved that if G is any group for which there is a positive integer n such that for every subgroup H of G there is a series of length n from H to G then G is nilpotent and the class of G is bounded above by a function of n. This theorem generalizes part of the result to be proved in 11.3.

7.61 Definition. For the purpose of proving the next major result, it is convenient to define *higher commutators*. If H, J, K are subgroups of G then we may have $[[H, J], K] \neq [H, [J, K]]$: see **393**. In order to simplify notation, we adopt the convention that $[H, J, K] = [[H, J], K]$. This is customary.

Let n be a positive integer and let $G_1, G_2, \ldots, G_n$ be subgroups of G (not necessarily distinct). Then we define

$$[G_1, G_2, \ldots, G_n] = \begin{cases} G_1 \text{ if } n = 1, \\ [[\ldots [[G_1, G_2], G_3], \ldots, G_{n-1}], G_n] \text{ if } n \geq 2. \end{cases}$$

For instance, with this notation, for each positive integer n,

$$\Gamma_n(G) = [G, G, \ldots, G],$$

where on the right G appears n times.

7.62 Lemma. *Let r, s be positive integers such that $r \leq s$, and let G_1, $G_2, \ldots, G_s, H, K$ be normal subgroups of G. Then*

$$[G_1, G_2, \ldots, G_{r-1}, HK, G_{r+1}, \ldots, G_s]$$
$$= [G_1, \ldots, G_{r-1}, H, G_{r+1}, \ldots, G_s][G_1, \ldots, G_{r-1}, K, G_{r+1}, \ldots, G_s].$$

Proof. The assertion is trivial if $r = s = 1$. Suppose first that $r = s > 1$, and let $J = [G_1, G_2, \ldots, G_{r-1}]$. Then

$$[G_1, G_2, \ldots, G_{r-1}, HK] = [J, HK] \text{ (by definition)}$$
$$= [J, H][J, K] \text{ (by 3.49 and } \textbf{169})$$
$$= [G_1, \ldots, G_{r-1}, H][G_1, \ldots, G_{r-1}, K]. \quad \text{(i)}$$

Next suppose that $r = 1 < s$. Then

$$
\begin{aligned}
[HK, G_2, G_3, \ldots, G_s] &= [[HK, G_2], G_3, \ldots, G_s] \text{ (by definition)} \\
&= [[H, G_2][K, G_2], G_3, \ldots, G_s] \text{ (by 169)} \\
&= [[H, G_2, G_3][K, G_2, G_3], \ldots, G_s] \text{ (similarly)} \\
&= \ldots \\
&= [H, G_2, \ldots, G_s][K, G_2, \ldots, G_s].
\end{aligned} \tag{ii}
$$

Now we consider the general case. By (i) and (ii), we may assume that $1 < r < s$. Then

$$
\begin{aligned}
&[G_1, G_2, \ldots, G_{r-1}, HK, G_{r+1}, \ldots, G_s] \\
&\quad = [[G_1, \ldots, G_{r-1}, HK], G_{r+1}, \ldots, G_s] \text{ (by definition)} \\
&\quad = [[G_1, \ldots, G_{r-1}, H][G_1, \ldots, G_{r-1}, K], G_{r+1}, \ldots, G_s] \text{ (by (i))} \\
&\quad = [[G_1, \ldots, G_{r-1}, H], G_{r+1}, \ldots, G_s] \\
&\qquad\quad [[G_1, \ldots, G_{r-1}, K], G_{r+1}, \ldots, G_s] \text{ (by (ii))} \\
&\quad = [G_1, \ldots, G_{r-1}, H, G_{r+1}, \ldots, G_s][G_1, \ldots, G_{r-1}, K, G_{r+1}, \ldots, G_s].
\end{aligned}
$$

The following important result was established by Fitting in 1938.

7.63 Theorem (H. Fitting [a27]). (i) *Suppose that H and K are nilpotent normal subgroups of G. Then HK is a nilpotent normal subgroup of G. Moreover, if H, K and HK have classes a, b and c, respectively, then $c \leqslant a + b$.*

(ii) *Every finite group has a nilpotent radical.*

Proof. (i) By hypothesis (see 7.54),

$$\Gamma_{a+1}(H) = 1 = \Gamma_{b+1}(K).$$

We wish to show that $\Gamma_{a+b+1}(HK) = 1$. By 7.62, for any positive integer n,

$$
\begin{aligned}
\Gamma_n(HK) &= [HK, HK, \ldots, HK] \\
&= [H, HK, \ldots, HK][K, HK, \ldots, HK] \\
&= \ldots .
\end{aligned}
$$

Thus, by repeated application of 7.62, $\Gamma_n(HK)$ can be expressed as a product of 2^n commutators $[L_1, L_2, \ldots, L_n]$, where, for each $i = 1, \ldots, n$, L_i is either H or K.

Let r be a positive integer. Since $H \trianglelefteq G$ and $\Gamma_r(H)$ is characteristic in $H, \Gamma_r(H) \trianglelefteq G$, by 3.15. Hence, by 3.53,

$$[\Gamma_r(H), K] \leqslant \Gamma_r(H).$$

Suppose that in a particular commutator $[L_1, L_2, \ldots, L_n]$, r of the L_i's are equal to H and $n - r$ are equal to K. Then it follows from the last inclusion that

$$[L_1, L_2, \ldots, L_n] \leqslant \Gamma_r(H).$$

Similarly, if $r < n$ then

$$[L_1, L_2, \ldots, L_n] \leqslant \Gamma_{n-r}(K).$$

Now we choose $n = a + b + 1$. Then, for any particular $[L_1, L_2, \ldots, L_n]$, either $r \geq a + 1$ or $n - r \geq b + 1$. In the former case, $\Gamma_r(H) = 1$, while in the latter case, $\Gamma_{n-r}(K) = 1$. Hence in any case

$$[L_1, L_2, \ldots, L_n] = 1.$$

This is true for every one of the 2^n commutators in the product expression for $\Gamma_n(HK) = \Gamma_{a+b+1}(HK)$. Hence

$$\Gamma_{a+b+1}(HK) = 1,$$

and therefore HK is nilpotent of class at most $a + b$.

(ii) This follows from (i) by an exactly similar argument to the deduction of (ii) from (i) in 7.48, with 'soluble' replaced by 'nilpotent'.

7.64 Definition. The nilpotent radical of a finite group G is denoted by $F(G)$ and called the *Fitting subgroup* of G. Note that $F(G)$ is a characteristic subgroup of G: see **160**.

391 Suppose that G is nilpotent.
   (i) If $H < G$ then $H < N_G(H)$ (cf. 5.6).
   (ii) If $1 < K \trianglelefteq G$ then $[K, G] < K$ and $K \cap Z(G) \neq 1$. (cf. 5.8. Hint. Consider the subgroups K_n, where $K_1 = K$ and, for each integer $n > 1$, $K_n = [K_{n-1}, G]$.)

392 Suppose that A is a maximal abelian normal subgroup of the nilpotent group G. Then A is a maximal abelian subgroup of G.
(See **235**, **251**. Hint. Use **391** (ii) in place of 5.8.)

*393 Let $G = \Sigma_3$. Show that there are subgroups H, J, K of G such that

$$[[H, J], K] \neq [H, [J, K]].$$

394 If G has abelian normal subgroups H and K such that $G = HK$ then G is nilpotent of class at most 2. (Remark. G need not be abelian: see **171**, **181**.)

395 Let G be a finite group.
   (i) If $K \trianglelefteq G$ then $F(K) \leq F(G)$.
   (ii) Show by an example that $F(G)$ need not contain $F(H)$ for every subgroup H of G.

396 Let G be a finite group.
   (i) Show that if G is soluble and $G \neq 1$ then $F(G) \neq 1$.
   (ii) Define subgroups $F_n(G)$ recursively as follows. Let $F_0(G) = 1$ and, for each positive integer n, let $F_n(G)/F_{n-1}(G) = F(G/F_{n-1}(G))$.

Then $1 = F_0(G) \leq F_1(G) \leq F_2(G) \leq \ldots,$

and this ascending sequence is called the *upper nilpotent series* (or *upper Fitting series*) of G. Prove that G is soluble if and only if $F_n(G) = G$ for some n.
   (iii) Suppose that G is soluble. The least integer n for which $F_n(G) = G$ is called the *nilpotent length* (or *Fitting height*) of G.

Let $1 = G_0 \trianglelefteq G_1 \trianglelefteq \ldots \trianglelefteq G_r = G$

be any series of G whose factors are all nilpotent. Prove that $G_i \leq F_i(G)$ for each $i = 0, 1, \ldots, r$. In particular, the derived length of G is not less than its nilpotent length. (Hint. Argue by induction on i. Note that by **336**, every nilpotent subnormal subgroup of a finite group H is contained in $F(H)$.)

We end this chapter with a few properties of $F(G)$.

7.65 Lemma. *Let G be a non-trivial finite group. Then $C_G(F(G))$ contains every minimal normal subgroup of G.*

Proof. Let $K = F(G)$ and $H = C_G(K)$. Let L be a minimal normal subgroup of G. If $L \not\leqslant K$ then, since $K \cap L < L$ and $K \cap L \trianglelefteq G, K \cap L = 1$. Then, by 3.53, $[K, L] = 1$. Hence $L \leqslant H$.

If on the other hand $L \leqslant K$ then, since $1 < L \trianglelefteq K$, there must be a minimal normal subgroup, M say, of K with $M \leqslant L$. Since K is nilpotent, $M \leqslant Z(K)$, by 7.58. Thus $Z(K) \cap L \neq 1$. But $Z(K) \trianglelefteq G$ **(121)**. Hence $Z(K) \cap L \trianglelefteq G$ and therefore, since L is minimal normal in $G, Z(K) \cap L = L$. Hence $L \leqslant Z(K) \leqslant H$. This completes the proof.

7.66 Theorem. *Let G be a non-trivial finite group. Then, for any chief series of G, say*

$$1 = G_0 < G_1 < \ldots < G_n = G,$$

$$F(G) = \bigcap_{i=1}^{n} C_G(G_i/G_{i-1}).$$

(See **390**.)

Proof. Let $K = F(G)$, $L = \bigcap_{i=1}^{n} C_G(G_i/G_{i-1})$.

Then, $L \trianglelefteq G$. Thus, in order to show that $L \leqslant K$ it is enough to show that L is nilpotent. For each $i = 1, \ldots, n, L \leqslant C_G(G_i/G_{i-1})$ and so, by **390**,

$$[L, G_i] \leqslant G_{i-1}.$$

Hence

$$[L \cap G_i, L] = [L, L \cap G_i] \leqslant L \cap G_{i-1}.$$

Therefore, by 7.45, the series

$$1 = (L \cap G_0) \trianglelefteq (L \cap G_1) \trianglelefteq \ldots \trianglelefteq (L \cap G_n) = L$$

is a central series of L, and so L is nilpotent. Hence $L \leqslant K$.

To complete the proof, it is enough to show that for each $i = 1, \ldots, n$, $K \leqslant C_G(G_i/G_{i-1})$. Now, by 3.30, 3.39, 3.40 and 7.46, $KG_{i-1}/G_{i-1} \trianglelefteq G/G_{i-1}$ and

$$KG_{i-1}/G_{i-1} \cong K/(K \cap G_{i-1}),$$

which is nilpotent. Hence

$$KG_{i-1}/G_{i-1} \leqslant F(G/G_{i-1}).$$

Since G_i/G_{i-1} is a minimal normal subgroup of G/G_{i-1} (7.36), it follows from 7.65 that $F(G/G_{i-1})$ centralizes G_i/G_{i-1}, hence that

$$KG_{i-1}/G_{i-1} \leqslant C_{G/G_{i-1}}(G_i/G_{i-1}).$$

Therefore (see **390**),

$$K \leqslant C_G(G_i/G_{i-1}).$$

This is true for each $i = 1, \ldots, n$. Hence $K = L$.

7.67 Theorem. *Let G be a finite soluble group. Then $C_G(F(G)) \leqslant F(G)$.*
Proof. Let $K = F(G)$ and $H = C_G(K)$. Then $H \trianglelefteq G$, by 4.36. Suppose, contrary to what we wish to show, that $H \not\leqslant K$. Then

$$H \cap K < H \text{ and } H \cap K \trianglelefteq G.$$

There is a chief series of G which includes both $H \cap K$ and H as terms. Let $J/(H \cap K)$ be a chief factor of G with $J \leqslant H$.

Since G is soluble, $J/(H \cap K)$ is abelian, by 7.56. Therefore (3.52),

$$J' \leqslant H \cap K \leqslant K.$$

Hence

$$\Gamma_3(J) = [J, J, J] = [J', J] \leqslant [K, H] = 1,$$

since $H = C_G(K)$. Thus, by 7.54, J is a nilpotent normal subgroup of G and therefore $J \leqslant K$. Hence $J \leqslant H \cap K$. This is a contradiction, since, by definition of $J, H \cap K < J$. Therefore we conclude that $H \leqslant K$.

*397 Let G be a finite group. Then $S(G)$, the *socle* of G, is defined to be the product of all the minimal normal subgroups of G, if $G \neq 1$; and $S(G) = 1$ if $G = 1$.

(i) $S(G)$ is a characteristic subgroup of G.

(ii) $F(G) \leqslant C_G(S(G))$.

(iii) Let $K \trianglelefteq G$. Then $C_G(K) = 1$ if and only if $Z(K) = 1$ and $S(G) \leqslant K$.

398 Let G be a non-trivial finite group.

(i) If

$$1 = G_0 < G_1 < \ldots < G_n = G$$

is a chief series of G, then $G/F(G)$ can be embedded in the direct product of the n groups $\mathrm{Aut}(G_i/G_{i-1}), i = 1, \ldots, n$. (Hint. Use 7.66, **390** and the obvious generalization of **109** to n normal subgroups of G.)

(ii) Suppose that G is supersoluble (**389**). Then $G/F(G)$ is abelian. (Hint. Apply (i) and 4.38(i). Remark. This gives an alternative proof for *finite* supersoluble groups of the result in **390** (iii).)

(iii) Suppose that G is supersoluble. Then for any prime divisor q of $|G/F(G)|$ there is a prime divisor p of $|G|$ such that $p \equiv 1 \bmod q$. Hence the largest prime divisor of $|G|$ does not divide $|G/F(G)|$. (Hint. Apply (i) and 4.38(ii). See also **609**.)

399 Let G be a finite soluble group. Then $|G|$ divides $|Z(F(G))| \cdot |\mathrm{Aut}(F(G))|$. (Hint. Apply 4.36 and 7.67.)

400 Suppose that G is a finite group such that $F(G)$ is abelian. Then $F(G)$ is the unique maximal abelian normal subgroup of G. Moreover, if G is soluble then $F(G)$ is a maximal abelian subgroup of G. (See **235**, **251**. See also **644**, **645**.)

401 (i) Suppose that G is a finite soluble group. If $F(G)$ is cyclic then G is supersoluble (**389**). (Hint. Apply 4.36, 4.38(i) and 7.67.)

(ii) Give an example of a finite supersoluble group G for which $F(G)$ is not cyclic.

8

DIRECT PRODUCTS AND THE STRUCTURE OF FINITELY GENERATED ABELIAN GROUPS

In considering possible programmes to classify groups, we may distinguish two related general problems. On the one hand, there is the problem of *construction*: starting from a collection of known groups, we want to build up other groups from them by explicit procedures. On the other hand, there is the problem of *decomposition*: we want to find out how any given group is built up by these procedures from 'simpler' components.

The easiest procedure is the direct product construction introduced in 2.31 and 2.36. We have obtained criteria in 2.34 and 3.54 for a group to be decomposable as a direct product of two groups. In the present chapter, we shall examine this procedure in further detail and eventually show that it is adequate for a description of the structure of finitely generated abelian groups.

We begin with a convention which simplifies notation.

8.1. Let $G = H \times K$. Then, by 2.33, we know that the map

$$h \mapsto (h, 1) \quad \text{(defined for all } h \in H)$$

is an isomorphism of H onto $H \times 1$, and the map

$$k \mapsto (1, k) \quad \text{(defined for all } k \in K)$$

is an isomorphism of K onto $1 \times K$. Providing that the groups H and K have only the identity element 1 in common, we *identify* H with the subgroup $H \times 1$ of G by identifying the elements h and $(h, 1)$ for all $h \in H$, and similarly we identify K with $1 \times K$ by identifying the elements k and $(1, k)$ for all $k \in K$. Then, by 2.33 and 3.11,

$$H \trianglelefteq G, \quad K \trianglelefteq G, \quad G = HK \quad \text{and} \quad H \cap K = 1.$$

Each element (h, k) of G is then identified with the product hk in G of the elements $h \in H$ and $k \in K$, and of course $hk = kh$. (Note that in making these identifications we are also identifying the groups $H \times K$ and $K \times H$; cf. 2.35.)

With this convention, the converse result contained in 3.54 can be stated as

8.2. *Suppose that H and K are normal subgroups of G such that $G = HK$ and $H \cap K = 1$. Then $G = H \times K$.*

We shall establish a generalization of this result to an arbitrary finite number of direct factors. First we introduce some notation.

8.3. If H and K are normal subgroups of G then it is easy to see that $HK = KH$ (cf. 3.38 and 95). It follows that if $G_1, G_2, \ldots, G_n$ are normal subgroups of G then the product $G_1 G_2 \ldots G_n$ does not depend on the ordering of the factors. We sometimes use the notation $\prod_{i=1}^{n} G_i$ (and similar expressions) for this product. Note that, by 3.39, $\prod_{i=1}^{n} G_i \trianglelefteq G$.

It is clear that the convention of 8.1 can be extended to the direct product of any finite number of groups. We sometimes denote the direct product of groups $G_1, G_2, \ldots, G_n$ by $\mathrm{Dr} \prod_{i=1}^{n} G_i$ (instead of $G_1 \times G_2 \times \ldots \times G_n$). If $n = 1$, then of course $\mathrm{Dr} \prod_i G_i = G_1$.

8.4 Theorem. *Suppose that* $G_1, G_2, \ldots, G_n$ *are normal subgroups of G, where n is a positive integer. Then the following three statements are equivalent:*

(i) $G = \mathrm{Dr} \prod_{i=1}^{n} G_i$.

(ii) *Every element g of G has a unique expression of the form $g = g_1 g_2 \cdots g_n$ with $g_i \in G_i$ for each $i = 1, \ldots, n$.*

(iii) $G = \prod_{i=1}^{n} G_i$ *and, for each integer m such that $1 < m \leqslant n$,*

$$\left(\prod_{i=1}^{m-1} G_i \right) \cap G_m = 1.$$

Proof. (i) $\Rightarrow$ (ii) If $G = \mathrm{Dr} \prod_{i=1}^{n} G_i$ then certainly each element of G is expressible in the form $g_1 g_2 \cdots g_n$, with $g_i \in G_i$ for each $i = 1, \ldots, n$. Moreover, the expression is unique, by definition of the direct product.

(ii) $\Rightarrow$ (iii) It is immediate from the hypothesis of (ii) that $G = \prod_{i=1}^{n} G_i$. Suppose that m is an integer such that $1 < m \leqslant n$, and let

$$g_m \in \left(\prod_{i=1}^{m-1} G_i \right) \cap G_m.$$

Then there are elements $g_1 \in G_1, g_2 \in G_2, \ldots, g_{m-1} \in G_{m-1}$ such that

$$g_m^{-1} = g_1 g_2 \cdots g_{m-1}.$$

Hence
$$1 = g_1 g_2 \cdots g_{m-1} g_m g_{m+1} \cdots g_n,$$

where, if $m < n, g_{m+1} = \ldots = g_n = 1$. Then $g_i \in G_i$ for every $i = 1, \ldots, n$, and so the hypothesis of uniqueness implies that $g_i = 1$ for every i; in particular, $g_m = 1$. Thus

$$\left(\prod_{i=1}^{m-1} G_i\right) \cap G_m = 1.$$

(iii) $\Rightarrow$ (i) For each integer $m = 1, \ldots, n$, let

$$J_m = \prod_{i=1}^{m} G_i.$$

Then $J_m \trianglelefteq G$ and, by hypothesis, $J_n = G$. We shall show by induction on m that, for each $m = 1, \ldots, n$,

$$J_m = \mathrm{Dr} \prod_{i=1}^{m} G_i.$$

This is trivial if $m = 1$. Suppose that $m > 1$ and, inductively, that

$$J_{m-1} = \mathrm{Dr} \prod_{i=1}^{m-1} G_i.$$

Then, by definition, $J_m = J_{m-1} G_m$.

Since J_{m-1} and G_m are normal in G, they are certainly normal in J_m; and, by hypothesis,

$$J_{m-1} \cap G_m = 1.$$

Therefore (8.2)

$$J_m = J_{m-1} \times G_m = \left(\mathrm{Dr} \prod_{i=1}^{m-1} G_i\right) \times G_m = \mathrm{Dr} \prod_{i=1}^{m} G_i.$$

Thus the induction argument goes through. Hence

$$G = J_n = \mathrm{Dr} \prod_{i=1}^{n} G_i.$$

8.5. Warning. Let the notation be as in 8.4. In order to establish that $G = \mathrm{Dr} \prod_{i=1}^{n} G_i$, it is *not* in general enough to prescribe that $G = \prod_{i=1}^{n} G_i$ and $G_i \cap G_j = 1$ whenever $i \neq j$. For instance, consider the group

$$G = C_2 \times C_2,$$

and let a and b be distinct non-trivial elements of G. Then $G = \{1, a, b, ab\}$. Let

$$A = \langle a \rangle, \qquad B = \langle b \rangle, \qquad C = \langle ab \rangle.$$

Then A, B, C are distinct subgroups of G of order 2. They are certainly normal in G. Moreover,

$$G = ABC \quad \text{and} \quad A \cap B = B \cap C = C \cap A = 1.$$

But $G \neq A \times B \times C$ since $|G| = 4$ while $|A \times B \times C| = 8$.

8.6 Theorem. *Let G be a non-trivial finite group and let $p_1, \ldots, p_s$ be the distinct prime divisors of $|G|$, where s is a positive integer. Suppose that, for each $i = 1, \ldots, s$, G has a normal Sylow p_i-subgroup P_i. Then $G = \mathrm{Dr} \prod_{i=1}^{s} P_i$ and G is nilpotent.*

Proof. For each $i = 1, \ldots, s$, let $|P_i| = p_i^{n_i}$. Then

$$|G| = \prod_{i=1}^{s} p_i^{n_i}.$$

For each $m = 1, \ldots, s$, let

$$J_m = \prod_{i=1}^{m} P_i \trianglelefteq G.$$

We show by induction on m that

$$|J_m| = \prod_{i=1}^{m} p_i^{n_i}.$$

This is trivial if $m = 1$, since $J_1 = P_1$. Suppose that $m > 1$ and, inductively, that

$$|J_{m-1}| = \prod_{i=1}^{m-1} p_i^{n_i}.$$

Then, by Lagrange's theorem, $J_{m-1} \cap P_m = 1$. Hence, by 3.40,

$$|J_m| = |J_{m-1} P_m| = |J_{m-1}| \, |P_m| = \prod_{i=1}^{m} p_i^{n_i}.$$

Thus the induction argument goes through. In particular,

$$|J_s| = \prod_{i=1}^{s} p_i^{n_i} = |G|,$$

so that

$$G = J_s.$$

Since also $J_{m-1} \cap P_m = 1$ whenever $1 < m \leqslant s$, it follows by 8.4 that

$$G = \mathrm{Dr} \prod_{i=1}^{s} P_i.$$

By 7.44, P_i is nilpotent for each $i = 1, \ldots, s$. Hence, by repeated application of 7.49(i), G is nilpotent.

Remarks. This result shows that a non-trivial finite abelian group is the direct product of its distinct Sylow subgroups. In 11.3, we shall prove that if G is any non-trivial finite nilpotent group then every Sylow subgroup of G is normal in G, hence, by 8.6, that G is the direct product of its distinct Sylow subgroups.

*402 Suppose that H and K are normal subgroups of G such that $G = H \times K$. If $H \leqslant J \leqslant G$ then $J = H \times (J \cap K)$; and $J \trianglelefteq G$ if and only if $J \cap K \trianglelefteq K$. (Hint. Apply Dedekind's rule 7.3.)

403 If H and K are normal subgroups of G such that $HK = G$ then

$$G/(H \cap K) = H/(H \cap K) \times K/(H \cap K) \cong (G/H) \times (G/K) \quad \text{(cf. 109).}$$

404 (i) Prove that G has a composition series if and only if G has only finitely many distinct subnormal subgroups. (Hints. To prove that if G has a composition series then G has only finitely many distinct subnormal subgroups, argue by induction on the composition length, s say, of G. For $s > 1$, note that every proper subnormal subgroup of G is contained in a maximal normal subgroup of G (see 363). Hence, by the induction hypothesis, it is enough to show that G has only finitely many distinct maximal normal subgroups. Let K be a maximal normal subgroup of G. Show, by means of the induction hypothesis, that there are only finitely many maximal normal subgroups of G which intersect K non-trivially. It only remains to consider the possibility that there is a maximal normal subgroup L of G such that $K \cap L = 1$, in which case $G = K \times L$ and K and L are both simple. Then consider $C_G(K)$ and $C_G(L)$ and see 3.6.)

(ii) Deduce from (i) that a group cannot have infinitely many distinct composition series.

(iii) Verify that the argument in (i) can be modified to prove that G has a chief series if and only if G has only finitely many distinct normal subgroups.

(Remark. A group G with an operator domain Ω can have an Ω-composition series but nevertheless have infinitely many distinct subnormal Ω-subgroups. For example, consider a vector space V of dimension 2 over an infinite field F, and regard F as an operator domain for V^+, as in 7.28 (5). Then the F-group V^+ has an F-composition series (of length 2): see 7.32 (4). The F-subgroups of V^+ are the subspaces of V, and these are normal subgroups of V^+, since V^+ is abelian. Since F is infinite, V has infinitely many distinct 1-dimensional subspaces.)

405 Let $G = \mathrm{Dr} \prod_{i=1}^{n} G_i$, where n is a positive integer. Suppose that for each $i = 1, \ldots, n, G_i$ has a composition series and the composition length of G_i is s_i. Then G has a composition series and the composition length of G is $\sum_{i=1}^{n} s_i$.

*406 Let $G = \mathrm{Dr} \prod_{i=1}^{n} G_i$, where n is a positive integer. Then $Z(G) = \mathrm{Dr} \prod_{i=1}^{n} Z(G_i)$.

*407 Suppose that n is a positive integer and $G_1, G_2, \ldots, G_n$ are normal subgroups of G such that $G = \mathrm{Dr} \prod_{i=1}^{n} G_i$. Let $\alpha \in \mathrm{Aut}\, G$. Then also $G = \mathrm{Dr} \prod_{i=1}^{n} G_i^\alpha$.

408 Let n be an integer, $n > 1$.

(i) Suppose that $K_1, K_2, \ldots, K_n$ are normal subgroups of G such that $\bigcap_{i=1}^{n} K_i = 1$ and, for each integer m with $1 < m \leqslant n, G = \left(\bigcap_{i=1}^{m-1} K_i \right) K_m$.

Then

$$G = \mathrm{Dr} \prod_{i=1}^{n} G_i,$$

where, for each $i = 1, \ldots, n$, $G_i = \bigcap_{\substack{j=1 \\ j \neq i}}^{n} K_j$.

(cf. 8.4. Hints. Apply 8.4 to show that $\prod_{i=1}^{n} G_i = \mathrm{Dr} \prod_{i=1}^{n} G_i$. Prove, by induction on m,

that for each integer $m = 2, \ldots, n$, $G = \prod_{i=1}^{m} \left(\bigcap_{\substack{j=1 \\ j \neq i}}^{m} K_j \right)$.)

(ii) Suppose that $G_1, G_2, \ldots, G_n$ are normal subgroups of G such that $G = \mathrm{Dr} \prod_{i=1}^{n} G_i$. For each $j = 1, 2, \ldots, n$, let

$$K_j = \prod_{\substack{i=1 \\ i \neq j}}^{n} G_i.$$

Then $K_1, K_2, \ldots, K_n$ fulfil the hypotheses of (i); and, for each $i = 1, \ldots, n$,

$$\bigcap_{\substack{j=1 \\ j \neq i}}^{n} K_j = G_i.$$

409 A group G can have normal subgroups G_1, G_2, G_3 such that
  (i) $G = G_1 G_2 G_3$,
  (ii) $G_1 G_2, G_2 G_3, G_3 G_1$ are proper subgroups of G,
  (iii) $G_1 \cap G_2 = G_2 \cap G_3 = G_3 \cap G_1 = 1$, and
  (iv) $G \neq G_1 \times G_2 \times G_3$ (cf. 8.4, 8.5).
Demonstrate this by considering an elementary abelian group G of order p^5 and three suitable subgroups of G of order p^2.

*410 Let G be a finite abelian group. Then G is cyclic if and only if every Sylow subgroup of G is cyclic.

411 Let $G_1, G_2, \ldots, G_n$ be normal subgroups of G such that $G = \prod_{i=1}^{n} G_i$, where n is a positive integer. Suppose that G has a composition series.
  (i) Every composition factor of G is isomorphic to a composition factor of one of $G_1, G_2, \ldots, G_n$.
  (ii) Suppose that whenever $i, j \in \{1, 2, \ldots, n\}$ with $i \neq j$, no composition factor of G_i is isomorphic to a composition factor of G_j. Then $G = \mathrm{Dr} \prod_{i=1}^{n} G_i$.

8.7 Theorem (R. Remak [a79], 1930). *Suppose that $K_1, K_2, \ldots, K_n$ are minimal normal subgroups of G, where n is a positive integer, and let*
$$K = \prod_{j=1}^{n} K_j. \text{ Then there is a subset } \{i_1, \ldots, i_m\} \text{ of } \{1, \ldots, n\} \text{ such that}$$
$$K = \mathrm{Dr} \prod_{j=1}^{m} K_{i_j}.$$

Proof. Let $\mathcal{S}$ denote the set of all non-empty subsets $\{i_1, \ldots, i_m\}$ of

$\{1, \ldots, n\}$ such that $i_1, \ldots, i_m$ are distinct and $\prod_{j=1}^{m} K_{i_j} = \mathrm{Dr} \prod_{j=1}^{m} K_{i_j}$. Trivially, $\{j\} \in \mathscr{S}$ for each $j = 1, \ldots, n$.

Now choose $\{i_1, \ldots, i_m\} \in \mathscr{S}$ with m as large as possible, and let

$$L = \prod_{j=1}^{m} K_{i_j} = \mathrm{Dr} \prod_{j=1}^{m} K_{i_j}.$$

By 3.39, K and L are normal subgroups of G and certainly $L \leqslant K$.

If $K \neq L$ then there is an integer $l \in \{1, \ldots, n\}$ such that $K_l \not\leqslant L$. Since K_l is minimal normal in G and $L \trianglelefteq G$, it follows that

$$K_l \cap L = 1.$$

Then $K_l L \trianglelefteq G$ and, by 8.2,

$$K_l L = K_l \times L.$$

Let $i_{m+1} = l$. Since $K_{i_j} \leqslant L$ for each $j = 1, \ldots, m$, i_{m+1} is distinct from $i_1, \ldots, i_m$ and

$$\prod_{j=1}^{m+1} K_{i_j} = LK_l = L \times K_l = \mathrm{Dr} \prod_{j=1}^{m+1} K_{i_j}.$$

Thus $\{i_1, \ldots, i_m, i_{m+1}\} \in \mathscr{S}$. But this is contrary to the choice of m.

Therefore we conclude that $K = L$ and this completes the proof of the theorem.

Remark. In particular, if G is a non-trivial finite group and $S(G)$ denotes the product of all the minimal normal subgroups of G (the so-called *socle* of G: see **397**) then $S(G)$ is the direct product of some of the minimal normal subgroups of G.

8.8 Definition. G is said to be *completely reducible* if either $G = 1$ or G is the direct product of a finite number of simple groups.

In particular, every simple group is completely reducible.

8.9 Lemma (Remak). *Suppose that G is a non-trivial finite completely reducible group: say $G = \mathrm{Dr} \prod_{j=1}^{n} K_j$, where, for each $j = 1, \ldots, n, K_j$ is a simple normal subgroup of G. If $Z(G) = 1$ then $K_1, \ldots, K_n$ are the only minimal normal subgroups of G and every non-trivial normal subgroup of G is a direct product of some of $K_1, \ldots, K_n$.*

Proof. For each $j = 1, \ldots, n, K_j$ is a simple normal subgroup of G, and therefore K_j is minimal normal in G.

Now assume that $Z(G) = 1$ and suppose, contrary to what we wish to show, that there is a minimal normal subgroup L of G distinct from $K_1, \ldots, K_n$. Then, for each $j = 1, \ldots, n$,

$$K_j \cap L = 1,$$

and so, by 3.53, $[K_j, L] = 1$.

Hence
$$C_G(L) \geqslant \prod_{j=1}^{n} K_j = G,$$

that is
$$L \leqslant Z(G) = 1,$$

a contradiction. Thus $K_1, \ldots, K_n$ are the only minimal normal subgroups of G.

Let $1 < K \trianglelefteq G$. We may choose the notation so that $K_1, \ldots, K_m$ are the minimal normal subgroups of G contained in K, while (if $m < n$) $K_{m+1}, \ldots,$ K_n are the minimal normal subgroups of G not contained in K. Let

$$H = \prod_{j=1}^{m} K_j \quad \text{and} \quad J = \prod_{j=m+1}^{n} K_j \quad \text{(with } J = 1 \text{ if } m = n\text{)}.$$

Then
$$H \leqslant K \leqslant G = H \times J,$$

from which it follows (**402**) that

$$K = H \times (J \cap K).$$

Now $J \cap K \trianglelefteq G$ and $J \cap K \leqslant J$, so that if $J \cap K \neq 1, J \cap K$ contains a minimal normal subgroup K_j of G with $j > m$. But then K_j is contained in K, in contradiction to the choice of m. Hence $J \cap K = 1$ and

$$K = H = \mathrm{Dr} \prod_{j=1}^{m} K_j.$$

Remarks. (1) With the notation of 8.9, since $Z\left(\mathrm{Dr} \prod_{j=1}^{n} K_j\right) = \mathrm{Dr} \prod_{j=1}^{n} Z(K_j)$ (**406**), we have $Z(G) = 1$ if and only if every K_j is a non-abelian simple group.

(2) Without the condition that $Z(G) = 1$, the result of 8.9 is not true in general. For instance, in the example of 8.5, $G = A \times B \cong C_2 \times C_2$, so that G is completely reducible; A and B are minimal normal subgroups of G, but there is also a third minimal normal subgroup C of G distinct from A and B.

(3) 8.9 remains true without the condition that G is finite: see **416**.

412 Let G be a finite group. Define $S_1(G)$ to be the product of all the *abelian* minimal normal subgroups of G (with $S_1(G) = 1$ if G has no abelian minimal normal subgroup); and $S_2(G)$ to be the product of all the *non-abelian* minimal normal subgroups of G (with $S_2(G) = 1$ if G has no non-abelian minimal normal subgroup). Let $S(G)$ denote the socle of G: see **397**. Then

(i) $S_1(G)$ and $S_2(G)$ are characteristic subgroups of G.
(ii) $S_1(G)$ is abelian (cf. **171**).

(iii) $Z(S_2(G)) = 1$, and if $S_2(G) \neq 1, S_2(G)$ is the direct product of all the minimal normal subgroups of G which it contains. (Hints. See 357. Follow part of the proof of 8.9.)

(iv) $S(G) = S_1(G) \times S_2(G)$.

413 Suppose that G is a finite completely reducible group.

(i) For any normal subgroup H of G, there is a completely reducible normal subgroup K of G such that $G = H \times K$. (Hints. If $H < G$, let $\{K_1, \ldots, K_m\}$ denote the set of all minimal normal subgroups of G not contained in H; note that this set is non-empty. Then let $\mathscr{S}$ denote the set of all non-empty subsets $\{i_1, \ldots, i_l\}$ of

$$\{1, \ldots, m\} \text{ such that } i_1, \ldots, i_l \text{ are distinct and } H\left(\prod_{j=1}^{l} K_{i_j}\right) = H \times \mathrm{Dr} \prod_{j=1}^{l} K_{i_j}. \text{ Note that}$$

$\{j\} \in \mathscr{S}$ for each $j = 1, \ldots, m$. Then use a similar argument to the one in the proof of 8.7 to show that there is a subset $\{i_1, \ldots, i_l\}$ of $\{1, \ldots, m\}$ such that $G = H \times \mathrm{Dr} \prod_{j=1}^{l} K_{i_j}$.

Finally, show that each K_{i_j} is simple, for $j = 1, \ldots, l$.)

(ii) Every quotient group and every normal subgroup of G is completely reducible. (Remark. These results remain true without the condition that G is finite: see 416.)

414 Suppose that G is finite and that L is any product of minimal normal subgroups of G. Let $H \trianglelefteq G$ with $H \leqslant L$. Then there is a normal subgroup K of G such that $K \leqslant L$ and $L = H \times K$. (Hint. Modify the argument in 413(i).)

415 Suppose that G is a finite group with the property that, for any normal subgroup H of G, there is a normal subgroup K of G such that $G = H \times K$. Then G is completely reducible. (Hints. Argue by induction. Use 402 to show that every normal subgroup of G has the same property as G. Remark. This is a converse to the result of 413(i).)

416 (i) Suppose that G is completely reducible (but not necessarily finite). Prove that G has a chief series.

(ii) Verify that the results of 8.9 and 413 remain true without the hypothesis that G is finite. Verify also that 415 remains true if the hypothesis that G is finite is replaced by the hypothesis that G has a chief series. (Hint. See 404(iii).)

417 Suppose that G is completely reducible. Then every non-trivial normal subgroup of G is a direct product of minimal normal subgroups of G. (Hint. See 8.7, 402, 404(iii), 413, 416.)

418 (i) Suppose that H is a non-trivial completely reducible normal subgroup of G such that $Z(H) = 1$. Then H is a direct product of minimal normal subgroups of G. (Hints. Argue by induction on the length of a chief series of H: see 416(i). Show that H contains a minimal normal subgroup, K say, of G and that $H = K \times C_H(K)$.)

(ii) The assertion in (i) is no longer true in general, without the hypothesis that $Z(H) = 1$. Demonstrate this by considering a suitable normal subgroup of the dihedral group D_8. (Hint. See 5.8 and 124.)

419 (i) Let H and K be completely reducible normal subgroups of G with $Z(H) = 1 = Z(K)$. Then HK is a completely reducible normal subgroup of G with $Z(HK) = 1$. (Hint. See 8.7, 8.9, 416, 418.)

(ii) Every finite group has an $\mathfrak{X}$-radical, where $\mathfrak{X}$ is the class of all completely reducible groups with trivial centre. (See also 426.)

(iii) A finite group need not have a $\mathfrak{Y}$-radical, where $\mathfrak{Y}$ is the class of all completely reducible groups. Demonstrate this by considering the dihedral group D_8.

420 Suppose that G is completely reducible. Then every subnormal subgroup H of

G is normal in G. (Hint. Argue by induction on the length of a series from H to G, and see **413** and **416**.)

421 (i) Suppose that G has a composition series, say

$$1 = G_0 \lhd G_1 \lhd \ldots \lhd G_n = G,$$

and suppose that there is no positive integer m such that $m < n$, $G_{m-1} \lhd G_{m+1}$ and G_{m+1}/G_{m-1} is the direct product of two isomorphic simple groups. Then every subnormal subgroup H of G is normal in G. (cf. **342**, **420**. Hints. Argue by induction on $j(G:H)$: see 7.20. Apply 7.24(i), **115**(i) and **403**.)

(ii) Suppose that G is finite and that all Sylow subgroups of G are cyclic. Then every subnormal subgroup of G is normal in G. (Hints. Show that all sections of G satisfy the same hypothesis as G, hence that no section of G is isomorphic to $C_p \times C_p$ for any prime p. See 5.11 and apply (i).

Remark. See also 10.26.)

We can now prove the main result on the structure of finite characteristically simple groups (see 7.37).

8.10 Theorem. *Let G be a non-trivial finite group. Then G is characteristically simple if and only if G is a direct product of finitely many isomorphic copies of a simple group.*

Proof. Suppose first that G is characteristically simple. Let K_1 be a minimal normal subgroup of G. For each $\alpha \in \text{Aut } G$, 3.29 shows that K_1^α is a minimal normal subgroup of G, and $K_1^\alpha \cong K_1$. Since G is finite, there are only finitely many distinct subgroups of G of the form K_1^α with $\alpha \in \text{Aut } G$, say n of them: let these be $K_1, K_2, \ldots, K_n$. Let

$$K = \prod_{j=1}^{n} K_j.$$

Now let $\gamma \in \text{Aut } G$. For each $j \in \{1, 2, \ldots, n\}$, $K_j = K_1^\alpha$ for some $\alpha \in \text{Aut } G$, and then, since $\alpha\gamma \in \text{Aut } G$, $K_j^\gamma = K_1^{\alpha\gamma} = K_l$ for some $l \in \{1, 2, \ldots, n\}$. Moreover, if $i, j \in \{1, 2, \ldots, n\}$ with $i \neq j$ then $K_i^\gamma \neq K_j^\gamma$. Hence

$$\{K_1^\gamma, \ldots, K_n^\gamma\} = \{K_1, \ldots, K_n\},$$

and therefore

$$K^\gamma = \prod_{j=1}^{n} K_j^\gamma = \prod_{j=1}^{n} K_j = K.$$

This is true for all $\gamma \in \text{Aut } G$, and so K is characteristic in G. Since $1 < K_1 \leqslant K$ and G is characteristically simple, it follows that

$$K = G.$$

By 8.7, it follows that G is the direct product of some of the subgroups $K_1, \ldots, K_n$. We may choose the notation so that, where $m \leqslant n$,

$$G = \text{Dr} \prod_{j=1}^{m} K_j.$$

Now any normal subgroup of K_1 is easily seen to be normal in G (cf. **111**). Therefore, since K_1 is minimal normal in G, it follows that K_1 is simple. Thus, since $K_j \cong K_1$ for each $j = 1, 2, \ldots, m$, G is the direct product of m isomorphic copies of the simple group K_1.

Suppose conversely that G is the direct product of m isomorphic copies of K_1, where m is a positive integer and K_1 a simple group: say

$$G = \mathrm{Dr} \prod_{j=1}^{m} K_j,$$

where, for each $j = 1, \ldots, m$, $K_j \cong K_1$.

If K_1 is abelian then $|K_1| = p$ for some prime p (3.6). Then $|K_j| = p$ for each $j = 1, \ldots, m$ and G is an elementary abelian group of order p^m. By 7.41, G is characteristically simple.

If K_1 is non-abelian then G is a direct product of non-abelian simple groups and so $Z(G) = 1$ (**406**). Let K be a non-trivial characteristic subgroup of G. Then K contains a minimal normal subgroup of G and so, by 8.9, $K \geqslant K_i$ for some $i \in \{1, \ldots, m\}$. Without loss of generality, we may suppose that

$$K \geqslant K_1.$$

If $m > 1$, let $j \in \{2, \ldots, m\}$. There is an isomorphism

$$\varphi : K_1 \to K_j.$$

Each element of G has a unique expression of the form $k_1 k_2 \ldots k_m$ with $k_i \in K_i$ for each $i = 1, \ldots, m$. Then we can define a map $\alpha : G \to G$ by

$$\alpha : k_1 k_2 \ldots k_{j-1} k_j k_{j+1} \ldots k_m \mapsto k_j^{\varphi^{-1}} k_2 \ldots k_{j-1} k_1^{\varphi} k_{j+1} \ldots k_m$$

for all $k_1 \in K_1, \ldots, k_m \in K_m$. It is easy to verify that α is an automorphism of G, and clearly

$$K_1^{\alpha} = K_j.$$

Since K is characteristic in G,

$$K = K^{\alpha} \geqslant K_1^{\alpha} = K_j.$$

Thus

$$K \geqslant K_j \quad \text{for every } j = 1, \ldots, m,$$

and so

$$K = G.$$

Hence K is characteristically simple. This completes the proof.

Remarks. (1) It follows in particular that every finite characteristically simple group is completely reducible.

(2) Without the condition that G is finite, the theorem fails: for instance, by **361** there are infinite abelian characteristically simple groups, and these cannot be direct products of finitely many isomorphic copies of

simple groups, since, by 3.6, abelian simple groups are finite. However, the theorem does remain true if the hypothesis that G is finite is replaced by the hypothesis that G has a chief series: see **423**.

8.11 Corollary. *Let G be a non-trivial finite group. Then every product of minimal normal subgroups of G is completely reducible.*

Proof. Let K be a product of minimal normal subgroups of G. Then, by 8.7, there are minimal normal subgroups $K_1, \ldots, K_m$ of G such that

$$K = \mathrm{Dr} \prod_{j=1}^{m} K_j.$$

By 7.38, K_j is characteristically simple for each $j = 1, \ldots, m$. Hence, by 8.10, K_j is completely reducible for each $j = 1, \ldots, m$. It follows that K is completely reducible.

422 Let G be a finite group.

(i) Suppose that there is a prime divisor p of $|G|$ such that p^2 does not divide $|G|$. Then, for any minimal normal subgroup K of G, either K is simple or p does not divide $|K|$. (Hint. Apply 8.10.)

(ii) Suppose that G has a subgroup H such that $|G:H| = p$ and $H_G = 1$. Then G has a unique minimal normal subgroup K, and K is simple. (See also **652**. Hint. Apply 4.14, (i) and **358**.)

423 Verify that 8.10 remains true if the hypothesis that G is finite is replaced by the hypothesis that G has a chief series. (Hint. See **404**(iii) and **416**.)

424 Suppose that G has a composition series, and let

$$\mathcal{K}(G, 1) = \{L_1, L_2, \ldots, L_s\},$$

where s is a positive integer. Note that (by **355** or **404**) G also has a chief series.

(i) Any chief factor of G is the direct product of finitely many isomorphic copies of L_j, for some $j \in \{1, 2, \ldots, s\}$. Moreover, for each $j \in \{1, 2, \ldots, s\}$ there is in any chief series of G at least one factor which is the direct product of finitely many isomorphic copies of L_j. (Hint. See **423**.)

(ii) Suppose that in a composition series of G there is just one factor isomorphic to L_1. Then in any chief series of G there is a factor which is isomorphic to L_1; and no other factor of the series is a direct product of isomorphic copies of L_1.

425 (a) Suppose that $L_1, L_2, \ldots, L_n$ are maximal normal subgroups of G (see **363**), where n is a positive integer, and let $L = \bigcap_{j=1}^{n} L_j$. Then G/L is completely reducible. (cf. 8.11. Hint. Argue by induction on n, and use **363** and **403**.)

(b) Let G be a finite group. We define $R(G)$ to be the intersection of all the maximal normal subgroups of G, if $G \neq 1$; and $R(G) = 1$ if $G = 1$. As before, $S(G)$ denotes the socle of G (**397**). Then the following three statements are equivalent:

(i) $R(G) = 1$.

(ii) G is completely reducible.

(iii) $S(G) = G$.

426 Let $\mathfrak{X}$ denote the class of all completely reducible groups with trivial centre. Let G be a finite group. Then the $\mathfrak{X}$-radical of G (see **419**) is the subgroup $S_2(G)$ of G, defined in **412**. (Hint. See **418**.)

8.12. We recall the *extension problem* for groups mentioned in chapter 1: given groups K and Q, find the groups G for which $K \trianglelefteq G$ and $G/K \cong Q$. This may be viewed as defining a construction procedure; though, as we have pointed out, unlike the direct product construction, in general this does not lead from K and Q to a unique type of group G. (Note that $K \times Q$ is one type of group obtained by this extension procedure.)

The corresponding decomposition procedure for a finite group G leads to the notion of a composition series of G: the group G is built up by this procedure from its composition factors, which are simple groups (7.2) and cannot be further decomposed.

For such decompositions, the Jordan–Hölder theorem (7.9) provides a *uniqueness* result. Although a finite group G may have several different composition series, any two of them have the same length and contain as factors simple groups of exactly the same types with the same multiplicities.

We may ask whether such a uniqueness result holds for decompositions of groups as direct products of indecomposable factors. (Recall **(81)** that G is said to be *decomposable* if it has *proper* subgroups H and K such that $G = H \times K$; and if not G is said to be *indecomposable*.) The answer is that there is such a result for *finite* groups; and also for infinite groups under certain conditions, but not in general. It is called the Krull–Remak–Schmidt theorem and, for a finite group G, asserts that if

$$G = H_1 \times \ldots \times H_m = K_1 \times \ldots \times K_n,$$

where $H_1, \ldots, H_m, K_1, \ldots, K_n$ are non-trivial and indecomposable, then $m = n$ and, by relabelling the suffices if necessary,

$$H_i \cong K_i \quad \text{for each } i = 1, \ldots, n.$$

In fact it provides even more information than this. We shall not in this book prove the general Krull–Remak–Schmidt theorem: for the proof see Huppert [b21] p. 60, theorem 1.12.3, or Rotman [b34] p. 80, theorem 4.36, or Scott [b36] p. 83, theorem 4.6.2, or Zassenhaus [b41] p. 114, theorem 7. However, we shall in 8.18 prove a special case which will be applied in chapter 9.

We need a few preliminary results. We begin with a result known as Fitting's lemma. Recall (2.18) that the endomorphisms of a group form a semigroup with respect to composition of maps. Thus, for each endomorphism φ of G and each positive integer k, there is a corresponding endomorphism φ^k of G.

8.13 Lemma (Fitting [a26], 1934). *Let G be a finite group. Regard G as an operator domain for G, as in 7.28 (3), and let φ be a G-endomorphism of G (see 7.29).*

  (i) *There is a positive integer k such that*

$$G = \operatorname{Ker} \varphi^k \times \operatorname{Im} \varphi^k.$$

(ii) *If G is indecomposable then either* $\varphi \in \operatorname{Aut} G$ *or, for some positive integer* k, φ^k *is the trivial endomorphism of G.*

Proof. For each positive integer j, let

$$K_j = \operatorname{Ker} \varphi^j.$$

Then clearly

$$K_1 \leqslant K_2 \leqslant K_3 \leqslant \ldots \leqslant G.$$

Since G is finite, it follows that there is a positive integer k such that

$$K_k = K_{k+1}.$$

By induction on l, we deduce that for every positive integer l,

$$K_k = K_{k+l}.$$

This is true for $l = 1$. Now suppose that $l > 1$ and inductively that $K_k = K_{k+l-1}$, and let $g \in K_{k+l}$. Then

$$(g\varphi^{l-1})\varphi^{k+1} = g\varphi^{k+l} = 1,$$

so that

$$g\varphi^{l-1} \in K_{k+1} = K_k.$$

Hence

$$g\varphi^{k+l-1} = 1,$$

so that

$$g \in K_{k+l-1} = K_k,$$

by the induction hypothesis. Since also $K_k \leqslant K_{k+l}$, this shows that

$$K_k = K_{k+l}.$$

This completes the induction argument.

Now let $K = K_k$ and $L = \operatorname{Im} \varphi^k$. Then $K \trianglelefteq G$ and, since φ^k is a G-endomorphism of G (353), L is a G-subgroup of G: that is, $L \trianglelefteq G$. Let $x \in K \cap L$. Then

$$x\varphi^k = 1 \text{ and } x = y\varphi^k$$

for some $y \in G$. Thus

$$y\varphi^{2k} = 1,$$

so that $y \in K_{2k} = K_k$, by the previous paragraph. Hence

$$x = y\varphi^k = 1.$$

Therefore $K \cap L = 1.$

It follows (3.40) that

$$L \cong KL/K.$$

But also, by the fundamental theorem on homomorphisms,

$$L \cong G/K.$$

Therefore, since $KL \leqslant G$ and G is finite,

$$G = KL.$$

Hence, by 8.2,

$$G = K \times L.$$

Now suppose that G is indecomposable. Then either $K = G$ and $L = 1$ or $K = 1$ and $L = G$. In the former case, φ^k is the trivial endomorphism of G. In the latter case, $\varphi^k \in \text{Aut } G$: this implies that φ is bijective, hence that $\varphi \in \text{Aut } G$.

8.14 Definition. For the statement of the next lemma, which is a deduction from Fitting's lemma, it is convenient to introduce *sums* of homomorphisms (cf. **33**).

Let φ and ψ be homomorphisms of G into H (arbitrary groups). We define a map

$$\varphi + \psi : G \to H$$

by $\qquad \varphi + \psi : g \mapsto (g\varphi)(g\psi) \quad$ for all $g \in G$.

In general, $\varphi + \psi$ is not a homomorphism, and $\varphi + \psi \neq \psi + \varphi$ (although if $\varphi + \psi$ is a homomorphism then $\varphi + \psi = \psi + \varphi$: see **430**). However, in 8.15 we shall be concerned with a special situation in which sums of homomorphisms are again homomorphisms.

The definition is extended in the natural way to arbitrary finite sums of homomorphisms. Let n be a positive integer and let $\varphi_1, \ldots, \varphi_n$ be homomorphisms of G into H. Then we define the map

$$\sum_{i=1}^{n} \varphi_i : G \to H$$

by $\qquad \displaystyle\sum_{i=1}^{n} \varphi_i : g \mapsto (g\varphi_1)(g\varphi_2)\ldots(g\varphi_n) \quad$ for all $g \in G$.

427 (i) Suppose that H and K are normal subgroups of G such that $G = H \times K$. Let π be the corresponding projection of G onto H (see 3.11) and let ι be the inclusion map of H into G. Let $\varphi = \pi\iota$. Then φ is a G-endomorphism of G and $\varphi^2 = \varphi$. Moreover, if H and K are proper subgroups of G (so that G is decomposable) then φ is not an automorphism of G and there is no positive integer k such that φ^k is the trivial endomorphism of G (cf. 8.13(ii)).

(ii) Suppose that φ is a G-endomorphism of G such that $\varphi^2 = \varphi$. Let $H = \text{Im } \varphi$ and $K = \text{Ker } \varphi$. Then $G = H \times K$ and $\varphi = \pi\iota$, where π is the corresponding projection of G onto H and ι is the inclusion map of H into G.

428 (i) Let φ be an endomorphism of G and let $J = \operatorname{Im} \varphi$. Then φ is a G-endomorphism of G if and only if, for every $g \in G, (g\varphi)g^{-1} \in C_G(J)$ (cf. **245, 353**(ii)).

(ii) Suppose that G is indecomposable and that $Z(G) = 1$. Then the only G-endomorphisms of G are the identity automorphism of G and the trivial endomorphism of G. (cf. 8.13(ii). Hint. Let φ be a G-endomorphism of G and let $J = \operatorname{Im} \varphi$. Use (i) and 8.2 to show that $G = C_G(J) \times J$.)

429 (i) Any endomorphism φ of G such that $\operatorname{Im} \varphi \leqslant Z(G)$ is a G-endomorphism of G. (See **428**(i).)

(ii) Let G be a finite non-abelian group. Suppose that G has no non-trivial abelian direct factor. Then, for any endomorphism φ of G such that $\operatorname{Im} \varphi \leqslant Z(G)$, there is a positive integer k such that φ^k is the trivial endomorphism of G.

430 (i) Let φ and ψ be homomorphisms of G into H. Then $\varphi + \psi$ is a homomorphism if and only if $[\operatorname{Im} \varphi, \operatorname{Im} \psi] = 1$. In particular, if $\varphi + \psi$ is a homomorphism then $\varphi + \psi = \psi + \varphi$.

(ii) If G is non-abelian and α is any automorphism of G then $\alpha + \alpha$ is not an endomorphism of G.

431 Let G be an abelian group. Then the set of all endomorphisms of G forms a ring when addition is defined as in 8.14 and multiplication is defined by composition of maps (see 2.18). We shall denote this ring by End G. The zero element of End G is the trivial endomorphism of G; and End G has a multiplicative identity element, namely the identity automorphism of G.

432 Let R be a ring with a multiplicative identity element 1. Then R is isomorphic (as a ring) to a subring of End R^+. (cf. **46**(i), 4.24. Hint. For each $a \in R$, let ρ_a be defined as in 2.11. Verify that the map $a \mapsto \rho_a$ is an injective ring homomorphism of R into End R^+.)

433 The rings $\mathbf{Z}$ and End $\mathbf{Z}^+$ are isomorphic, and, for every positive integer n, the rings $\mathbf{Z}_n$ and End $\mathbf{Z}_n^+$ are isomorphic (cf. **46**(ii)).

8.15 Lemma. *Let G be a non-trivial finite indecomposable group. Suppose that $\varphi_1, \varphi_2, \ldots, \varphi_n$ are G-endomorphisms of G such that, for each $j = 1, \ldots, n$,*

$$\sum_{i=1}^{j} \varphi_i \text{ is a } G\text{-endomorphism of } G, \text{ and } \sum_{i=1}^{n} \varphi_i = 1, \text{ the identity automorphism}$$

of G. Then at least one of $\varphi_1, \ldots, \varphi_n$ is an automorphism of G.

Proof. We argue by induction on n. The result is trivial if $n = 1$. Now suppose that $n = 2$. Then $\varphi_1 + \varphi_2 = 1$, so that

$$\varphi_1 = \varphi_1(\varphi_1 + \varphi_2) = \varphi_1^2 + \varphi_1\varphi_2$$
$$= (\varphi_1 + \varphi_2)\varphi_1 = \varphi_1^2 + \varphi_2\varphi_1,$$

by definition of composition of maps and since φ_1 is an endomorphism of G. Hence, since G is a group,

$$\varphi_1\varphi_2 = \varphi_2\varphi_1.$$

Let ζ denote the trivial endomorphism of G. If neither φ_1 nor φ_2 is an automorphism of G then, by Fitting's lemma (8.13), there are positive integers k_1, k_2 such that

$$\varphi_1^{k_1} = \varphi_2^{k_2} = \zeta.$$

Then, since $\varphi_1 + \varphi_2 = 1$ and $\varphi_1\varphi_2 = \varphi_2\varphi_1$,

$$1 = (\varphi_1 + \varphi_2)^{k_1+k_2} = \sum_{i=0}^{k_1+k_2} \binom{k_1 + k_2}{i} \varphi_1^{k_1+k_2-i}\varphi_2^i$$

(where $\varphi_1^0 = \varphi_2^0 = 1$).

When $0 \leqslant i \leqslant k_2, k_1 + k_2 - i \geqslant k_1$ and so $\varphi_1^{k_1+k_2-i} = \zeta$. Hence also, since φ_2^i is an endomorphism of $G, \varphi_1^{k_1+k_2-i}\varphi_2^i = \zeta$. When $k_2 < i \leqslant k_1 + k_2$, $\varphi_2^i = \zeta$ and so also $\varphi_1^{k_1+k_2-i}\varphi_2^i = \zeta$. Therefore the equation above gives

$$1 = \zeta.$$

This is a contradiction since $G \neq 1$. We conclude that either φ_1 or φ_2 must be an automorphism of G.

Finally, suppose that $n > 2$. Let

$$\psi = \sum_{i=1}^{n-1} \varphi_i.$$

By hypothesis, ψ and φ_n are G-endomorphisms of G and $\psi + \varphi_n = 1$. Hence, by what we have proved above, either ψ or φ_n is an automorphism of G. If $\varphi_n \in \operatorname{Aut} G$, we are done. Suppose $\psi \in \operatorname{Aut} G$. Then $\psi^{-1} \in \operatorname{Aut} G$ and

$$1 = \sum_{i=1}^{n-1} \varphi_i\psi^{-1}.$$

It is easy to check (see 353) that $\varphi_1\psi^{-1}, \ldots, \varphi_{n-1}\psi^{-1}$ are G-endomorphisms of G and, for each $j = 1, \ldots, n-1$, $\sum_{i=1}^{j} \varphi_i\psi^{-1}$ is a G-endomorphism of G. Then, by the induction hypothesis, $\varphi_i\psi^{-1} \in \operatorname{Aut} G$ for some $i \in \{1, \ldots, n-1\}$. Then $\varphi_i = (\varphi_i\psi^{-1})\psi \in \operatorname{Aut} G$. This completes the induction argument.

8.16. Suppose that $G_1, G_2, \ldots, G_n$ are normal subgroups of G such that $G = \operatorname{Dr} \prod_{j=1}^{n} G_j$, where n is a positive integer. Then every element of G is uniquely expressible in the form $g_1g_2 \ldots g_n$, with $g_j \in G_j$ for each $j = 1, \ldots, n$. Therefore, for each $i = 1, \ldots, n$, we can define a map

$$\pi_i : G \to G_i$$

by $\pi_i : g_1g_2 \ldots g_n \mapsto g_i$ (for all $g_1 \in G_1, g_2 \in G_2, \ldots, g_n \in G_n$).

Then π_i is called the *projection* of G onto G_i relative to the decomposition $G = G_1 \times \ldots \times G_n$ (cf. 3.11).

Now G_i is a G-subgroup of G, and it is easy to verify that π_i is a G-homomorphism. Let γ_i denote the inclusion map of G_i into G; this is obviously also a G-homomorphism. Then $\pi_i\gamma_i$ is a G-endomorphism of G, and for each $j = 1, \ldots, n$, the map $\sum_{i=1}^{j} \pi_i\gamma_i$ is defined:

$$\sum_{i=1}^{j} \pi_i\gamma_i : g_1 g_2 \cdots g_j g_{j+1} \cdots g_n \mapsto g_1 g_2 \cdots g_j \text{ (for all } g_1 \in G_1, g_2 \in G_2, \ldots g_n \in G_n).$$

Then $\sum_{i=1}^{j} \pi_i\gamma_i$ is a G-endomorphism of G, and

$$\sum_{i=1}^{n} \pi_i\gamma_i = 1,$$

the identity automorphism of G.

8.17. *Let G be any non-trivial finite group. Then there are non-trivial indecomposable normal subgroups $G_1, G_2, \ldots, G_n$ of G such that*

$$G = \text{Dr} \prod_{j=1}^{n} G_j.$$

Proof. We argue by induction on $|G|$. If G is itself indecomposable, we set $n = 1$ and $G_1 = G$: then there is nothing more to prove. Suppose that G is decomposable: then there are proper subgroups H and K of G such that

$$G = H \times K.$$

Then H and K are non-trivial and $|H| < |G|, |K| < |G|$. Hence, by the induction hypothesis, there are non-trivial indecomposable normal subgroups $G_1, \ldots, G_m$ of H and $G_{m+1}, \ldots, G_n$ of K such that $H = \text{Dr} \prod_{j=1}^{m} G_j$ and $K = \text{Dr} \prod_{j=m+1}^{n} G_j$. Then $G_1, G_2, \ldots, G_n$ are non-trivial indecomposable normal subgroups of G and

$$G = \text{Dr} \prod_{j=1}^{m} G_j \times \text{Dr} \prod_{j=m+1}^{n} G_j = \text{Dr} \prod_{j=1}^{n} G_j.$$

This completes the induction argument.

434 Let $\alpha \in \text{Aut } G$.

(i) If α is a G-automorphism of G then there is a unique G-endomorphism φ of G such that

$$\alpha + \varphi = 1, \text{ the identity automorphism of } G.$$

Moreover, if G is non-abelian then $\varphi \notin \text{Aut } G$ (cf. 8.15).

(ii) If α is not a G-automorphism of G then there is no endomorphism φ of G such that

$$\alpha + \varphi = 1.$$

(Hint. See **245**, **353**(ii) **429**(i).)

435 (i) Verify that 8.17 remains true if the hypothesis that G is finite is replaced by the hypothesis that G has a chief series.

(ii) Give an example of an indecomposable group which does not have a chief series.

We shall now prove a special case of the Krull–Remak–Schmidt theorem, namely that a non-trivial finite group with trivial centre has just one decomposition as a direct product of non-trivial indecomposable normal subgroups (i.e. unique, apart from ordering of the factors).

8.18 Theorem. *Let G be a non-trivial finite group with $Z(G) = 1$. Suppose that*

$$G = H_1 \times \dots \times H_m = K_1 \times \dots \times K_n,$$

where m, n are positive integers and $H_1, \dots, H_m, K_1, \dots, K_n$ non-trivial indecomposable normal subgroups of G. Then $m = n$ and, by relabelling the suffices if necessary, $H_i = K_i$ for each $i = 1, \dots, n$.

Proof. We argue by induction on n. If $n = 1$ then, since K_1 is indecomposable and $H_1, \dots, H_m$ are non-trivial, $m = 1$ and $H_1 = K_1$. Therefore we may assume that $n > 1$. This implies also that $m > 1$, for a similar reason.

Let π_1 denote the projection of G onto H_1, relative to the decomposition $G = H_1 \times \dots \times H_m$, and, for each $i = 1, \dots, n$, let ρ_i denote the projection of G onto K_i, relative to the decomposition $G = K_1 \times \dots \times K_n$. Further, let κ_i denote the inclusion map of K_i into G, and let

$$\pi_i^* = \kappa_i \pi_1 = \pi_1|_{K_i} : K_i \to H_1 \text{ and } \rho_i^* = \rho_i|_{H_1} : H_1 \to K_i.$$

Each $\rho_i^* \pi_i^*$ is a G-endomorphism of H_1, hence also an H_1-endomorphism of H_1.

Moreover, for each $j = 1, \dots, n$, $\sum_{i=1}^{j} \rho_i \kappa_i$ is defined and is a G-endomorphism of G (see 8.16). Therefore, since π_1 is a G-homomorphism of G into H_1, so also is $\left(\sum_{i=1}^{j} \rho_i \kappa_i \right) \pi_1$, and clearly

$$\left(\sum_{i=1}^{j} \rho_i \kappa_i \right) \pi_1 = \sum_{i=1}^{j} \rho_i \pi_i^*.$$

The restriction of this to H_1 is $\sum_{i=1}^{j} \rho_i^* \pi_i^*$, which is thus a G-endomorphism of H_1, hence also an H_1-endomorphism of H_1. For all $h \in H_1$,

$$h = h\pi_1 = ((h\rho_1)(h\rho_2)\dots(h\rho_n))\pi_1$$
$$= (h\rho_1^*\pi_1^*)(h\rho_2^*\pi_2^*)\dots(h\rho_n^*\pi_n^*),$$

so that

$$\sum_{i=1}^{n} \rho_i^* \pi_i^* = 1,$$

the identity automorphism of H_1. Therefore, since H_1 is non-trivial and indecomposable, 8.15 shows that for some $i, \rho_i^* \pi_i^* \in \text{Aut } H_1$. We may suppose the notation chosen so that $\rho_1^* \pi_1^* \in \text{Aut } H_1$. It follows in particular that ρ_1^* is injective.

Let $J = H_2 \times \ldots \times H_m$ and $L = K_2 \times \ldots \times K_n$. Then $G = H_1 \times J = K_1 \times L$.

Since $Z(G) = 1$, it follows (**406**) that

$$Z(H_1) = Z(J) = Z(K_1) = Z(L) = 1.$$

Now

$$J \leqslant C_G(H_1) \leqslant G = H_1 \times J,$$

and so (by **402**)

$$C_G(H_1) = (H_1 \cap C_G(H_1)) \times J = J, \text{ since } Z(H_1) = 1.$$

By exactly similar arguments,

$$C_G(J) = H_1, C_G(K_1) = L \quad \text{and} \quad C_G(L) = K_1.$$

Now $L = \text{Ker } \rho_1$. Therefore, since ρ_1^* is injective,

$$1 = \text{Ker } \rho_1^* = H_1 \cap \text{Ker } \rho_1 = H_1 \cap L.$$

Hence, by 3.53, $H_1 \leqslant C_G(L)$. Thus

$$H_1 \leqslant K_1 \leqslant G = H_1 \times J,$$

and so (again by **402**)

$$K_1 = H_1 \times (K_1 \cap J).$$

Since K_1 is indecomposable and $H_1 \neq 1$, it follows that

$$K_1 = H_1.$$

Hence

$$J = C_G(H_1) = C_G(K_1) = L.$$

Thus

$$J = H_2 \times \ldots \times H_m = K_2 \times \ldots \times K_n.$$

Since $Z(J) = 1$, the induction hypothesis now implies that $m = n$ and, by suitable choice of the notation,

$$H_i = K_i$$

for each $i = 2, \ldots, m$. This completes the induction argument.

436 Let G be a non-trivial finite group with $Z(G) = 1$. Then, by 8.17 and 8.18, there is a unique decomposition, say

$$G = G_1 \times \ldots \times G_n,$$

such that $G_1, \ldots, G_n$ are non-trivial indecomposable normal subgroups of G. Suppose that no two of the groups $G_1, \ldots, G_n$ are isomorphic. Then $G_1, \ldots, G_n$ are characteristic subgroups of G and

$$\text{Aut } G \cong \text{Aut } G_1 \times \ldots \times \text{Aut } G_n.$$

(cf. **342**; also see **94**.)

437 Give an example of a finite abelian group G such that $G = A \times B$, where A and B are non-isomorphic non-trivial indecomposable subgroups of G, and such that G has subgroups A^* and B^*, distinct from A and B, and with $G = A^* \times B^*$ (cf. 8.18).

We shall now prove a result about subgroups of the direct product of two groups. In chapter 9 we shall apply this result to the extension problem: see 9.28.

8.19 Theorem (Remak [a80], Klein, Fricke [b26]). *Let H and K be normal subgroups of G such that $G = H \times K$, and let π and ρ be the corresponding projections of G onto H and K, respectively. Let $L \leqslant G$. Then*

(i) $(H \cap L) \trianglelefteq L\pi \leqslant H, (K \cap L) \trianglelefteq L\rho \leqslant K$ *and* $L\pi/(H \cap L) \cong L\rho/(K \cap L)$.

(ii) $L = (H \cap L) \times (K \cap L)$ *if and only if* $L\pi = H \cap L$ *(or if and only if* $L\rho = K \cap L$*).*

Proof. (i) We know that π and ρ are homomorphisms (3.11). Since $H \trianglelefteq G, (H \cap L) \trianglelefteq L \leqslant G$. Therefore (**87**)

$$(H \cap L)\pi \trianglelefteq L\pi \leqslant G\pi = H.$$

By definition, $\pi|_H$ is the identity map on H.
Therefore $\qquad\qquad (H \cap L)\pi = H \cap L.$

Hence $\qquad\qquad (H \cap L) \trianglelefteq L\pi \leqslant H.$

Similarly $\qquad\qquad (K \cap L) \trianglelefteq L\rho \leqslant K.$

We now define a map

$$\varphi : L\pi \to L\rho/(K \cap L).$$

For each element $h \in L\pi$, there is an element $k \in K$ such that $hk \in L$. Then $k \in L\rho$, and we define

$$h\varphi = k(K \cap L).$$

The element k is not necessarily uniquely determined by h, and so we must check that this definition of $h\varphi$ does not depend on the choice of k. If also $k' \in K$ with $hk' \in L$ then

$$k^{-1}k' = (hk)^{-1}(hk') \in K \cap L,$$

and so

$$k'(K \cap L) = k(K \cap L).$$

Thus φ is well defined.

Let $h_1, h_2 \in L\pi$ and let $k_1, k_2 \in K$ with $h_1 k_1, h_2 k_2 \in L$. Then $h_1 h_2 \in L\pi$, $k_1 k_2 \in K$ and, since $[H, K] = 1$,

$$(h_1 h_2)(k_1 k_2) = (h_1 k_1)(h_2 k_2) \in L.$$

Therefore

$$(h_1 h_2)\varphi = k_1 k_2 (K \cap L) = (h_1 \varphi)(h_2 \varphi).$$

Thus φ is a homomorphism. It is surjective because, for any $k \in L\rho$, there is an element $h \in H$ such that $hk \in L$, and then $h \in L\pi$ and $h\varphi = k(K \cap L)$. Moreover,

$$\begin{aligned}
\text{Ker } \varphi &= \{h \in L\pi : hk \in L \text{ for some element } k \in K \cap L\} \\
&= \{h \in L\pi : h \in L\} \\
&= H \cap L \text{ (since } (H \cap L)\pi = H \cap L).
\end{aligned}$$

Therefore, by the fundamental theorem on homomorphisms,

$$L\pi/(H \cap L) = L\pi/\text{Ker } \varphi \cong \text{Im } \varphi = L\rho/(K \cap L).$$

(ii) Clearly

$$(H \cap L) \times (K \cap L) \leqslant L \leqslant L\pi \times L\rho.$$

If $L\pi = H \cap L$ then it follows from (i) that $L\rho = K \cap L$. Then the inclusions above imply that

$$L = (H \cap L) \times (K \cap L).$$

If, conversely, $L = (H \cap L) \times (K \cap L)$ then it is clear from the definitions of π and ρ that

$$L\pi = H \cap L \quad \text{and} \quad L\rho = K \cap L.$$

8.20 Corollary. *Let* $G = H \times K$. *Suppose that* G *is finite and that* $(|H|, |K|) = 1$. *Then, for every subgroup* L *of* G,

$$L = (H \cap L) \times (K \cap L).$$

Proof. Let $L \leqslant G$ and let π, ρ be defined as in 8.19. Then $L\pi \leqslant H$ and $L\rho \leqslant K$. Hence, by hypothesis,

$$(|L\pi|, |L\rho|) = 1.$$

Since, by 8.19(i), $L\pi/(H \cap L) \cong L\rho/(K \cap L)$, this implies that $|L\pi/(H \cap L)| = 1$, hence that $L\pi = H \cap L$. Thus, by 8.19(ii),

$$L = (H \cap L) \times (K \cap L).$$

Remark. This result would of course fail in general without the condition that $(|H|, |K|) = 1$. For instance, let $G = \langle a \rangle \times \langle b \rangle$ with $o(a) = o(b) = 2$. Then $\langle ab \rangle$ is a subgroup of G of order 2, but $\langle a \rangle \cap \langle ab \rangle = 1 = \langle b \rangle \cap \langle ab \rangle$.

438 Let H and K be normal subgroups of G such that $G = H \times K$, and let π and ρ be the corresponding projections of G onto H and K, respectively. Suppose that

$$H_2 \trianglelefteq H_1 \leqslant H, \qquad K_2 \trianglelefteq K_1 \leqslant K \quad \text{and} \quad H_1/H_2 \cong K_1/K_2.$$

Let θ be any isomorphism of H_1/H_2 onto K_1/K_2, and let

$$L = \{hk : h \in H_1, k \in K_1 \text{ and } (hH_2)\theta = kK_2\}.$$

Then $L \leqslant G$ and

$$H \cap L = H_2, \qquad L\pi = H_1, \qquad K \cap L = K_2, \qquad L\rho = K_1.$$

439 Let H and K be normal subgroups of G such that $G = H \times K$, and let π be the corresponding projection of G onto H. Let $L \leqslant G$ and let $J = (H \cap L) \times (K \cap L)$. Then $J \trianglelefteq L$ and

$$L/J \cong L\pi/(H \cap L).$$

(See 8.19. Hint. Let $\pi_1 : L \to L\pi$ be defined by restriction of π, and let $\nu : L\pi \to L\pi/(H \cap L)$ be the natural homomorphism. Consider the map $\pi_1 \nu$.)

440 (Remak [a80]). Let H and K be normal subgroups of G such that $G = H \times K$, and let π and ρ be the corresponding projections of G onto H and K, respectively. Let $L \leqslant G$. Then the following two statements are equivalent:
  (i) $L \trianglelefteq G$.
  (ii) $(H \cap L) \trianglelefteq H$, $(K \cap L) \trianglelefteq K$, $L\pi/(H \cap L) \leqslant Z(H/(H \cap L))$ and $L\rho/(K \cap L) \leqslant Z(K/(K \cap L))$.
(Hint. To prove that (ii) $\Rightarrow$ (i), let $J = (H \cap L) \times (K \cap L)$. Note that $J \trianglelefteq G$ and use **151** to show that $L/J \leqslant Z(G/J)$.)

441 Let H and K be normal subgroups of G such that $G = H \times K$, and let π and ρ be the corresponding projections of G onto H and K, respectively. A subgroup L of G is said to be *a subdirect product of H and K* if $L\pi = H$ and $L\rho = K$.
  (i) Let $L \leqslant G$. Then L is a subdirect product of H and K if and only if $HL = G = KL$.
  (ii) Let L be a subdirect product of H and K. Then $L \trianglelefteq G$ if and only if $G' \leqslant L$. (Hint. Apply **165** and **440**.)
  (iii) Suppose that G is finite and that $(|H/H'|, |K/K'|) = 1$. Then no proper normal subgroup of G is a subdirect product of H and K. (Hint. Apply (i) and 8.19.)

442 Let $H \trianglelefteq G$ and $K \trianglelefteq G$. Verify that the homomorphism ψ defined in **109** maps $G/(H \cap K)$ onto a subdirect product of G/H and G/K (see **441**).

443 Let H and K be normal subgroups of G such that $G = H \times K$. Then the following two statements are equivalent:
  (i) L is a subdirect product of H and K (**441**).
  (ii) For some group J, there are surjective homomorphisms $\varphi : H \to J$ and $\psi : K \to J$ such that

$$L = \{hk : h \in H, k \in K \text{ and } h\varphi = k\psi\}.$$

(Hint. To prove that (i) $\Rightarrow$ (ii), see the proof of 8.19.)

It is convenient to regard the direct product of a finite number of copies of a group G as a group of maps from a suitable set into G. We introduce this group of maps here; we shall return to it in chapter 9. The definition can also be generalized to arbitrary direct products: see **444, 445**.

8.21 Lemma. *Let X be a non-empty finite set and let G^X denote the set of all maps of X into the group G. For any $f_1, f_2 \in G^X$, let $f_1 f_2 \in G^X$ be defined, for all $x \in X$, by*

$$(f_1 f_2)(x) = f_1(x) f_2(x).$$

(N.B. This operation of multiplication is *not* composition of maps, and in the present case we write the maps on the *left* of the elements to which they apply. In his book, Scott uses the notation $f_1 + f_2$ for the map which we denote here by $f_1 f_2$: see [b36] p. 14, example 11. This would be consistent with 8.14, but we adopt the notation which is more usual in the present context.) *With respect to this operation of multiplication, G^X acquires the structure of a group which we shall denote by* Dr G^X.

For each $x \in X$, let

$$G_x = \{ f \in G^X : f(y) = 1 \text{ whenever } x \neq y \in X \}.$$

Then

$$G \cong G_x \trianglelefteq \text{Dr } G^X$$

and

$$\text{Dr } G^X = \text{Dr} \prod_{x \in X} G_x.$$

Thus Dr G^X *is the direct product of $|X|$ isomorphic copies of G.*

Proof. Certainly G^X is non-empty and is closed with respect to the multiplication defined above. Since multiplication in G is associative, so also is this multiplication in G^X. There is an identity element for G^X, namely the map

$$e : X \to G$$

defined, for all $x \in X$, by $\qquad e(x) = 1.$

Moreover, every element $f \in G^X$ has an inverse $f^{-1} \in G^X$, defined for all $x \in X$, by

$$f^{-1}(x) = f(x)^{-1}.$$

Hence G^X is a group with respect to the multiplication defined above. We denote this group by Dr G^X.

Now let $x \in X$ and let $G^* = \text{Dr } G^X$. Then we define a map

$$\varphi_x : G \to G^*,$$

as follows. For each $g \in G$,

$$\varphi_x : g \mapsto g_x,$$

where g_x is the map of X into G defined, for all $y \in X$, by

$$g_x(y) = \begin{cases} 1 \text{ if } y \neq x, \\ g \text{ if } y = x. \end{cases}$$

Then, for all $g, g' \in G$,

$$(gg')_x = g_x g'_x,$$

so that φ_x is a homomorphism. Moreover,

$$\text{Ker } \varphi_x = \{g \in G : g_x = e\} = 1,$$

and, by definition,

$$\text{Im } \varphi_x = G_x.$$

Thus $G \cong G_x \leqslant G^*.$

If $g \in G$ and $f \in G^*$ then, whenever $x \neq y \in X$,

$$\begin{aligned}(f^{-1}g_x f)(y) &= f(y)^{-1}g_x(y)f(y) \\ &= f(y)^{-1}f(y) \text{ (since } x \neq y) \\ &= 1.\end{aligned}$$

Therefore $f^{-1}g_x f \in G_x$. Hence $G_x \trianglelefteq G^*$.

Finally, we want to show that

$$G^* = \text{Dr} \prod_{x \in X} G_x.$$

This is obvious if $|X| = 1$. Suppose that $|X| > 1$. Then, for each $x \in X$, any element of $\prod_{y \neq x} G_y$ maps x to 1, and so

$$G_x \cap \prod_{y \neq x} G_y = 1.$$

Moreover, $G^* = \prod_{x \in X} G_x :$

for if $f \in G^*$ then we can express f in the form

$$f = \prod_{x \in X} (f(x))_x,$$

(where the ordering of elements in the product on the right is immaterial, since any two such elements commute). Now the result follows, by 8.4.

444 Let X be a non-empty finite set. To each $x \in X$ let there be associated an arbitrary group G^x. (These groups G^x need not be distinct.) Let D denote the set of all maps f of X into the *set* $\bigcup_{x \in X} G^x$ which satisfy the condition

$$f(x) \in G^x$$

for all $x \in X$. For any $f_1, f_2 \in D$, we may define a product map $f_1 f_2 \in D$, for all $x \in X$, by

$$(f_1 f_2)(x) = f_1(x)f_2(x).$$

Then D acquires the structure of a group with respect to this operation of multiplication.

For each $x \in X$, let

$$G_x = \{f \in D : f(y) = 1 \text{ whenever } x \neq y \in X\}.$$

Then $G^x \cong G_x \trianglelefteq D.$

Moreover, $D = \text{Dr} \prod_{x \in X} G_x.$

(Remarks. 8.21 is the special case in which $G^x = G$ for all $x \in X$. The representation of the direct product of a *finite* collection of groups given above is an appropriate basis for a generalization to a definition of direct products of possibly infinite collections of groups: see also **445**.)

*445 Let X be a non-empty set (possibly infinite) and, as in 8.21, let G^X denote the set of all maps of X into the group G.

(i) Let multiplication of elements of G^X be defined as in 8.21. Show that with respect to this operation of multiplication, G^X acquires the structure of a group: the group is called the *cartesian power* (or *unrestricted direct power*) of G *with index set* X, and will be denoted by $\text{Cr } G^x$. Let $C = \text{Cr } G^x$.

For each $f \in C$, the *support* of f is defined to be the set

$$s(f) = \{x \in X : f(x) \neq 1\} \subseteq X.$$

Let $f, f' \in C$. Show that (cf. **110**)

(ii) $s(f^{-1}) = s(f)$,

(iii) $s(ff') \subseteq s(f) \cup s(f')$,

(iv) $s(f^{-1}f'f) = s(f')$.

(v) If $s(f) \cap s(f') = \emptyset$ prove that $ff' = f'f$.

(vi) Let $D = \{f \in C : |s(f)| < \infty\}$. Prove that $D \trianglelefteq C$. The group D is called the *direct power* (or *restricted direct power*) of G *with index set* X, and is denoted by $\text{Dr } G^X$. Note that this notation is consistent with 8.21: in fact, if $G \neq 1$ then $D = C$ if and only if $|X| < \infty$.

(vii) For each $x \in X$, let

$$G_x = \{f \in C : f(y) = 1 \text{ whenever } x \neq y \in X\}.$$

Prove that $G \cong G_x \trianglelefteq C$, $G_x \leqslant D$ and $[G_x, G_y] = 1$ whenever x and y are distinct elements of X. Moreover, every element of D is expressible in the form $\prod_{x \in X} f_x$, where $f_x \in G_x$ for all $x \in X$, and $f_x = e$, the identity element of C, for all but finitely many values of x; and the expression is unique apart from ordering of the factors.

(When it is non-trivial, the 'product' $\prod_{x \in X} f_x$ is of course interpreted as the product of the *finitely* many f_x distinct from e. Products of infinitely many elements are not defined in general.)

(viii) Suppose that Y is a non-empty set such that there is an injective map of Y into X. Show that $\text{Cr } G^Y$ can be embedded in $\text{Cr } G^X$ and that $\text{Cr } G^Y / \text{Dr } G^Y$ can be embedded in $\text{Cr } G^X / \text{Dr } G^X$.

(ix) Suppose that X is infinite and $G \neq 1$. Prove that $\text{Cr } G^X / \text{Dr } G^X$ is infinite. (Hint. It may be assumed that there is an injective map of the set **N** of all positive integers into X. Hence, by (viii), it is enough to prove that $\text{Cr } G^N / \text{Dr } G^N$ is infinite.)

In the remainder of this chapter, we turn our attention to abelian groups. Starting from cyclic groups, with whose structure we are already familiar (see 3.25, 3.31, 3.32), we can form many other abelian groups simply by using the direct product construction. Any group which is a direct product of finitely many cyclic groups is certainly abelian; and it is also finitely generated (see **108**). We shall prove the fundamental structure theorem which asserts, conversely, that every finitely generated abelian group is the direct product of finitely many cyclic subgroups. This result is one of the outstanding achievements of the classical period of group theory: for finite groups it was partially known to Gauss, and proved completely in that case in 1879 by Frobenius and L. Stickelberger [a33]. We shall also show that we can decide whether or not two finitely generated abelian groups are of the same type by comparing certain systems of integers associated with the groups.

Before establishing these results, we make some remarks on notation.

8.22. Abelian groups appear in a natural way in the general framework of algebra as the additive groups of rings (see 2.11). Perhaps for this reason, it is conventional in developing the theory of abelian groups to write the group operation as addition rather than multiplication. This convention has various notational consequences. The identity element of an abelian group is called the *zero* element and denoted by 0. One refers to the *sum* of two subgroups H and K of an abelian group G instead of to their product, and writes $H + K$ instead of HK. If G is the *direct sum* of H and K (that is, if $G = H + K$ and $H \cap K = 0$) then one writes $G = H \oplus K$. Illogically, perhaps, one nevertheless speaks of *quotient* groups of an abelian group and denotes them as before. So, for instance, the isomorphism theorem of 3.40 would, for an abelian group G, be expressed as follows: if H and K are subgroups of G then $H/(H \cap K) \cong (H + K)/K$.

However, in this book we shall *not* adopt these conventions. Since the theory of abelian groups forms only a small part of the subject matter of the book, it seems more natural and economical to retain the notational conventions already established in the preceding pages.

We come now to the proof of the structure theorem for finitely generated abelian groups. Various different methods of proof appear in the literature. From a wider point of view, it is illuminating to place the structure theorem in the more general setting of results on modules over rings: see for instance Hartley and Hawkes [b18] and Rotman [b34] chapters 4 and 9. For the limited aims of the present chapter, we follow a brief and ingenious proof due to R. Rado [a77]. An alternative method for finite abelian groups is outlined in **448–452**: this approach is based on results in Fuchs [b11] vol. 1.

We begin with a lemma. In 8.23 and 8.24, we use the notation

$$(m_1, m_2, \ldots, m_s)$$

to denote the greatest common divisor of a sequence of integers $m_1, m_2, \ldots, m_s$ which are not all 0, where s is a positive integer. Note that

$$(m_1, m_2, \ldots, m_s) = (m_{i_1}, m_{i_2}, \ldots, m_{i_k}),$$

where $m_{i_1}, \ldots, m_{i_k}$ are those of the integers $m_1, \ldots, m_s$ which are not 0.

8.23 Lemma. *Let H be a finitely generated abelian group. Suppose that $\{x_1, x_2, \ldots, x_s\}$ is a set of generators of H, where s is a positive integer.*

Let $m_1, m_2, \ldots, m_s$ be non-negative integers, not all 0, such that $(m_1, m_2, \ldots, m_s) = 1$. Then there is a set of generators $\{y_1, y_2, \ldots, y_s\}$ of H such that

$$y_1 = \prod_{i=1}^{s} x_i^{m_i}.$$

Proof. Let $m = \sum_{i=1}^{s} m_i$, a positive integer. We argue by induction on m. If $m = 1$ then $m_i \neq 0$ for only one value of i; we may assume without loss of generality that $m_1 \neq 0$, and then $m_1 = 1$. In this case the result is trivial.

Now suppose that $m > 1$. Then, since $(m_1, m_2, \ldots, m_s) = 1$, $m_i \neq 0$ for at least two values of i. We may assume that

$$m_1 \geqslant m_2 > 0.$$

Then $m_1 - m_2, m_2, m_3, \ldots, m_s$ are non-negative integers, not all 0, and

$$(m_1 - m_2, m_2, m_3, \ldots, m_s) = 1.$$

Moreover, $\{x_1, x_1 x_2, x_3, \ldots, x_s\}$ is a set of generators of H, since $x_2 = x_1^{-1}(x_1 x_2)$. We may apply the induction hypothesis to any suitable set of generators of H. Then, since

$$m_1 - m_2 + \sum_{i=2}^{s} m_i = m - m_2 < m,$$

the induction hypothesis implies that there is a set of generators $\{y_1, y_2, \ldots, y_s\}$ of H such that

$$y_1 = x_1^{m_1 - m_2}(x_1 x_2)^{m_2} x_3^{m_3} \ldots x_s^{m_s}$$

$$= \prod_{i=1}^{s} x_i^{m_i} \text{ (since } H \text{ is abelian).}$$

This completes the induction argument.

446 Show, by considering the group Σ_3, that 8.23 does not remain true in general for non-abelian groups.

447 Show, by considering the group C_6, that in 8.23 we cannot in general choose $\{y_2, \ldots, y_s\}$ to be a subset of $\{x_1, x_2, \ldots, x_s\}$.

8.24 Structure theorem for finitely generated abelian groups. *Let r be a positive integer and let G be an r-generator abelian group. Then there are elements $x_1, x_2, \ldots, x_r$ of G such that*

$$G = \mathrm{Dr} \prod_{i=1}^{r} \langle x_i \rangle.$$

Proof. If $r = 1$ then G is cyclic and there is nothing more to prove. Therefore we may assume that $r > 1$. We consider the set $\mathcal{R}$ of all ordered sets

$$(x_1, x_2, \ldots, x_r)$$

of elements of G such that

$$o(x_1) \leqslant o(x_2) \leqslant \ldots \leqslant o(x_r) \qquad \text{(i)}$$

and

$$\langle x_1, x_2, \ldots, x_r \rangle = G.$$

Here the elements $x_1, x_2, \ldots, x_r$ need not all be distinct; and in the inequalities (i), we treat ∞ as a 'number' greater than every positive integer. Clearly any set of r generators of G can be ordered (possibly in several different ways) so that it becomes a member of $\mathcal{R}$. Thus $\mathcal{R} \neq \emptyset$.

We choose a member of $\mathcal{R}$ which satisfies certain minimality conditions on orders of elements. For all members $(x_1, x_2, \ldots, x_r)$ of $\mathcal{R}$, let N_1 be the smallest value of $o(x_1)$; thus N_1 is either a positive integer or ∞. Then, for all members $(x_1, x_2, \ldots, x_r)$ of $\mathcal{R}$ such that $o(x_1) = N_1$, let N_2 be the smallest value of $o(x_2)$. Then, for all members $(x_1, x_2, \ldots, x_r)$ of $\mathcal{R}$ such that $o(x_1) = N_1$ and $o(x_2) = N_2$, let N_3 be the smallest value of $o(x_3)$. And so on.

Now we choose some member

$$(x_1, x_2, \ldots, x_r) \in \mathcal{R}$$

with, for each $i = 1, \ldots, r$, $\qquad o(x_i) = N_i$.

Then $\{x_1, x_2, \ldots, x_r\}$ is a set of r generators of G with the following property: whenever $\{y_1, y_2, \ldots, y_r\}$ is a set of r generators of G and j is a positive integer such that (if $j > 1$)

$$o(x_i) = o(y_i) \quad \text{for all } i < j,$$

then

$$o(x_j) \leqslant o(y_j) \quad \text{for all } i \geqslant j.$$

Since G is abelian, every element of G is expressible in the form

$$\prod_{i=1}^{r} x_i^{n_i}$$

with suitable integers $n_1, n_2, \ldots, n_r$ (**69**); and $\langle x_i \rangle \trianglelefteq G$ for each $i = 1, \ldots, r$. Hence

$$G = \prod_{i=1}^{r} \langle x_i \rangle.$$

We claim that

$$G = \text{Dr} \prod_{i=1}^{r} \langle x_i \rangle.$$

Assume to the contrary that this is false. Then it follows easily from

8.4 that there are integers $n_1, n_2, \ldots, n_r$ such that

$$\prod_{i=1}^{r} x_i^{n_i} = 1 \quad \text{and} \quad x_i^{n_i} \neq 1$$

for some i. We may suppose without loss of generality that $n_1, n_2, \ldots, n_r$ are all non-negative: for if, say, $n_j < 0$ then, in the argument above, we may replace x_j by x_j^{-1} and n_j by $-n_j$. (Since $o(x_j^{-1}) = o(x_j)$, this replacement does not alter the properties prescribed above.)

We define integers $l_1, l_2, \ldots, l_r$ such that, for each $i = 1, \ldots, r$,

$$0 \leqslant l_i < o(x_i) \quad \text{and} \quad x_i^{l_i} = x_i^{n_i}.$$

This may be done as follows.

If $o(x_i) < \infty$ then, by the division algorithm, there are integers q_i and s_i such that $n_i = q_i o(x_i) + s_i$ and $0 \leqslant s_i < o(x_i)$. Then $x_i^{n_i} = x_i^{s_i}$, and we define $l_i = s_i$.

If $o(x_i) = \infty$ then we define $l_i = n_i$. (We have arranged above that n_i is non-negative.)

Since, by assumption, $x_i^{n_i} \neq 1$ for some $i, l_i > 0$ for some i. Let j be the least positive integer such that $l_j > 0$: thus, if $j > 1, l_i = 0$ for every $i < j$.

Now let

$$d = (l_1, l_2, \ldots, l_r),$$

and, for each $i = 1, \ldots, r$, let

$$m_i = l_i / d.$$

Then $m_1, m_2, \ldots, m_r$ are non-negative integers such that

$$(m_1, m_2, \ldots, m_r) = 1.$$

Since $m_i = 0$ for every $i < j$,

$$(m_j, m_{j+1}, \ldots, m_r) = 1.$$

Let

$$H = \langle x_j, x_{j+1}, \ldots, x_r \rangle \leqslant G.$$

Then, by 8.23, there is a set of generators $\{y_j, y_{j+1}, \ldots, y_r\}$ of H such that

$$y_j = \prod_{i=j}^{r} x_i^{m_i}.$$

Hence

$$y_j^d = \prod_{i=j}^{r} x_i^{l_i}$$

$$= \prod_{i=1}^{r} x_i^{l_i} \text{ (since } l_i = 0 \text{ whenever } i < j)$$

$$= \prod_{i=1}^{r} x_i^{n_i} \text{ (by definition of } l_1, \ldots, l_r)$$

$$= 1.$$

But now $\quad G = \langle x_1, \ldots, x_{j-1}, y_j, y_{j+1}, \ldots, y_r \rangle$

and $\qquad o(y_j) \leqslant d \leqslant l_j < o(x_j).$

This contradicts the choice of $(x_1, x_2, \ldots, x_r)$. Therefore we conclude that

$$G = \operatorname{Dr} \prod_{i=1}^{r} \langle x_i \rangle.$$

448 (i) Let G be a finite abelian group and let x be an element of G of largest possible order, say n. Then $g^n = 1$ for every $g \in G$.
(Hint. Use **6** to show that if there were an element $g \in G$ with $g^n \neq 1$ then, for suitable positive integers j and k, $o(g^j x^k) > n$.)
  (ii) Show by an example that the statement in (i) would not be true in general without the condition that G is abelian.

449 Let $H \leqslant G$, a finite abelian group. Let K be a subgroup of G maximal subject to $H \cap K = 1$. Suppose that $g \in G$ and $g^p \in K$ for some prime p. Then $g \in HK$ ($= H \times K$, by 8.2). (Hint. If $g \notin K$, show that there are elements $h \in H$ and $k \in K$ such that $h = kg^r$, where r is an integer not divisible by p.)

450 Let $H \leqslant G$, a finite abelian group. Let K be a subgroup of G maximal subject to $H \cap K = 1$. Then the following two statements are equivalent:
  (i) $G = H \times K$.
  (ii) For any prime p and any elements $g \in G, h \in H, k \in K$ such that $g^p = hk$, there is an element $h' \in H$ such that $h = (h')^p$.
(Hint. To prove that (ii) $\Rightarrow$ (i), suppose that $(H \times K) < G$, consider an element of prime order in $G/(H \times K)$ and apply **449**.)

451 Let G be a finite abelian group and let x be an element of G of largest possible order, say n. Then $\langle x \rangle$ is a direct factor of G.
(Hints. Let $H = \langle x \rangle \leqslant G$, and let K be defined as in **450**. Use **448** and **450** to show that $G = H \times K$. Deal separately with the cases in which p divides n and p does not divide n.)

452 Let G be a non-trivial finite abelian group. Then there are non-trivial elements $x_1, x_2, \ldots, x_r$ of G such that $G = \operatorname{Dr} \prod_{i=1}^{r} \langle x_i \rangle$ and (if $r > 1$) $o(x_i)$ is divisible by $o(x_{i+1})$ for all $i = 1, \ldots, r-1$. (Hint. Argue by induction on $|G|$ and use **448** and **451**. Remark. In **463**, we shall see that the sequence of positive integers $r, o(x_1), o(x_2), \ldots, o(x_r)$ is uniquely determined by G.)

453 Let G be a non-trivial finite group and let n be a positive integer. Then the following two statements are equivalent:
  (i) G is abelian and $g^n = 1$ for every $g \in G$.
  (ii) G is a direct product of cyclic subgroups each of which has order dividing n.

454 (i) Suppose that $G = H \times K$ and let A be an abelian group. Then

$$\operatorname{Hom}(G, A) \cong \operatorname{Hom}(H, A) \times \operatorname{Hom}(K, A) \quad \text{(see 33)}.$$

(ii) If J is a finite cyclic group then $\mathrm{Hom}(J, \mathbf{C}^\times) \cong J$.

(iii) Deduce that if G is a finite abelian group then $\mathrm{Hom}(G, \mathbf{C}^\times) \cong G$ (cf. **41**(ii)).

8.25. It is natural to ask whether there is a *uniqueness* theorem for decompositions of a finitely generated abelian group as a direct product of cyclic groups (cf. 8.12). The proof of 8.24 allows the possibility that one or more of the elements $x_1, \ldots, x_r$ may be equal to the identity element 1. More significantly, the direct factors $\langle x_i \rangle$ in 8.24 need not be indecomposable: for instance, if $o(x_i) = 6$ then $\langle x_i \rangle = \langle x_i^2 \rangle \times \langle x_i^3 \rangle$ (see **81**).

Even if the direct factors in 8.24 are non-trivial and indecomposable, we cannot expect to obtain such a strong uniqueness theorem as in 8.18. For instance, consider again the group $G = C_2 \times C_2$, and let A, B, C be the three distinct subgroups of G of order 2 (see 8.5). Then

$$G = A \times B = B \times C = C \times A,$$

and these are essentially different decompositions of G as a direct product of non-trivial indecomposable subgroups. However, we shall prove that, as in this example, in any two decompositions of a finitely generated abelian group G as a direct product of non-trivial indecomposable subgroups, the number of factors in both decompositions is the same and the factors in one decomposition can be paired isomorphically with the factors of the other decomposition.

We shall establish this by means of several intermediate results. First, we note that any non-trivial finitely generated abelian group does have a decomposition as a direct product of finitely many indecomposable subgroups.

8.26 (cf. **81**, **132**). *Let G be a finitely generated abelian group. Then G is indecomposable if and only if G is cyclic of prime power or infinite order.*
Proof. Suppose that G is indecomposable. Then, by 8.24, G is cyclic. If G is finite then it follows from 8.6 that G has prime power order.

Suppose conversely that G is cyclic, of prime power or infinite order. If $|G| = p^m$, where m is a positive integer, then, by 3.32, every non-trivial subgroup of G contains the unique subgroup of G of order p, and so any two non-trivial subgroups of G have non-trivial intersection; thus G is indecomposable. If G is infinite then it follows from 3.25 that any two non-trivial subgroups of G again have non-trivial intersection, so that G is indecomposable.

8.27 Corollary. *Let G be a non-trivial finitely generated abelian group. Then G is the direct product of finitely many indecomposable subgroups.*
Proof. By 8.24, there are subgroups H and K of G such that

$$G = H \times K,$$

where H is finite and either $K = 1$ or K is the direct product of a finite number of infinite cyclic subgroups of G. Now the result follows from 8.17 and 8.26.

8.28 Definitions. (i) G is said to be *periodic* (or to be a *torsion* group) if every element of G has finite order. Every finite group is periodic; and there are also infinite periodic groups, such as $\mathbf{Q}^+/\mathbf{Z}^+$ (**195**).

(ii) G is said to be *torsion-free* (or *aperiodic* or *locally infinite*) if every non-trivial element of G has infinite order. For example, the groups $\mathbf{Z}^+, \mathbf{Q}^+, \mathbf{R}^+, \mathbf{C}^+, \mathbf{Q}^\times_{\text{pos}}, \mathbf{R}^\times_{\text{pos}}$ are torsion-free. The only group which is both periodic and torsion-free is the trivial group (of order 1).

(iii) In general, an infinite group may have non-trivial elements of finite orders and also elements of infinite orders. Such a group is said to be *mixed*. For instance, the groups $\mathbf{Q}^\times, \mathbf{R}^\times, \mathbf{C}^\times$ are mixed.

8.29. We mention here a famous problem of Burnside. Clearly any finite group is both finitely generated and periodic. In 1902, Burnside asked whether, conversely, a group which is both finitely generated and periodic is necessarily finite. This question has a positive answer for *soluble* groups (see **455**). The general question remained unanswered until 1964, when E. S. Golod and I. R. Shafarevich ([a41], [a42]) showed that for any prime p, there is a 3-generator infinite group in which every element has order a power of p; thus answering Burnside's question in the negative. (In this connexion, see Herstein [b20] chapter 8.)

In the Golod–Shafarevich examples, there is no finite upper bound on the orders of elements. On the other hand, in a finite group G, every element x satisfies the equation $x^{|G|} = 1$. Therefore one may ask whether a group G is necessarily finite if it is finitely generated and there is a positive integer n such that $x^n = 1$ for every $x \in G$. This question has been the subject of a great deal of study since Burnside first formulated his problem. It is easy to show that G must be finite if $n = 2$ (see **3** and **69**); and it is also known that G is finite if $n = 3$ (Burnside [a10] and F. W. Levi and B. L. van der Waerden [a70]; see Huppert [b21] p. 290, theorem 3.6.6) or if $n = 4$ (I. N. Sanov [a82]) or if $n = 6$ (M. Hall [a46]). The solution in the case $n = 6$ uses ideas from a seminal paper of P. Hall and G. Higman [a52]: this paper has also had a profound influence on later investigations on finite simple groups.

However, in 1968, P. S. Novikov and S. I. Adyan ([a75]) established that for every odd integer $n \geqslant 4381$, there is a 2-generator infinite group G such that $x^n = 1$ for every $x \in G$.

455 A finitely generated periodic soluble group is necessarily finite. (Hint. Argue by induction on derived length, and apply **108**, **195** and **196**.)

We shall now show that any *abelian* group has a 'periodic radical' (see 3.45).

8.30 Theorem. *Let G be an abelian group and let H be the set of all elements of G which have finite orders. Then H is a periodic subgroup of G, and G/H is torsion-free. We call H the* torsion subgroup *of G and write $H = T(G)$.*
Proof. Certainly $1 \in H$, so that $H \neq \emptyset$. Let $h_1, h_2 \in H$, and let

$$o(h_1) = n_1, \qquad o(h_2) = n_2.$$

Then n_1, n_2 are positive integers and, since G is abelian,

$$(h_1 h_2^{-1})^{n_1 n_2} = (h_1^{n_1})^{n_2} (h_2^{n_2})^{-n_1} = 1.$$

Therefore, since $n_1 n_2$ is a positive integer, $o(h_1 h_2^{-1}) < \infty$, and so $h_1 h_2^{-1} \in H$. Thus $H \leqslant G$. By definition, H is periodic.

Since G is abelian, we can form the quotient group G/H. Let $g \in G$, and suppose that gH is an element of finite order n in G/H. Then

$$g^n H = (gH)^n = H,$$

by hypothesis, so that $\qquad\qquad g^n \in H.$

Hence $o(g^n) < \infty$: say $o(g^n) = m$. Then

$$g^{nm} = 1,$$

and nm is a positive integer. Therefore $o(g) < \infty$, and so $g \in H$. Thus

$$gH = H.$$

Hence G/H is torsion free.
Remark. In a non-abelian group G, the set of all elements whose orders are finite need not form a subgroup of G: see **45, 142**. See also **458**.

8.31 Lemma. *Let G be an abelian group. Suppose that G_0 is a periodic subgroup of G such that G/G_0 is torsion-free. Then $G_0 = T(G)$.*
Proof. Since G_0 is periodic, $G_0 \leqslant T(G)$. Then $T(G)/G_0$ is a periodic subgroup of G/G_0. Since G/G_0 is torsion-free, this implies that $G_0 = T(G)$.

8.32 Lemma. *Let G and H be abelian groups. If $G \cong H$ then $T(G) \cong T(H)$ and $G/T(G) \cong H/T(H)$.*
Proof. Suppose that φ is an isomorphism of G onto H. If $x \in T(G)$ then $x^n = 1$ for some positive integer n. Hence $(x\varphi)^n = x^n \varphi = 1$, so that $x\varphi \in T(H)$. Thus $T(G)\varphi \leqslant T(H)$. Every element of H is expressible in the form $y\varphi$, with $y \in G$. If $y\varphi \in T(H)$ then $(y\varphi)^n = 1$ for some positive integer n. Hence $y^n \varphi = 1$ and so, since φ is an isomorphism, $y^n = 1$. Therefore $y \in T(G)$ and $y\varphi \in T(G)\varphi$. Thus $T(G)\varphi = T(H)$, so that φ maps $T(G)$ isomorphically onto $T(H)$. Now it follows by 3.29 that $G/T(G) \cong H/T(H)$.

456 Let $H \leqslant G$, an abelian group. Show that
   (i) $T(H) = H \cap T(G)$,
   (ii) $T(G)/T(H) \cong HT(G)/H \leqslant T(G/H)$.

Show by an example that we may have $HT(G)/H < T(G/H)$.

457 Show that $T(\mathbf{Q}^{\times}) = C_2 = T(\mathbf{R}^{\times})$, $T(\mathbf{C}^{\times}) = V$, the multiplicative group of all complex roots of 1 (**131**), and

$$\mathbf{Q}^{\times}/T(\mathbf{Q}^{\times}) \cong \mathbf{Q}_{\text{pos}}^{\times}, \quad \mathbf{R}^{\times}/T(\mathbf{R}^{\times}) \cong \mathbf{R}_{\text{pos}}^{\times},$$

$$\mathbf{C}^{\times}/T(\mathbf{C}^{\times}) \cong \mathbf{R}_{\text{pos}}^{\times} \times (\mathbf{R}^{+}/\mathbf{Q}^{+}).$$

458 Let G be a (not necessarily abelian) group in which the set of all elements which have finite orders forms a subgroup H of G. Then H is a periodic characteristic subgroup of G, and G/H is torsion-free.

459 Let $U(G) = \{x \in G : \text{either } x = 1 \text{ or } o(x) = \infty\}$.

   (i) Suppose that G is abelian. Show that $U(G)$ is a subgroup of G if and only if either $U(G) = 1$ or $U(G) = G$; that is, if and only if G is either periodic or torsion-free.

   (ii) Show by an example that if G is non-abelian, $U(G)$ can be a non-trivial proper subgroup of G.

8.33 Lemma. *Let n be a positive integer, and let G and H be abelian groups.*

   (i) *Let $G^n = \{g^n : g \in G\}$. Then $G^n \leqslant G$.*

   (ii) *If $G \cong H$ then $G^n \cong H^n$ and $G/G^n \cong H/H^n$.*

Proof. Since G is abelian, the map $\lambda_n : g \mapsto g^n$, defined for all $g \in G$, is an endomorphism of G. Then $G^n = \text{Im } \lambda_n \leqslant G$.

Now suppose that φ is an isomorphism of G onto H. For any $x \in G$, $x^n\varphi = (x\varphi)^n \in H^n$. Thus $G^n\varphi \leqslant H^n$. Every element of H is expressible in the form $y\varphi$, with $y \in G$, and $(y\varphi)^n = y^n\varphi \in G^n\varphi$. Thus $G^n\varphi = H^n$, so that φ maps G^n isomorphically onto H^n. It follows by 3.29 that $G/G^n \cong H/H^n$.

8.34 Lemma. *Let n be a positive integer. Let G be a finitely generated abelian group, so that, by 8.24, there are elements $x_1, \ldots, x_r$ of G such that*

$$G = \text{Dr} \prod_{i=1}^{r} \langle x_i \rangle. \text{ Then}$$

$$G^n = \text{Dr} \prod_{i=1}^{r} \langle x_i^n \rangle.$$

Proof. Let $g \in G$. Then there are integers $m_1, \ldots, m_r$ such that

$$g = \prod_{i=1}^{r} x_i^{m_i}.$$

Then

$$g^n = \prod_{i=1}^{r} x_i^{m_i n} \in \langle x_1^n, \ldots, x_r^n \rangle.$$

Thus

$$G^n = \langle x_1^n, \ldots, x_r^n \rangle.$$

Since, for each $i = 1, \ldots, r$,

$$\langle x_i^n \rangle \leqslant \langle x_i \rangle,$$

it follows that

$$G^n = \mathrm{Dr} \prod_{i=1}^{r} \langle x_i^n \rangle.$$

8.35 Definition. Let r be a positive integer. A group which is the direct product of r infinite cyclic subgroups is said to be *free abelian of rank r*. The trivial group (of order 1) is said to be *free abelian of rank* 0.

Note that any free abelian group is torsion-free.

8.36 Lemma. *Let n and r be positive integers. Let G be a free abelian group of rank r. Then G/G^n is the direct product of r cyclic subgroups, all of order n.*

Proof. By hypothesis, there are r elements $x_1, \ldots, x_r$ of G, all of infinite order, such that

$$G = \mathrm{Dr} \prod_{i=1}^{r} \langle x_i \rangle.$$

Then, by 8.34,

$$G^n = \mathrm{Dr} \prod_{i=1}^{r} \langle x_i^n \rangle.$$

Hence (see **111**)

$$G/G^n \cong (\langle x_1 \rangle / \langle x_1^n \rangle) \times \ldots \times (\langle x_r \rangle / \langle x_r^n \rangle).$$

By 3.25, $\langle x_i \rangle / \langle x_i^n \rangle$ is cyclic of order n, for each $i = 1, \ldots, n$.

We deduce that the rank of a free abelian group is uniquely determined by the group.

8.37 Corollary. *Let G and H be free abelian groups of rank r and s, respectively, where r and s are non-negative integers. Then $G \cong H$ if and only if $r = s$.*

Proof. It is clear from the definition that if $r = s$ then $G \cong H$. Suppose, conversely, that $G \cong H$. We may assume that G and H are non-trivial, so that r and s are both positive. By 8.33, $G/G^2 \cong H/H^2$; and, by 8.36, G/G^2 and H/H^2 are elementary abelian groups, of orders 2^r and 2^s, respectively. Since $|G/G^2| = |H/H^2|$, it follows that $r = s$.

460 Let n be a positive integer, G an abelian group and $G_n = \{ g \in G : g^n = 1 \}$.

  (i) Then $G_n \leqslant G$, and $G/G_n \cong G^n$.

  (ii) Suppose that G is finite, so that (by 8.24 or **452**) there are elements $x_1, \ldots, x_r$ of G such that

$$G = \mathrm{Dr} \prod_{i=1}^{r} \langle x_i \rangle.$$

For each $i = 1, \ldots, r$, let $m_i = o(x_i)$ and $k_i = m_i/(m_i, n)$. Then

$$G_n = \mathrm{Dr} \prod_{i=1}^{r} \langle x_i^{k_i} \rangle.$$

(iii) If G is finite then $G/G^n \cong G_n$.

(iv) Show by an example that if G is infinite, we need not have $G/G^n \cong G_n$ for any $n > 1$.

(v) Show by an example that, even when G is finite, for some $n > 1$ we need not have $G = G_n \times G^n$.

461 Let n be a positive integer and let G and H be abelian groups. If $G \cong H$ then $G_n \cong H_n$ (see **460**).

462 Let G be a non-trivial finite abelian group. Then, by **452**, there are non-trivial elements $x_1, x_2, \ldots, x_r$ of G such that $G = \text{Dr} \prod_{i=1}^{r} \langle x_i \rangle$ and (if $r > 1$) $o(x_i)$ is divisible by $o(x_{i+1})$, for all $i = 1, \ldots, r-1$.

(i) If p is a prime divisor of $o(x_r)$, then G_p is elementary abelian of order p^r (where G_p is defined as in **460**).

(ii) The integer r is the least positive integer n such that G is an n-generator group. (Hint. Apply (i), 2.30(i) and **388**.)

463 Let G and H be non-trivial finite abelian groups. Then, by **452**, there are positive integers r and s and non-trivial elements $x_1, x_2, \ldots, x_r$ of G and $y_1, y_2, \ldots, y_s$ of H such that $G = \text{Dr} \prod_{i=1}^{r} \langle x_i \rangle$, $H = \text{Dr} \prod_{j=1}^{s} \langle y_j \rangle$, and (if $r > 1$) $o(x_i)$ is divisible by $o(x_{i+1})$ for all $i = 1, \ldots, r-1$ and (if $s > 1$) $o(y_j)$ is divisible by $o(y_{j+1})$ for all $j = 1, \ldots, s-1$.

Prove that if $G \cong H$ then $r = s$ and $o(x_i) = o(y_i)$ for all $i = 1, \ldots, r$. (Remark. This shows that if G is a non-trivial finite abelian group then the positive integers $r, o(x_1), o(x_2), \ldots, o(x_r)$, given by **452**, are uniquely determined by G; and, conversely, they obviously determine uniquely the type of G. The integers $o(x_1), o(x_2), \ldots, o(x_r)$ are sometimes called the *invariants* of G. Hints. By **462**(ii), $r = s$. Argue by induction on $|G|$ to show that $o(x_i) = o(y_i)$ for all $i = 1, \ldots, r$. Let p be a prime divisor of $o(x_r)$. Then p divides $o(x_i)$ for all $i = 1, \ldots, r$ and, by **460**(ii), **461** and **462**(i), p divides $o(y_i)$ for all $i = 1, \ldots, r$. If G and H are not elementary, consider G^p and H^p and apply 8.33, 8.34 and the induction hypothesis.)

464 Find the invariants (**463**) of the finite abelian groups $C_4 \times C_6, C_6 \times C_{15} \times C_{21}$ and $C_2 \times C_{10} \times C_{15}$.

465 Let r be a positive integer, and let G be a free abelian group of rank r. Then, by hypothesis, there are elements $x_1, \ldots, x_r$ of G, all of infinite order, such that

$$G = \text{Dr} \prod_{i=1}^{r} \langle x_i \rangle.$$

(i) Let $g \in G$. Then there are integers $n_1, \ldots, n_r$, *uniquely determined by* g, such that

$$g = \prod_{i=1}^{r} x_i^{n_i}.$$

(ii) Let H be any r-generator abelian group. Then there is a homomorphism of G onto H.

466 Let r be a positive integer, and let G be a free abelian group of rank r. Let $H \leqslant G$. Then H is free abelian of rank s for some $s \leqslant r$. (Hint. Apply **388** and 8.24.)

467 Let $K \leqslant G$, an abelian group. Suppose that G/K is free abelian of rank r for some positive integer r. Then there is a subgroup H of G such that $G = H \times K$. (Hint. There are r elements $x_1, \ldots, x_r$ of G such that $G/K = \text{Dr} \prod_{i=1}^{r} \langle x_i K \rangle$ and the elements $x_1 K, \ldots, x_r K$ of G/K all have infinite order. Let $H = \langle x_1, \ldots, x_r \rangle \leqslant G$.)

468 Let r be a positive integer, and let G be a free abelian group of rank r. Then
  (i) G can be embedded in $\mathbf{Q}_{\text{pos}}^{\times}$.
  (ii) If $r > 1$, G cannot be embedded in $\mathbf{Q}^{+}$.

We now note a consequence of the structure theorem 8.24.

8.38 Lemma. *Let G and H be finitely generated abelian groups.*
  (i) *There is a non-negative integer r and a free abelian subgroup K of G of rank r such that $G = T(G) \times K$; and $T(G)$ is finite.*
  (ii) *Let $G = T(G) \times K$ and $H = T(H) \times L$, where K is a free abelian subgroup of G of rank r, L a free abelian subgroup of H of rank s, and r and s are non-negative integers. Then $G \cong H$ if and only if $T(G) \cong T(H)$ and $r = s$.*

Proof. (i) By 8.24, there are subgroups G_0 and K of G such that

$$G = G_0 \times K,$$

with G_0 finite and K free abelian of rank r, for some non-negative integer r. Then, since $G/G_0 \cong K$, which is torsion-free, it follows, by 8.31, that $G_0 = T(G)$.

  (ii) If $r = s$ then $K \cong L$. If also $T(G) \cong T(H)$ then, since $G = T(G) \times K$ and $H = T(H) \times L$, it follows that $G \cong H$.

Suppose, conversely, that $G \cong H$. Then, by 8.32, $T(G) \cong T(H)$ and $K \cong G/T(G) \cong H/T(H) \cong L$. Then also, by 8.37, $r = s$.

Remark. Let G be an abelian group. It has been proved by R. Baer [a4] (1936) and S. V. Fomin [a28] (1937) that if there is a positive integer n such that $x^n = 1$ for every $x \in T(G)$ (and, in particular, if $T(G)$ is finite) then $T(G)$ is a direct factor of G. See Fuchs [b11] vol. 2, p. 187, theorem 100.1. However, in general $T(G)$ need not be a direct factor of G; see for example Macdonald [b30] p. 223, example 11.14.

The last lemma reduces the problem of finding conditions for two finitely generated abelian groups to be isomorphic to the corresponding problem for finite abelian groups. We now make a further reduction.

8.39 Lemma. *Let G and H be finite abelian groups. Then $G \cong H$ if and only if, for every prime p, G and H have isomorphic Sylow p-subgroups.*
Proof. Suppose that φ is an isomorphism of G onto H. Let p be any prime, and let P be the Sylow p-subgroup of G. Then $P \cong P\varphi \leqslant H$. Moreover, since $|P| = |P\varphi|$ and $|G| = |H|$, $P\varphi$ is the Sylow p-subgroup of H.

Suppose, conversely, that for every prime p, G and H have isomorphic Sylow p-subgroups. If $G = 1$, it follows immediately that $H = 1$. Suppose that $G \neq 1$, and let the distinct prime divisors of $|G|$ be $p_1, \ldots, p_s$, where s is a positive integer. For each $i = 1, \ldots, s$, let P_i, Q_i denote the Sylow p_i-subgroups of G, H respectively. Then, by hypothesis, for each $i = 1, \ldots, s$,

$$P_i \cong Q_i,$$

and $|H|$ is divisible by no prime distinct from $p_1, \ldots, p_s$. By 8.6,

$$G = \mathrm{Dr} \prod_{i=1}^{s} P_i \quad \text{and} \quad H = \mathrm{Dr} \prod_{i=1}^{s} Q_i.$$

Hence
$$G \cong H.$$

It remains to consider the question of isomorphism of finite abelian p-groups. We do this next.

8.40 Lemma. *Let G and H be non-trivial finite abelian p-groups. By 8.24, we may decompose G and H as direct products of non-trivial cyclic subgroups: say*

$$G = \mathrm{Dr} \prod_{i=1}^{t} \langle x_i \rangle \quad \text{and} \quad H = \mathrm{Dr} \prod_{j=1}^{u} \langle y_j \rangle,$$

where t and u are positive integers. For each $i = 1, \ldots, t$ and each $j = 1, \ldots, u$, let

$$o(x_i) = p^{m_i} \quad \text{and} \quad o(y_j) = p^{n_j}.$$

We may suppose the notation chosen so that

$$m_1 \geqslant m_2 \geqslant \ldots \geqslant m_t \quad \text{and} \quad n_1 \geqslant n_2 \geqslant \ldots \geqslant n_u.$$

Then $G \cong H$ if and only if $t = u$ and $m_i = n_i$ for each $i = 1, \ldots, t$.

Proof. It is clear that if $t = u$ and $m_i = n_i$ for each $i = 1, \ldots, t$, then $G \cong H$.

Suppose, conversely, that $G \cong H$. We argue by induction on m_1. If $m_1 = 1$ then, since, by hypothesis, $m_t > 0$, it follows that $m_i = 1$ for every $i = 1, \ldots, t$. Thus G is an elementary abelian p-group, and so H is also an elementary abelian p-group. Hence $n_j = 1$ for every $j = 1, \ldots, u$. Then also $p^t = |G| = |H| = p^u$, so that $t = u$.

Now suppose that $m_1 > 1$. Then also $n_1 > 1$, by the case already established. Let k be the largest integer such that $m_k > 1$ and let l be the largest integer such that $n_l > 1$. Then $1 \leqslant k \leqslant t$, and if $k < t$ then $m_i = 1$ for all $i > k$. Similarly, $1 \leqslant l \leqslant u$, and if $l < u$ then $n_j = 1$ for all $j > l$. Hence

$$|G| = p^{m_1 + \ldots + m_k + t - k} \quad \text{and} \quad |H| = p^{n_1 + \ldots + n_l + u - l}.$$

Therefore, since $G \cong H$,

$$\sum_{i=1}^{k} m_i + t - k = \sum_{j=1}^{l} n_j + u - l. \tag{i}$$

By 8.33 and 8.34,

$$\mathrm{Dr} \prod_{i=1}^{t} \langle x_i^p \rangle = G^p \cong H^p = \mathrm{Dr} \prod_{j=1}^{u} \langle y_j^p \rangle.$$

For each $i = 1, \ldots, t$ and each $j = 1, \ldots, u$,

$$o(x_i^p) = p^{m_i-1} \quad \text{and} \quad o(y_j^p) = p^{n_j-1}.$$

Hence also

$$G^p = \text{Dr} \prod_{i=1}^{k} \langle x_i^p \rangle \quad \text{and} \quad H^p = \text{Dr} \prod_{j=1}^{l} \langle y_j^p \rangle,$$

with

$$o(x_i^p) = p^{m_i-1}, o(y_j^p) = p^{n_j-1},$$

$$m_1 - 1 \geqslant m_2 - 1 \geqslant \ldots \geqslant m_k - 1 > 0 \quad \text{and} \quad n_1 - 1 \geqslant n_2 - 1 \geqslant \ldots \geqslant n_l - 1 > 0.$$

Since $G^p \cong H^p$ and $o(x_1^p) = p^{m_1-1}$, it follows, by the induction hypothesis, that

$$k = l \quad \text{and, for each } i = 1, \ldots, k, m_i - 1 = n_i - 1.$$

Then it follows, by equation (i), that $t = u$; and also, since $m_i = 1 = n_i$ whenever $i > t, m_i = n_i$ for each $i = 1, \ldots, t$. This completes the induction argument.

469 Let n and m be positive integers, and let G be an abelian group. Then
   (i) $G^{nm} = (G^n)^m$.
   (ii) If m divides n then $G^n \leqslant G^m$.
   (iii) If $H \leqslant G$ then $(G/H)^n = G^n H/H$.
   (iv) If $G = H \times K$ then $G^n = H^n \times K^n$; and $G^n = G$ if and only if $H^n = H$ and $K^n = K$.
   (v) If G is periodic and, for every $g \in G, (o(g), n) = 1$, then $G^n = G$.
   (vi) Suppose that $n > 1$ and that G is finitely generated. Then $G^n = G$ if and only if G is finite and $(|G|, n) = 1$.
   (vii) Suppose that $n > 1$. Then $G^n = G$ if and only if $G^p = G$ for every prime divisor p of n.

470 Let G be an abelian group. Then G is said to be *divisible* (or *radicable* or *complete*) if $G^n = G$ for every positive integer n. (We shall keep to the most frequently used term, *divisible*, even though it derives from the customary additive notation for abelian groups and the term *radicable* would be more appropriate to our multiplicative notation.)
   (i) The groups $\mathbf{Q}^+, \mathbf{R}^{\times}_{\text{pos}}, \mathbf{C}^{\times}, C_{p^\infty} (144)$ are all divisible.
   (ii) If G is divisible, then so is every quotient group of G.
   (iii) If $G = H \times K$, then G is divisible if and only if H and K are both divisible.
   (iv) If G is divisible and non-trivial, then G cannot be finitely generated.

471 Let G be an abelian group and suppose that H is a divisible subgroup of G (**470**). Assume that K is a subgroup of G which is maximal subject to $H \cap K = 1$. (Remark. Zorn's lemma, which is a version of the so-called axiom of choice for sets, guarantees the existence of such a subgroup K: see, for instance, Fuchs [b11] vol. 1, pp. 1–2 and p. 48.)
   Prove that $G = H \times K$. (Hints. Let $J = HK$. Suppose, contrary to what we wish to prove, that $J < G$. Then there is an element $x \in G$ and a positive integer m such that $x \notin J$ and $x^m \in J$. Let n be the least such integer. By the divisibility of H, there is an element $h \in H$ such that $(xh^{-1})^n \in K$. Moreover, there are elements $h' \in H, k' \in K$ and an integer r such that $1 \neq h' = (xh^{-1})^r k'$; and r is not divisible by n. Use the division algorithm to derive a contradiction to the choice of n.)

We have now reached a very satisfactory position with regard to finitely generated abelian groups. Any such group G determines a certain system of integers (see 8.41) and the type of G is uniquely determined by this system.

8.41 Uniqueness theorem for finitely generated abelian groups. *Any finitely generated abelian group G determines a system of non-negative integers, as follows. There are non-negative integers r and s; and if $s > 0$, there are distinct primes $p_1, \ldots, p_s$, positive integers $t_1, \ldots, t_s$ and positive integers m_{ij} $(i = 1, \ldots, s, j = 1, \ldots, t_i)$; such that $G = T(G) \times K$, where K is a free abelian group of rank r, and if $s = 0$, $T(G) = 1$, while if $s > 0$, $T(G)$ is the direct product of $\sum\limits_{i=1}^{s} t_i$ cyclic subgroups whose orders are $p_i^{m_{ij}}(i = 1, \ldots, s; j = 1, \ldots, t_i)$. Moreover, two finitely generated abelian groups are isomorphic if and only if they determine the same system of integers.*
Proof. Let G be a finitely generated abelian group. By 8.38, $G = T(G) \times K$, where K is a free abelian group of rank r for some non-negative integer r, and $T(G)$ is finite. If $T(G) = 1$, let $s = 0$. If $T(G) \neq 1$, let s denote the number of distinct prime divisors of $|T(G)|$, and let these primes be $p_1, \ldots, p_s$. In this case, by 8.6, $T(G) = \mathrm{Dr} \prod\limits_{i=1}^{s} P_i$, where P_i is the unique Sylow p_i-subgroup of $T(G)$. Finally, by 8.24, for each $i = 1, \ldots, s$, P_i is the direct product of a finite number, t_i say, of non-trivial cyclic subgroups; let the orders of these cyclic subgroups be $p_i^{m_{ij}}(j = 1, \ldots, t_i)$.
The fact that two finitely generated abelian groups are isomorphic if they determine the same system of integers follows immediately from the definition of the system of integers. Suppose, conversely, that G and H are isomorphic finitely generated abelian groups. Then the fact that they determine the same system of integers follows from 8.38, 8.39 and 8.40.
Remark. Every system of non-negative integers of the kind specified in the theorem does arise from some finitely generated abelian group. This is clear: we need only form a suitable direct product of cyclic groups of prime power and infinite orders.

8.42 Corollary. *Let G be a non-trivial finitely generated abelian group, and let*

$$G = H_1 \times \ldots \times H_m = K_1 \times \ldots \times K_n,$$

where m, n are positive integers and $H_1, \ldots, H_m, K_1, \ldots, K_n$ are non-trivial indecomposable subgroups of G. Then $m = n$ and, by relabelling the suffices if necessary, $H_i \cong K_i$ for each $i = 1, \ldots, n$.
Proof. For each $i = 1, \ldots, m$, H_i is isomorphic to a quotient group of G, and is therefore finitely generated. Similarly, K_j is finitely generated for each $j = 1, \ldots, n$. Therefore, by 8.26, each H_i and each K_j is cyclic

of prime power or infinite order. Now the result follows, by 8.31 and 8.41.

8.43 Corollary (1.5). *Let* $n = p_1^{m_1} p_2^{m_2} \dots p_s^{m_s}$, *where* $s, m_1, \dots, m_s$ *are positive integers and* $p_1, \dots, p_s$ *distinct primes. Then, if* $v_a(n)$ *denotes the number of distinct types of abelian groups of order* n,

$$v_a(n) = v_a(p_1^{m_1}) v_a(p_2^{m_2}) \dots v_a(p_s^{m_s}),$$

and for each $j = 1, \dots, s$, $v_a(p_j^{m_j})$ *is the number of partitions of* m_j.
Proof. The first assertion follows from 8.6 and 8.39, and the second assertion from 8.40.

8.44. There is an extensive theory of infinite abelian groups. Some remarkable phenomena occur. For instance, A. L. S. Corner [a18] has shown that there is a torsion-free abelian group G for which $G \cong G \times G \times G$ but $G \not\cong G \times G$.

For further information about infinite abelian groups, see Fuchs [b11], Griffith [b15], Kaplansky [b24], Kurosh [b27] vol. 1, part 2, and Rotman [b34] chapters 9 and 10.

472 (i) For each positive integer m, let $\rho(m)$ denote the number of partitions of m. Verify that $\rho(1) = 1, \rho(2) = 2, \rho(3) = 3, \rho(4) = 5, \rho(5) = 7, \rho(6) = 11$ and $\rho(7) = 15$.
(ii) For each positive integer $k \leqslant 12$, find the least positive integer n_k such that $v_a(n_k) = k$.
Show that there is no positive integer n such that $v_a(n) = 13$.

473 (i) Let G be any non-trivial finitely generated abelian group, let n be any integer with $n > 1$, and let G^* be the direct product of n isomorphic copies of G. Then $G \not\cong G^*$.
(ii) Let G be any non-trivial group, let N denote the set of all positive integers, and let $G^* = \mathrm{Dr}\, G^N$, the direct power of G with index set N (**445**). Then G^* is an infinite group, and $G^* \cong G^* \times G^*$.

9

GROUP ACTIONS ON GROUPS

In chapters 4 and 5, we have discussed and applied the idea of a group action on a set. A group can also act on other mathematical systems. When this occurs, we add to the axioms of 4.1 further axioms to ensure that the action respects the structure of the particular system on which the group acts. There is in particular a very highly developed theory of group actions on *vector spaces*: this is usually called *representation theory*. This theory provides powerful tools for proving results about abstract groups: for instance, the theorems of Burnside and Frobenius mentioned in 4.29 and **248** are most easily (or, in the case of the theorem of Frobenius, only) proved by means of representation theory. We shall not in this book discuss representation theory further, but refer for introductory treatments to Lang [b28] chapter 18 and Serre [b37]. The most comprehensive account available is by Curtis and Reiner [b7].

In this chapter, we shall discuss the idea of a group action on a *group*. Note that in the following definition the axioms are those of 4.1 together with one extra axiom of structure preservation. As usual, H and K always denote groups.

9.1 Definition. We say that H *acts on* K (*as a group*) if, to each $h \in H$ and each $k \in K$, there corresponds a unique element $k^h \in K$ such that, for all $k, k_1, k_2 \in K$ and $h, h_1, h_2 \in H$,

$$(k^{h_1})^{h_2} = k^{h_1 h_2}, \qquad k^1 = k,$$

and
$$(k_1 k_2)^h = k_1^h k_2^h.$$

For the theory of group actions on groups, the 'exponential' notation k^h is a convenient one, which, as we shall see, fits in well with the previous notation for conjugates.

9.2 Examples. (i) Let R be a ring with a multiplicative identity element 1. Then the group $R^\times$ acts on the group R^+ (see 2.11 and 2.12) when we define, for each $a \in R^+$ and each $b \in R^\times$,

$$a^b = ab,$$

the product in the ring R. Here the conditions for the action are that for all $a, a_1, a_2, \in R^+$ and $b, b_1, b_2 \in R^\times$,

$$(ab_1)b_2 = a(b_1 b_2), \qquad a1 = a,$$

and $$(a_1 + a_2)b = a_1 b + a_2 b.$$

These are satisfied, by the associative law of multiplication and a distributive law in R, and the defining property of the identity element of R.

(ii) Let $H \leqslant \text{Aut } K$. Then H acts on K. In this case, each $h \in H$ is an automorphism of K and, for $k \in K$, k^h is the image of k under h. The conditions of 9.1 are clearly satisfied. This action is the *natural action* of H on K.

(iii) Let $K \trianglelefteq G$. Then G acts on K (as a group) *by conjugation*: to each $k \in K$ and each $g \in G$ there corresponds the element

$$k^g = g^{-1}kg \in K,$$

in accordance with previous notation (4.25). Here the conditions for the action are that for all $k, k_1, k_2 \in K$ and $g, g_1, g_2 \in G$,

$$(k^{g_1})^{g_2} = k^{g_1 g_2}, \qquad k^1 = k$$

and $$(k_1 k_2)^g = k_1^g k_2^g,$$

which we have already verified in 4.25 and 4.36 (or 2.19). In particular, we see that the action defined in 4.25 is an action of G on itself *as a group*.

On the other hand, the action of G on itself by right multiplication discussed in 4.23 is an action on G as a set but *not* as a group, unless $|G| = 1$.

We now note the analogues of 4.3 and 4.4.

9.3 Theorem. *Let H act on K. Then, to each $h \in H$ there corresponds a map $\varphi_h : K \to K$, defined by $\varphi_h : k \mapsto k^h$, and this is an automorphism of K. Moreover, the map $\varphi : H \to \text{Aut } K$, defined by $\varphi : h \mapsto \varphi_h$, is a homomorphism. We call φ the* automorphism representation *of H corresponding to the action; or, more frequently, for brevity, we simply call φ the* action.
Proof. Let $h \in H$. Since the action of H on K is in particular an action on K as a set, 4.3 shows that $\varphi_h \in \Sigma_K$. Then, for $k_1, k_2 \in K$,

$$(k_1 k_2)\varphi_h = (k_1 k_2)^h = k_1^h k_2^h = (k_1 \varphi_h)(k_2 \varphi_h),$$

and so $\varphi_h \in \text{Aut } K$. Hence the map $\varphi : h \mapsto \varphi_h$ (defined for all $h \in H$) is a map of H into $\text{Aut } K$, and, by 4.3, it is a homomorphism.

9.4 Theorem. *Let φ be a homomorphism of H into $\text{Aut } K$. Then H acts on K when we define, for each $h \in H$ and $k \in K$,*

$$k^h = k(h\varphi),$$

and the corresponding action is φ.

Proof. Since Aut $K \leqslant \Sigma_K$, it is clear from 4.4 that the equation above defines an action of H on K *as a set.* It is an action on K *as a group* because, for $h \in H$ and $k_1, k_2 \in K$,

$$\begin{aligned}
(k_1 k_2)^h &= (k_1 k_2)(h\varphi) \\
&= (k_1(h\varphi))(k_2(h\varphi)) \quad \text{(since } h\varphi \in \text{Aut } K) \\
&= k_1^h k_2^h.
\end{aligned}$$

Finally, it is clear from 4.3, 4.4 and 9.3 that the corresponding action is φ.

9.5. In 4.36 we proved that, whenever $H \leqslant G, C_G(H) \trianglelefteq N_G(H)$ and $N_G(H)/C_G(H)$ can be embedded in Aut H. We see now the true context of this result. Since $H \trianglelefteq N_G(H), N_G(H)$ acts on H *as a group* by conjugation. The action has $C_G(H)$ as its kernel, and so the result of 4.36 follows from 9.3 and the fundamental theorem on homomorphisms.

474 Formulate the appropriate axioms for a group action on a vector space.
  Show that if a group G acts on a vector space $V \neq 0$, there is a corresponding homomorphism $G \to \text{GL}(V)$ (the *linear representation* of G corresponding to the action). Show, conversely, that for each homomorphism $\theta : G \to \text{GL}(V)$ there is an action of G on V with corresponding linear representation θ.

475 Let R be a ring with a multiplicative identity element 1. Then the action of $R^\times$ on R^+ defined in 9.2(i) is faithful.

*****476** Let $K \trianglelefteq G$, and let the action of G on K by conjugation be φ (see 9.2(iii)). Then Ker $\varphi = C_G(K)$.

477 Let H act on K. If $K \neq 1$, then the action is intransitive.

  Suppose that H acts on K, say with action φ. Because H and K are both groups, a construction is available which would not make sense for group actions on general sets. This construction embeds both H and K in a group G in such a way that the action φ is preserved within G. This generalizes the direct product construction (2.31).

9.6 Theorem. *Let H act on K. Then the set of all ordered pairs (h, k) with $h \in H$ and $k \in K$ acquires the structure of a group G when we define, for all $h_1, h_2 \in H$ and $k_1, k_2 \in K$,*

$$(h_1, k_1)(h_2, k_2) = (h_1 h_2, k_1^{h_2} k_2).$$

Proof. Closure is immediate from the definition of multiplication. Let $h_1, h_2, h_3 \in H$ and $k_1, k_2, k_3 \in K$. Then, using the associativity of multiplication in H and in K,

$$\begin{aligned}
((h_1, k_1)(h_2, k_2))(h_3, k_3) &= (h_1 h_2, k_1^{h_2} k_2)(h_3, k_3) \\
&= (h_1 h_2 h_3, (k_1^{h_2} k_2)^{h_3} k_3) \\
&= (h_1 h_2 h_3, k_1^{h_2 h_3} k_2^{h_3} k_3),
\end{aligned}$$

by 9.1, and

$$(h_1,k_1)((h_2,k_2)(h_3,k_3)) = (h_1,k_1)(h_2h_3,k_2^{h_3}k_3)$$
$$= (h_1h_2h_3,k_1^{h_2h_3}k_2^{h_3}k_3).$$

By 9.1, $k^1 = k$ for every $k \in K$. Also by 9.3, for every $h \in H$ the map $k \mapsto k^h$ is an automorphism of K. Therefore $1^h = 1$ and $(k^{-1})^h = (k^h)^{-1}$ (2.9). It follows by the rule of multiplication that $(1,1)$ is the identity element of G and that any $(h,k) \in G$ has an inverse element $(h^{-1},(k^{-1})^{h^{-1}}) \in G$.

9.7 Definition. Before naming the group G of 9.6 we make a notational convention. In forming such a group G from H and K, we shall from now on assume that H and K have only the identity element 1 in common. (This is not a serious restriction, because if H and K have common non-trivial elements, we may replace H or K by an isomorphic copy in which the elements are denoted by new symbols so that only the identity elements of the groups bear the same symbol.) Then in the group G of 9.6 we replace each ordered pair (h, k) by the symbol hk. This convention simplifies notation considerably. It corresponds to the convention which we have already adopted in 8.1 for direct products: see also 9.9. The rule of multiplication in G is then

$$(h_1k_1)(h_2k_2) = (h_1h_2)(k_1^{h_2}k_2)$$

for all $h_1, h_2 \in H$ and $k_1, k_2 \in K$.

Let the action of H on K be φ. Then we call the group G of all juxtaposed symbols hk, with $h \in H$ and $k \in K$ and multiplication given by the equation above, *the semidirect product of K by H with action φ*. We shall denote this group by $H_\varphi \times K$. (Warning. This definition and notation differ from the corresponding ones in Rotman [b34] pp. 135–8. The difference arises from the fact that Rotman makes maps operate on the left of elements. When the appropriate translations are made, the definitions given here are equivalent.)

9.8. For any groups H and K, there is a *trivial action* of H on K: we define, for each $h \in H$ and $k \in K$,

$$k^h = k.$$

The conditions of 9.1 are obviously satisfied. The corresponding automorphism representation of H is the trivial homomorphism $\zeta : H \to \operatorname{Aut} K$, namely

$$\zeta : h \mapsto 1 \quad \text{for all } h \in H.$$

The semidirect product $H_\zeta \times K$ of K by H with trivial action consists of all symbols hk, with $h \in H$ and $k \in K$, and multiplication given by

$$(h_1k_1)(h_2k_2) = (h_1h_2)(k_1k_2),$$

for all $h_1, h_2 \in H$ and $k_1, k_2 \in K$. Clearly this group is just the direct product $H \times K$ (with the typical element (h, k) replaced by hk, as in 8.1).

When we form semidirect products with non-trivial actions, we usually get groups other than direct products. In particular, note that if H and K are abelian groups and H acts on K with non-trivial action, φ say, then the group $H_\varphi \times K$ is non-abelian. This construction process is of considerable importance.

9.9 Theorem. *Let H act on K, say with action φ. Let $G = H_\varphi \times K$. For each $h \in H$, we identify h with the element $h1 \in G$ and, for each $k \in K$, we identify k with the element $1k \in G$ (cf. 8.1). Then $H \leqslant G$, $K \trianglelefteq G$, $G/K \cong H$, $G = HK$ and $H \cap K = 1$. Moreover, the action of H on K is the restriction to H of the action by conjugation of G on K (see 9.2(iii)).*

Proof. First, note that the map

$$h \mapsto h1,$$

defined for all $h \in H$, is an injective homomorphism of H into G. This justifies the identification of h with $h1$. Then also H is identified with the image of the injective homomorphism above, and so $H \leqslant G$. Similarly, by identification of k with $1k$ for all $k \in K$, $K \leqslant G$. By definition of multiplication in G,

$$hk = (h1)(1k).$$

Hence, when h has been identified with $h1$ and k with $1k$, hk becomes the *product* in G of $h \in G$ and $k \in G$ (not just a juxtaposition of two unrelated symbols). Thus $G = HK$. Moreover, $H \cap K = 1$, since the only element of G which has both the forms $h1$ and $1k$ is 11, which we now denote by 1. (Remark. It was in order to ensure that the identifications made here would be unambiguous that we demanded in 9.7 that groups H and K should have only the element 1 in common.)

Now consider the map $\quad \psi : G \to H,$

defined by $\quad \psi : hk \mapsto h$

for all $h \in H$ and $k \in K$. By the definition of multiplication in G given in 9.7, ψ is a surjective homomorphism and $\operatorname{Ker} \psi = K$. Hence, by the fundamental theorem on homomorphisms, $K \trianglelefteq G$ and $G/K \cong H$. Finally, let $h \in H$ and $k \in K$. Then in G,

$$h^{-1}kh = h^{-1}hk^h = k^h.$$

Thus k^h is now the conjugate in G of k by h, and so the original action of H on K is the same as the action defined by restriction to H of the action by conjugation of G on K.

9.10. We now point out some important examples.

(i) Let K be any group, and consider the natural action of Aut K on K. Then the action is the identity map ε on Aut K: for every $\alpha \in$ Aut K,

$$\varepsilon : \alpha \mapsto \alpha.$$

This defines a semidirect product

$$(\text{Aut } K)_\varepsilon \times K,$$

which is called *the holomorph* of K and denoted by Hol K.

By 9.9, $K \trianglelefteq$ Hol K and, for every $\alpha \in$ Aut K and $k \in K$, $\alpha^{-1}k\alpha = k^\alpha$ (where the product on the left is defined in Hol K). Thus every automorphism of K is obtained by restriction from an *inner* automorphism of Hol K.

(ii) More generally, let $H \leqslant$ Aut K and consider the natural action of H on K. Then the action is the inclusion map $\iota : H \to$ Aut K. The corresponding semidirect product

$$H_\iota \times K$$

is said to be *a relative holomorph* of K. Clearly, by definition,

$$K \leqslant H_\iota \times K = HK \leqslant \text{Hol } K.$$

(iii) Let A be an abelian group such that $b^2 \neq 1$ for some $b \in A$. Let

$$\eta : A \to A$$

be the map $\eta : a \mapsto a^{-1}$ for every $a \in A$. Then $\eta \in$ Aut A, $\eta^2 = 1$ and $\eta \neq 1$ since $b^{-1} \neq b$.

The relative holomorph $\langle \eta \rangle A$ of A is a non-abelian group, called a *generalized dihedral* group, and denoted by Dih A. Then $|\text{Dih } A : A| = 2$. It is easy to see that for each integer $n \geqslant 3$, Dih $C_n \cong D_{2n}$, the dihedral group of order $2n$, and Dih $C_\infty \cong D_\infty$, the infinite dihedral group (57).

478 Let H act on K, say with action φ, and let $G = H_\varphi \times K$.
(i) For each $J \leqslant H$, define

$$C_K(J) = \{k \in K : k^j = k \text{ for all } j \in J\}$$

and, for each $L \leqslant K$, define

$$C_H(L) = \{h \in H : l^h = l \text{ for all } l \in L\}.$$

Then $C_K(J) = K \cap C_G(J)$ and $C_H(L) = H \cap C_G(L)$ (where $C_G(J), C_G(L)$ are the usual centralizers in G). Note that $C_K(H) = \text{Fix}_K(H)$ and $C_H(K) = \text{Ker } \varphi$; in particular, $\text{Fix}_K(H) \leqslant K$.
(ii) $\text{Ker } \varphi \trianglelefteq G$, and in fact $\text{Ker } \varphi = H_G$, the core of H in G.
(iii) For each $J \leqslant H$, $N_G(J) = N_H(J)C_K(J)$.
(iv) For each $J \leqslant H$, $N_G(J) \leqslant N_G(C_K(J))$.

479 Suppose that H is a finite p-group which acts on the finite group K. If p divides $|K|$ then $|\text{Fix}_K(H)| > 1$.

480 Let the non-trivial group H act on the non-trivial group K, say with action φ, and let $G = H_\varphi \times K$. The following two statements are equivalent:

(i) For every non-trivial element h of H, $h\varphi$ is a fixed-point-free automorphism of K (see **54**).

(ii) G is a Frobenius group and H is a Frobenius complement in G (see **248**).

481 Let p and q be primes such that $p > q$.

(i) The only action of a group of order p on a group of order q is the trivial action.

(ii) The only action of a group of order p on a group of order q^2 is the trivial action, unless $p = 3$ and $q = 2$. (Hint. Use 2.16, 2.17, 2.36, **40**, **46**, **47**, **222**.)

482 (i) $\mathrm{Aut}(C_2 \times C_2) \cong \Sigma_3$.

(ii) $\mathrm{Hol}(C_2 \times C_2) \cong \Sigma_4$.

(Hint. Let $K = C_2 \times C_2$. For (ii), consider the action of $\mathrm{Hol}\,K$ by right multiplication on the set of right cosets of $\mathrm{Aut}\,K$ in $\mathrm{Hol}\,K$, and use 3.53.)

*483 If $K_1 \cong K_2$ then $\mathrm{Hol}\,K_1 \cong \mathrm{Hol}\,K_2$, and every relative holomorph of K_1 is isomorphic to a relative holomorph of K_2 (cf. **48**).

484 Let V be a vector space $\neq 0$.

(i) Then $\mathrm{GL}(V)$ acts naturally on the additive group V^+ of V (see **47**). Let G denote the corresponding relative holomorph $\mathrm{GL}(V)V^+$ of V^+.

(ii) For each linear map $\lambda : V \to V$ and each vector $v \in V$, let $(\lambda; v)$ be the map $V \to V$ defined, for all $x \in V$. by

$$(\lambda; v) : x \mapsto x\lambda + v.$$

Then $(\lambda; v) \in \Sigma_V$ if and only if $\lambda \in \mathrm{GL}(V)$. Moreover, the set $\{(\lambda; v) : \lambda \in \mathrm{GL}(V), v \in V\}$ is a subgroup $\mathscr{A}(V)$ of Σ_V, called the *affine group* of V, and $\mathscr{A}(V) \cong G$.

485 (i) Let n be an integer, $n \geqslant 3$. Then

$$\mathrm{Dih}\,C_n \cong D_{2n}.$$

(ii) $\mathrm{Dih}\,C_\infty \cong D_\infty$.

*486 Let A be an abelian group and let $L = \mathrm{Hol}\,A$. Then

(i) $C_L(A) = A$.

(ii) If $H \leqslant \mathrm{Aut}\,A$ and G is the relative holomorph HA of A then $Z(G) = \mathrm{Fix}_A(H)$. In particular, if $H \neq 1$ then G is non-abelian.

487 Let A be a cyclic group and let $G = \mathrm{Hol}\,A$.

(i) Then G is supersoluble. (See **389**. Hint. See 4.38. If $|A| = \infty$, use **46**.)

(ii) If either A has odd finite order or A is infinite then $Z(G) = 1$. (Hint. Use **486**.)

(iii) G is nilpotent if and only if $|A| = 2^n$ for some non-negative integer n. (Hints. See **243**. If $|A|$ is finite but not a power of 2, find a subgroup of G which is not nilpotent.)

488 Let A be an abelian group with an element b such that $b^2 \neq 1$, and let $D = \mathrm{Dih}\,A$.

(i) Every element of $D \backslash A$ has order 2. (cf. **59**, **142**. For a converse result see **504**).

(ii) $Z(D) = \{a \in A : a^2 = 1\}$.

(iii) A is a characteristic subgroup of D.

(iv) $\mathrm{Aut}\,D \cong \mathrm{Hol}\,A$.

489 (i) If H_1 and H_2 are conjugate subgroups of $\mathrm{Aut}\,K$, then the relative holomorphs $H_1 K$ and $H_2 K$ are conjugate subgroups of $\mathrm{Hol}\,K$.

(ii) $\mathrm{Aut}\,K$ can have isomorphic subgroups H_1 and H_2 such that the relative holomorphs $H_1 K$ and $H_2 K$ are not isomorphic. (Hint. Consider $K = C_3 \times C_3$ and suitable subgroups H_1, H_2 of $\mathrm{Aut}\,K$ of order 2.)

490 Let $\mathbf{R}$ be identified with the euclidean line E^1 in the usual way: see **56**. Then Isom $\mathbf{R} \cong \mathrm{Dih}\,\mathbf{R}^+$.

9.11 Definition. Let $K \trianglelefteq G$. We say that G *splits over* K if there is a subgroup H of G such that $G = HK$ and $H \cap K = 1$. Any such subgroup H is said to be a *complement to K in G*.

Note that a subgroup H of G is a complement to K in G if and only if every element of G is *uniquely* expressible in the form hk with $h \in H$, $k \in K$. Then also, by 3.40, $G/K = HK/K \cong H/(H \cap K) \cong H$. Moreover, for every $g \in G$, H^g is a complement to K in G: for

$$G = G^g = H^g K^g = H^g K,$$

since $K \trianglelefteq G$, and $\quad H^g \cap K = H^g \cap K^g = (H \cap K)^g = 1$.

9.12 Lemma. *Let $K \trianglelefteq G$ and let $K \leqslant J \leqslant G$. If G splits over K then J splits over K.*
Proof. Certainly $K \trianglelefteq J$. Let H be a complement to K in G. Then $K \leqslant J \leqslant G = HK$, so that, by Dedekind's rule (7.3),

$$J = (H \cap J)K.$$

Moreover, $\qquad (H \cap J) \cap K = H \cap K = 1$.

Therefore $H \cap J$ is a complement to K in J.

In chapter 10 we shall establish some important sufficient conditions for splitting. We show now that there is an intimate connexion between splitting and semidirect products.

9.13 Theorem. (i) *Suppose that H acts on K, say with action φ, and let $G = H_\varphi \times K$. Then G splits over K, and H is a complement to K in G.*

(ii) *Let $K \trianglelefteq G$. Suppose that G splits over K, and let H be a complement to K in G. Let φ be the action of H on K defined by restriction of the action of G on K by conjugation. Then $G = H_\varphi \times K$.*
Proof. (i) This is immediate from 9.9.

(ii) Let $K \trianglelefteq G$, and suppose that H is a complement to K in G. Let $\varphi : H \to \mathrm{Aut}\,K$ be the action of H on K defined by restriction of the action of G on K by conjugation: thus, for every $h \in H$,

$$h\varphi : k \mapsto k^h = h^{-1}kh$$

for every $k \in K$. Every element of G is uniquely expressible in the form hk, with $h \in H$ and $k \in K$. Moreover, multiplication in G is given by the rule

$$(h_1 k_1)(h_2 k_2) = h_1 h_2 h_2^{-1} k_1 h_2 k_2 = h_1 h_2 k_1^{h_2} k_2$$

for all $h_1, h_2 \in H$ and $k_1, k_2 \in K$. The elements of G are thus the same as the elements of $H_\varphi \times K$ and the rule of multiplication is the same. Hence $G = H_\varphi \times K$.

9.14 Lemma. *Let H act on K, say with action φ, and let $J = \operatorname{Im} \varphi \leqslant \operatorname{Aut} K$. If the action is faithful then the group $H_\varphi \times K$ is isomorphic to the relative holomorph JK of K.*

Proof. Suppose that the given action is faithful; thus the homomorphism $\varphi : H \to \operatorname{Aut} K$ is injective. Let $G = H_\varphi \times K$, and define a map

$$\varphi^* : G \to \operatorname{Hol} K$$

by
$$\varphi^* : hk \mapsto (h\varphi)k \in \operatorname{Hol} K$$

for every $h \in H$ and $k \in K$. (This map is well defined since every element of G is uniquely expressible in the form hk with $h \in H, k \in K$.) For all $h_1, h_2 \in H$ and $k_1, k_2 \in K$,

$$\begin{aligned}
((h_1 k_1)(h_2 k_2))\varphi^* &= (h_1 h_2 k_1^{h_2 \varphi} k_2)\varphi^* \quad (\text{since } k_1^{h_2} = k_1^{h_2 \varphi}) \\
&= (h_1 h_2)\varphi k_1^{h_2 \varphi} k_2 \\
&= (h_1 \varphi)(h_2 \varphi)(h_2 \varphi)^{-1} k_1 (h_2 \varphi) k_2 \quad (\text{in Hol } K) \\
&= (h_1 \varphi) k_1 (h_2 \varphi) k_2 \\
&= ((h_1 k_1)\varphi^*)((h_2 k_2)\varphi^*).
\end{aligned}$$

Thus φ^* is a homomorphism. Moreover,

$$\begin{aligned}
\operatorname{Ker} \varphi^* &= \{hk : h \in \operatorname{Ker} \varphi, k = 1\} \\
&= 1 \quad (\text{since } \varphi \text{ is injective}).
\end{aligned}$$

Therefore φ^* is injective, and so

$$G \cong \operatorname{Im} \varphi^* = JK \leqslant \operatorname{Hol} K.$$

491 Let n be a positive integer and F a field. Then
(i) $\operatorname{GL}_n(F)$ splits over $\operatorname{SL}_n(F)$.
(ii) If $n > 1, \Sigma_n$ splits over A_n.

492 Let n be a positive integer and s a divisor of n. Then the cyclic group C_n splits over C_s if and only if $(s, n/s) = 1$.

493 Let $1 < K < G = C_\infty$. Then G does not split over K.

494 Let n be a positive integer and let V be a vector space of dimension n over the field $\mathbf{Z}_p$. Then V^+ splits over every subgroup.

495 Let $X = E^2$, the euclidean plane, $G = \operatorname{Isom} X, T = \operatorname{Tr} X$ and $H = T \cup \bigcup_{s \in X} \operatorname{Rot}(X; s)$: see 2.23. Then
(i) G splits over H.
(ii) G splits over T, and the complements to T in G are isomorphic to $\operatorname{Dih} U$, where U is the circle group (see 2.32(iv), **61**, **103**, **112**).

496 Let $K \leqslant Z(G)$ (so that, in particular, $K \trianglelefteq G$). If G splits over K, and H is a complement to K in G, then $G = H \times K$.

497 Let P be a finite non-abelian p-group. Then P does not split over $Z(P)$. (Hint. Apply **406** and **496**.)

498 Let $K \leqslant G \leqslant \operatorname{Hol} K$. Then G is a relative holomorph of K.

499 Suppose that G splits over an abelian normal subgroup A, and let H be a complement to A in G. Then $C_G(A) = H_G \times A$ (cf. **478**(ii), **486**).

500 Let J and K be normal subgroups of G with $K \leqslant J$.

(i) If G splits over J then G/K splits over J/K. (Hint. Apply 7.3.)

(ii) Suppose that G splits over K and let H be a complement to K in G. Then G splits over J if and only if H splits over $H \cap J$. (Hint. Apply 7.3 again.)

501 Let K and L be normal subgroups of G, and let $J = KL \trianglelefteq G$. Suppose that G/L splits over J/L, and let H/L be a complement to J/L in G/L. Suppose also that H splits over $H \cap K$. Then G splits over K.

502 Let $K \trianglelefteq G$, a finite group. Suppose that G/K is a p-group and let P be a Sylow p-subgroup of G. Then G splits over K if and only if P splits over $P \cap K$.

503 Let G be a non-trivial finite soluble group, and suppose that $|G|$ is neither a prime nor the square of a prime. Prove that G has a proper normal subgroup K such that $|G| < |K|^2$. (See the remarks at the end of chapter 6; see also **664**. Hints. Assume the result false, and suppose that G is a group of least possible order which violates the result. Let L be a minimal normal subgroup of G. Show that $|G/L| = p$, for some prime p, and that $|L| < p$. Let P be a Sylow p-subgroup of G. Apply 9.13 to show that $G = P_\varphi \times L$ for some action φ of P on L. If φ is non-trivial, use 7.56, 7.40, **47**, 2.17 and 2.16 to derive a contradiction.)

504 Let G be a group with a subgroup K of index 2 such that every element of $G \backslash K$ has order 2. Then K is abelian, and if $k^2 \neq 1$ for some $k \in K$, $G \cong \text{Dih } K$. (This is a converse to **488**(i). Hint. Use 9.13 and **52**(i).)

505 (i) Suppose that $J \leqslant \bar{G}$, $K \trianglelefteq \bar{G}$ and $\bar{G} = JK$ (where possibly $J \cap K \neq 1$). Let H be a group isomorphic to J such that $H \cap K = 1$, and let θ be an isomorphism of H onto J. Then H acts on K when we define, for every $h \in H$ and $k \in K$,

$$k^h = k^{h\theta},$$

the conjugate in $\bar{G}$ of k by $h\theta$.

Let this action be $\varphi : H \to \text{Aut } K$, and let $G = H_\varphi \times K$. Then the map

$$\theta^* : G \to \bar{G},$$

defined by
$$\theta^* : hk \mapsto (h\theta)k$$

for every $h \in H$ and $k \in K$, is a surjective homomorphism, $H \cap \text{Ker } \theta^* = 1 = K \cap \text{Ker } \theta^*$, and $\text{Ker } \theta^* \cong J \cap K$.

(ii) Let H act on K, say with action φ, and let $G = H_\varphi \times K$. If $L \trianglelefteq G$, with $H \cap L = 1 = K \cap L$, then there are subgroups S of H and T of K and an isomorphism θ of S onto T such that $S \trianglelefteq H$; for all $s \in S$, $h \in H$ and $k \in K$, $s^h\theta = (s\theta)^h$ and $k^s = k^{s\theta}$; and $L = \{s(s^{-1}\theta) : s \in S\} \cong S$.

Conversely, if $S \trianglelefteq H$, $T \leqslant K$ and there is an isomorphism θ of S onto T such that for all $s \in S$, $h \in H$ and $k \in K$, $s^h\theta = (s\theta)^h$ and $k^s = k^{s\theta}$, then, if $L = \{s(s^{-1}\theta) : s \in S\}$, $S \cong L \trianglelefteq G$ and $H \cap L = 1 = K \cap L$.

(iii) Let H act on K, with action φ, and let $G = H_\varphi \times K$. Suppose that $L \trianglelefteq G$, with $H \cap L = 1 = K \cap L$, and let $\bar{G} = G/L$, $\bar{H} = HL/L$ and $\bar{K} = KL/L$. Then $H \cong \bar{H} \leqslant \bar{G}$, $K \cong \bar{K} \trianglelefteq \bar{G}$, $\bar{G} = \bar{H}\bar{K}$ and $\bar{H} \cap \bar{K} \cong L$.

506 Let $H = \langle h \rangle$ be a cyclic group of order 4 and $K = \langle k \rangle$ a cyclic group of order $2n$, where n is an integer greater than 1.

(i) There is a unique action φ of H on K for which $k^h = k^{-1}$.
Let $G = H_\varphi \times K$.

(ii) Then $Z(G) = \langle h^2 \rangle \times \langle k^n \rangle \cong C_2 \times C_2$, and $G/Z(G) \cong D_{2n}$ for $n \geqslant 3$, while for $n = 2$, $G/Z(G) \cong C_2 \times C_2$.

(iii) Let $L = \langle h^2 k^n \rangle \trianglelefteq G$, and let $\bar{G} = G/L$, $\bar{H} = HL/L$, $\bar{K} = KL/L$. Then

$C_4 \cong \bar{H} \leqslant \bar{G}, C_{2n} \cong \bar{K} \trianglelefteq \bar{G}, \bar{G} = \bar{H}\bar{K}$ and $|\bar{H} \cap \bar{K}| = 2$. The group $\bar{G}$ is called *the dicyclic group of order* $4n$.

(iv) Let $\overline{Z(G)} = Z(G)/L$. Then $Z(\bar{G}) = \overline{Z(G)}$, of order 2 (cf. **150**). Hence $\bar{G}/Z(\bar{G}) \cong G/Z(G)$.

(v) $\bar{G}$ has just one element of order 2, and $\bar{G}$ does not split over $\bar{K}$.

(vi) $\bar{G} \not\cong D_{4n}$.

(vii) If $n = 2$ then $\bar{G} \cong Q_8$ (see **181**).

507 Let H act on K with action φ. Let $G = H_\varphi \ltimes K, J = \operatorname{Im} \varphi \leqslant \operatorname{Aut} K$ and $L = \operatorname{Ker} \varphi \trianglelefteq H$. Then

(i) $L \trianglelefteq G$ and G/L is isomorphic to the relative holomorph JK of K. (This generalizes 9.14. We know from **478** that $L \trianglelefteq G$.)

(ii) For each $\alpha \in \operatorname{Aut} H, H$ also acts on K with action $\alpha\varphi$. Then $H_{\alpha\varphi} \ltimes K \cong G$ and $\operatorname{Ker}(\alpha\varphi) = L^{\alpha^{-1}} = \{l^{\alpha^{-1}} : l \in L\}$.

508 (i) Let $K \trianglelefteq G$, and let ν be the natural homomorphism of G onto G/K. Then G splits over K if and only if there is a homomorphism $\theta : G/K \to G$ such that $\theta\nu$ is the identity map on G/K.

(ii) Suppose that $K \trianglelefteq G$, with G/K infinite cyclic. Then G splits over K (cf. **467**).

(iii) Use (ii), together with **133**(i) and (ii), to give another proof that $\mathbf{Q}^+$ has no non-trivial cyclic quotient group.

*509 (i) Let $K \trianglelefteq G$, and let φ be the action of G on K by conjugation. Let $g \in G$. Then $g\varphi \in \operatorname{Inn} K$ if and only if $g \in C_G(K)K$.

(ii) (Hölder[a60]) A group K is said to be *complete* if $Z(K) = 1$ and $\operatorname{Aut} K = \operatorname{Inn} K$. If $K \trianglelefteq G$ and K is complete then $G = C_G(K) \times K$. Thus a complete group K is a direct factor of every extension of K.

(iii) Σ_3 is complete. (Remark. In fact, Σ_n is complete for every integer $n \geqslant 3$ with $n \neq 6$. For a proof of this result, see Kurosh [b27] vol. 1, pp. 92–5 or Rotman [b34] pp. 132–4. Σ_6 is not complete.)

We shall illustrate these results by applying them to obtain more information about groups of order pq, where p and q are distinct primes: cf. 5.16, 5.17, 5.18. In order to do this, we need some further information about the automorphism group of a group of order p (cf. 4.38).

Recall that a cyclic group G of finite order n has a unique subgroup G_s of order s for each divisor s of n (3.32); moreover, G_s is cyclic and it follows that $G_s = \{x \in G : x^s = 1\}$ (**139**). We prove a converse result and use this to show that the automorphism group of a group of order p is cyclic.

9.15 Lemma. (i) *Let G be a group of finite order n such that, for every divisor s of n, $|\{x \in G : x^s = 1\}| \leqslant s$. Then G is cyclic* (cf. **139**(iii)).

(ii) *Let F be any field. Then every finite subgroup of $F^\times$ is cyclic.*

(iii) *If $|G| = p$ then $\operatorname{Aut} G$ is cyclic of order $p - 1$.*

Proof. (i) Let $g \in G, o(g) = s$, and $H = \langle g \rangle \leqslant G$. Then $|H| = s$, s is a divisor of n, and $h^s = 1$ for every $h \in H$. Hence, by hypothesis, $H = \{x \in G : x^s = 1\}$, and so every element of order s in G lies in H. Let G^* be a cyclic group of order n, and let H^* be the unique subgroup of G^* of order s (3.32). Since H and H^* are cyclic groups of the same order,

$H \cong H^*$ (2). Therefore, since every element of order s in G lies in H and every element of order s in G^* lies in H^*, it follows that G and G^* have exactly the same number of elements of order s. This is true for every divisor s of n for which G has an element of order s. Hence, since $|G| = |G^*|$ and G^* has an element of order n, G must have an element of order n, and so G is cyclic.

(ii) Let $G \leqslant F^\times$ with $|G| = n < \infty$. It is well known that for any positive integer s and any polynomial $f(x)$ of degree s with coefficients in a field, the equation $f(x) = 0$ has at most s roots in the field. In particular, there are at most s distinct elements x of F which satisfy the equation $x^s - 1 = 0$. Hence also, for every divisor s of n, $|\{x \in G : x^s = 1\}| \leqslant s$. Therefore, by (i), G is cyclic.

(iii) Since $|G| = p$,

$$G \cong \mathbf{Z}_p^+ \quad \textbf{(40)}.$$

Hence
$$\mathrm{Aut}\, G \cong \mathbf{Z}_p^\times \quad \textbf{(46)}.$$

Since p is prime, $\mathbf{Z}_p$ is a field. Therefore, by (ii), Aut G is cyclic.

We now prove

9.16 Theorem *Let p and q be primes such that $p > q$. If $p \not\equiv 1$ mod q then $v(pq) = 1$, while if $p \equiv 1$ mod q then $v(pq) = 2$.* (This includes the result of 5.18. Recall that for any positive integer n, $v(n)$ denotes the number of types of groups of order n.)

Proof. Since $p > q$, $q \not\equiv 1$ mod p. Let G be a group of order pq, P a Sylow p-subgroup of G, and Q a Sylow q-subgroup of G. Then $P \cong C_p$ and $Q \cong C_q$. By 5.16, $P \lhd G$. Moreover, $PQ = G$ and $P \cap Q = 1$. Thus G splits over P, and Q is a complement to P in G. Hence, by 9.13, $G = Q_\varphi \times P$, where $\varphi : Q \to \mathrm{Aut}\, P$ is defined by restriction of the action of G on P by conjugation. If φ is trivial then, by 9.8, $G = Q \times P \cong C_q \times C_p \cong C_{pq}$ **(78)**. Suppose that φ is non-trivial. Since Q has prime order, it follows that Ker $\varphi = 1$: that is, the action is faithful. Let $J = \mathrm{Im}\, \varphi \leqslant \mathrm{Aut}\, P$. Then $|J| = q$ and, since $|\mathrm{Aut}\, P| = p - 1$, it follows that $p \equiv 1$ mod q. Moreover, by 9.14, $G \cong JP \leqslant \mathrm{Hol}\, P$. By 9.15, Aut P is cyclic and therefore (3.32) J is the *unique* subgroup of Aut P of order q. Therefore JP is the unique relative holomorph of P of order pq. Thus we have proved that if $p \not\equiv 1$ mod q then $G \cong C_{pq}$ and so $v(pq) = 1$; while if $p \equiv 1$ mod q then G is isomorphic to either C_{pq} or the unique relative holomorph of C_p of order pq **(483)**; hence, since these two groups are non-isomorphic **(486)**, $v(pq) = 2$.

We shall establish next an interesting alternative characterization of the holomorph of a group.

9.17 Theorem. *Let K be any group and let $\rho^1 : K \to \Sigma_K$ be the right regular permutation representation of K (4.23). Let $K^* = \mathrm{Im}\, \rho^1 \leqslant \Sigma_K$. Then* Aut $K \leqslant \Sigma_K$, (Aut $K) \cap K^* = 1$ *and*

$$N_{\Sigma_K}(K^*) = (\text{Aut } K)K^* \cong \text{Hol } K.$$

Proof. For each $k \in K$, let $k^* = k\rho^1 \in \Sigma_K$: then, for all $x \in K$,

$$xk^* = xk.$$

Certainly Aut $K \leqslant \Sigma_K$ (2.18). Let $k \in K$. Then $k = 1k^*$: and if $k^* \in$ Aut K then $1k^* = 1$, hence $k = 1$. Thus (Aut $K) \cap K^* = 1$.

Let $\alpha \in$ Aut K. Then, for all $k, x \in K$,

$$\alpha^{-1}k^*\alpha : x \mapsto (x^{\alpha^{-1}}k)^{\alpha} = xk^{\alpha}.$$

Hence

$$\alpha^{-1}k^*\alpha = (k^{\alpha})^* \in K^*. \tag{i}$$

Thus $$\text{Aut } K \leqslant N_{\Sigma_K}(K^*) = L, \text{ say.}$$

Now let $\sigma \in L$. We want to show that $\sigma \in (\text{Aut } K)K^*$. Suppose that σ maps $1 \in K$ to $t \in K$. Let

$$\tau = \sigma(t^{-1})^* \in L.$$

Then τ fixes $1 \in K$. Moreover, since $K^* \trianglelefteq L$, the map

$$k^* \mapsto \tau^{-1}k^*\tau$$

(defined for all $k^* \in K^*$) is an automorphism of K^*. Since the map $k \mapsto k^*$ is an isomorphism of K onto K^*, it is clear that the map Aut $K \to$ Aut K^* defined by $\alpha \mapsto \alpha^*$, where, for each $\alpha \in$ Aut K,

$$\alpha^* : k^* \mapsto (k^{\alpha})^*$$

(for all $k \in K$) is an isomorphism of Aut K onto Aut K^* (cf. **48**). Hence there is a unique $\alpha \in$ Aut K such that, for all $k \in K$,

$$\tau^{-1}k^*\tau = (k^{\alpha})^*.$$

Now, for all $x \in K$,

$$x\tau = 1x^*\tau = 1\tau(x^{\alpha})^* = 1(x^{\alpha})^* = x^{\alpha}.$$

Hence $$\tau = \alpha.$$

Thus $$\sigma = \alpha t^* \in (\text{Aut } K)K^*.$$

Therefore $$L = (\text{Aut } K)K^*.$$

Finally, define a map

$$\theta : \text{Hol } K \to \Sigma_K$$

by $\theta : \alpha k \mapsto \alpha k^*$, for every $\alpha \in$ Aut K and $k \in K$.

Then, for all $\beta, \gamma \in$ Aut K and $x, y \in K$,

$$(\beta x \gamma y)\theta = (\beta \gamma x^\gamma y)\theta$$
$$= \beta \gamma (x^\gamma y)^*$$
$$= \beta \gamma (x^\gamma)^* y^*$$
$$= \beta \gamma \gamma^{-1} x^* \gamma y^* \quad \text{(by (i))}$$
$$= \beta x^* \gamma y^*$$
$$= (\beta x)\theta \cdot (\gamma y)\theta.$$

Thus θ is a homomorphism. Moreover, if $\alpha k \in \mathrm{Ker}\ \theta$ then, for all $x \in K$,

$$x^\alpha k = x,$$

hence (choosing $x = 1$) $k = 1$, and so also $\alpha = 1$. Thus $\mathrm{Ker}\ \theta = 1$, so that θ is injective. Hence $\mathrm{Hol}\ K \cong \mathrm{Im}\ \theta = (\mathrm{Aut}\ K)K^*$.

510 (i) Let K and L be normal subgroups of G such that $K \cap L = 1$. If G/L splits over KL/L then G splits over K.

(ii) Suppose that $Z(K) = 1$. Then every extension of K splits over K if and only if $\mathrm{Aut}\ K$ splits over $\mathrm{Inn}\ K$ (see 9.26).

(iii) Let p be an odd prime. Then the dihedral group D_{2p} of order $2p$ is complete (**509**) if and only if $p = 3$. Every extension of D_{2p} splits over D_{2p} if and only if $p \not\equiv 1 \bmod 4$. (Hints. Use **117, 124,** 9.12, 9.15, **476, 488, 492, 500, 509**.)

511 Use **452** and the fact that $\mathbf{Z}_p$ is a field to prove that $\mathbf{Z}_p^\times$ is cyclic. (Remark. This gives an alternative method of proving 9.15(iii).)

512 Let A be a group of prime order. Then $\mathrm{Hol}\ A$ is metacyclic (see **152**; cf. **487**).

513 No two of the following groups of order 30 are isomorphic and every group of order 30 is isomorphic to one of them: $C_{30}, C_5 \times D_6, C_3 \times D_{10}, D_{30}$. Hence $\nu(30) = 4$. (Hint. Let G be a group of order 30. Show that G has a cyclic normal subgroup K of order 15 and that G splits over K. Use **78** and **94**.)

514 (i) Find five groups of order 12 no two of which are isomorphic. (Hint. See **506**.)

(ii) Show that every group of order 12 is isomorphic to one of these groups, and hence that $\nu(12) = 5$. (Hints. Any group G of order 12 has a normal subgroup of order either 3 or 4 over which G splits. Use **489**(i) and **507**(ii).)

515 (i) Let G be a non-cyclic group of order 8 with a cyclic subgroup K of order 4. Show that if G splits over K then G is isomorphic to either $C_2 \times C_4$ or D_8, while if G does not split over K then $G \cong Q_8$. (Hint. See **505**(i) and **506**.)

(ii) Hence show that $\nu(8) = 5$.

516 (i) Let A be a cyclic group of order 8, say $A = \langle a \rangle$. Show that there is a unique automorphism α of A such that $a^\alpha = a^3$, and that $o(\alpha) = 2$.

Let T be the relative holomorph $\langle \alpha \rangle A$ of A: then T is called the *semidihedral group of order* 16. Show that $T' = \langle a^2 \rangle$, of order 4. Deduce that T has just three subgroups of index 2, and prove that of these three subgroups, one is isomorphic to C_8, another to D_8, and the third to Q_8. Deduce that each of these three subgroups of index 2 is characteristic in T, and that $T \not\cong D_{16}$. (Hint. See **59** and **515**.)

(ii) Let $G = \mathrm{GL}_2(\mathbf{Z}_3)$: then $|G| = 48$ (see 2.16 and 2.17). Let $x, y \in G$ be defined as

$$x = \begin{pmatrix} 0 & 1 \\ 1 & -1 \end{pmatrix} \quad \text{and} \quad y = \begin{pmatrix} 1 & 0 \\ -1 & -1 \end{pmatrix}.$$

Show that $o(x) = 8, o(y) = 2$ and $x^y = x^3$. Hence prove that the Sylow 2-subgroups of G are isomorphic to T. (cf. **193**. Hint. Apply 9.14.)

517 Let p be a prime, $p \geqslant 5$. If $p \not\equiv 1 \bmod 4$, $\nu(4p) = 4$, while if $p \equiv 1 \bmod 4$, $\nu(4p) = 5$. (Hint. Use 5.19 and **481**; cf. **514**.)

518 Let K be any group and $\rho^1 : K \to \Sigma_K$ the right regular permutation representation of K. Let $K^* = \operatorname{Im} \rho^1 \leqslant \Sigma_K$. For each $k \in K$, let $k_* : K \to K$ be defined by $k_* : x \mapsto kx$, for all $x \in K$, and let $K_* = \{k_* : k \in K\}$. Then $k_* \in \Sigma_K$, $K_* \leqslant \Sigma_K$, $C_{\Sigma_K}(K^*) = K_* \cong K$, $K^* \cap K_* \cong Z(K)$, $(\operatorname{Aut} K) \cap K_* = 1$,
$$N_{\Sigma_K}(K_*) = (\operatorname{Aut} K)K_* = (\operatorname{Aut} K)K^* = N_{\Sigma_K}(K^*) \qquad C_{\Sigma_K}(K_*) = K^*$$
and
$$N_{\Sigma_K}(K^*)/C_{\Sigma_K}(K^*) \cong \operatorname{Aut} K \cong N_{\Sigma_K}(K_*)/C_{\Sigma_K}(K_*).$$

519 (i) Suppose that θ is an isomorphism of G_1 onto G_2, $K_1 \leqslant G_1$ and θ maps K_1 to $K_2 \leqslant G_2$. Then θ maps $C_{G_1}(K_1)$ to $C_{G_2}(K_2)$.

(ii) For any group $K, C_{\operatorname{Hol} K}(K) \cong K$ (cf. **486**).

520 Let K be any group, $\rho^1 : K \to \Sigma_K$ the right regular permutation representation of K, and $K^* = \operatorname{Im} \rho^1 \leqslant \Sigma_K$. Then $N_{\Sigma_K}(K^*)$ consists of all $\sigma \in \Sigma_K$ such that, for all $x, y, z \in K$,
$$(xy^{-1}z)\sigma = (x\sigma)(y\sigma)^{-1}(z\sigma).$$

We shall now associate to any group G and any action of a group H on a finite *set* X an action of H on the *group* $\operatorname{Dr} G^X$ (see 8.21). This leads to a useful construction of groups known as *wreath products*.

9.18 Lemma. *Suppose that H acts on the finite set X. Let G be any group and let $G^* = \operatorname{Dr} G^X$. Then H acts on G^* (as a group) when, for each $h \in H$ and each $f \in G^*$, we define $f^h \in G^*$, for all $x \in X$, by*
$$f^h(x) = f(xh^{-1}).$$

(*Note*. Here, as in 8.21, the elements of G^* are maps of X into G which are written on the *left* of the elements of X to which they apply.)
Proof. Let $f, f_1, f_2 \in G^*, h, h_1, h_2 \in H$ and $x \in X$. Then, using the axioms of 4.1, and the definition of multiplication in G^*, we find that
$$\begin{aligned}(f^{h_1})^{h_2}(x) &= f^{h_1}(xh_2^{-1}) = f((xh_2^{-1})h_1^{-1}) \\ &= f(x(h_1h_2)^{-1}) = f^{h_1h_2}(x),\end{aligned}$$
$$f^1(x) = f(x1^{-1}) = f(x),$$
and
$$\begin{aligned}(f_1f_2)^h(x) &= (f_1f_2)(xh^{-1}) = f_1(xh^{-1})f_2(xh^{-1}) \\ &= f_1^h(x)f_2^h(x) = (f_1^hf_2^h)(x).\end{aligned}$$
Thus
$$(f^{h_1})^{h_2} = f^{h_1h_2},$$
$$f^1 = f,$$
and
$$(f_1f_2)^h = f_1^hf_2^h.$$
This verifies the axioms of 9.1.

9.19 Definitions. Let H, X, G and G^* be as in 9.18. Let φ denote the action of H on G^* defined in 9.18. Then the corresponding semidirect product $H_\varphi \times G^*$ of G^* by H is said to be *a wreath product of G by H*, often denoted by $G \wr H$. The normal subgroup G^* is sometimes called the *base group* of the wreath product.

We emphasize that a group $G \wr H$ is determined by G, H *and an action of H on a set*. Different actions of H may lead to different wreath products of G by H, so that the notation $G \wr H$ is ambiguous. (However, see 9.20 (3) below.)

9.20 Remarks. (1) Note that if G and H are finite groups, then a wreath product $G \wr H$ determined by an action of H on a finite set X is a finite group of order $|G|^{|X|} . |H|$.

(2) Suppose that $|X| = 1$. Then we can evidently identify G and $G^* = \operatorname{Dr} G^X$: we need only identify each $g \in G$ with the element of G^* which maps the unique element of X to g. The only action of H on X in this case is trivial, and, by 9.18, the corresponding action of H on G^* ($= G$) is trivial. Hence, by 9.8, the corresponding wreath product of G by H in this case is $H \times G = G \times H$, the direct product of G and H.

(3) Suppose that H is finite. In the absence of an explicit specification of the relevant action of H, the notation $G \wr H$ is conventionally taken to denote the wreath product of G by H corresponding to the action of H on itself by right multiplication (4.23). This is called the *regular* wreath product of G by H.

Regular wreath products are the most frequently occurring examples in the literature. We shall have reason to consider also *natural* wreath products; namely wreath products determined by actions as in 4.2(i).

(4) Suppose that H acts on an infinite set X. For any group G, there are corresponding actions of H on both the groups $\operatorname{Cr} G^X$ and $\operatorname{Dr} G^X$ (**445**), defined in exactly the same way as in 9.18. The corresponding semidirect products of $\operatorname{Cr} G^X$ by H and $\operatorname{Dr} G^X$ by H are called *unrestricted* and *restricted* wreath products of G by H, respectively. Most of the wreath products which we shall consider in this book will be defined by actions on finite sets, when this distinction between 'unrestricted' and 'restricted' does not arise.

9.21. In working with wreath products, it is convenient to adopt the notation of the proof of 8.21. Suppose that H acts on the finite set X, let G be any group, W the corresponding wreath product $G \wr H$ of G by H, and $G^* = \operatorname{Dr} G^X$, the base group of W. According to 8.21,

$$G^* = \operatorname{Dr} \prod_{x \in X} G_x,$$

where, for each $x \in X$,

$$G_x = \{f \in G^* : f(y) = 1 \text{ whenever } x \neq y \in X\}.$$

For each $x \in X$ and each $g \in G$, let $g_x \in G^*$ be defined, for all $y \in X$, by

$$g_x(y) = \begin{cases} 1 & \text{if } y \neq x \\ g & \text{if } y = x. \end{cases}$$

Recall that the map

$$g \mapsto g_x$$

is an injective homomorphism of G into G^*, with image

$$\{g_x : g \in G\} = G_x.$$

Now for all $g \in G, h \in H$ and $x, y \in X$,

$$\begin{aligned} g_x^h(y) &= g_x(yh^{-1}) \\ &= \begin{cases} 1 & \text{if } yh^{-1} \neq x \\ g & \text{if } yh^{-1} = x \end{cases} \\ &= g_{xh}(y). \end{aligned}$$

Thus, for all $g \in G, h \in H$ and $x \in X$,

$$g_x^h = g_{xh}.$$

These last equations completely determine the action of H on G^* in the group W.

We make one more remark on notation. Suppose that H is a finite cyclic group, say $H = \langle h \rangle$, of order n; and let W be the *regular* wreath product $G \wr H$. Then

$$G^* = \mathrm{Dr} \prod_{i=1}^{n} G_{h^i}$$

and for all $g \in G$ and all $i = 1, \ldots, n$,

$$g_{h^i}^h = g_{h^{i+1}}.$$

It is then more economical to suppress the appearances of h in the suffices of elements of G^*. We write

$$G^* = \mathrm{Dr} \prod_{i=1}^{n} G_i,$$

where

$$G_i = \{g_i : g \in G\}$$

and the action of H on G^* in W is determined by the equations

$$g_i^h = g_{i+1},$$

where the suffices are now interpreted modulo n: thus $g_{n+1} = g_1$.

521 Suppose that group H acts on the finite set X and let ρ be the corresponding permutation representation of H. Let G be any non-trivial group, $G^* = \operatorname{Dr} G^X$, and let φ be the action of H on G^* defined in 9.18. Then $\operatorname{Ker} \varphi = \operatorname{Ker} \rho$.

In particular, if the action of H on X is faithful then the corresponding action of H on G^* is faithful.

522 Let groups H, J act on finite sets X, Y, respectively, and suppose that these actions are equivalent (4.19). Let G be any group. Then the corresponding wreath products $G \wr H, G \wr J$ are isomorphic.

523 Suppose that group H acts on the finite set X. Let G be any group and let W be the corresponding wreath product of G by H. Suppose also that G acts on the set Y.

(i) Then W acts on the product set $X \times Y$ when we define

$$(x, y) hf = (xh, yf(xh))$$

for all $x \in X, y \in Y, h \in H$ and $f \in G^* = \operatorname{Dr} G^X$.

(ii) If the actions of H on X and G on Y are both transitive then the action of W on $X \times Y$ defined in (i) is transitive.

(iii) If the actions of H on X and G on Y are both faithful then the action of W on $X \times Y$ defined in (i) is faithful.

(iv) For any positive integers n and m, $(nm)!$ is divisible by $(m!)^n (n!)$.

524 Suppose that group H acts *transitively* on the finite set X. Let G be any group and let W be the corresponding wreath product $G \wr H$.

(i) Let $x \in X$. Then $W = \langle G_x, H \rangle$, where G_x is defined as in 9.21. Moreover, if $|X| > 1$ and $|G| > 1$ then neither G_x nor H is normal in W.

(ii) If H is an n-generator group and G an m-generator group, where n and m are positive integers, then W is an $(n + m)$-generator group.

525 $C_2 \wr C_2 \cong D_8$, the dihedral group of order 8. (Here, as elsewhere when no prescription of the relevant action is made, the wreath product in question is the *regular* one.)

*526 Suppose that H acts on the finite set X. Let G be any group, W the corresponding wreath product of G by H, and G^* the base group of W. Let $K \leqslant G$ and let

$$K^* = \{ f \in G^* : f(x) \in K \text{ for all } x \in X \}.$$

(i) Then $K^* \leqslant G^*, K^* \cong \operatorname{Dr} K^X, H \leqslant N_W(K^*)$ and $HK^* \cong K \wr H$, the wreath product corresponding to the given action of H on X.

(ii) If $K \trianglelefteq G$ then $K^* \trianglelefteq W$ and

$$W/K^* \cong (G/K) \wr H,$$

the wreath product corresponding to the given action of H on X.

Wreath product constructions provide a very useful source of examples: they yield relatively complicated groups in which it is nevertheless practicable to perform calculations. As an illustration, we shall now show that there are soluble groups of derived length n for every positive integer n (see 7.52).

9.22 Lemma. *Let G be any soluble group, say of derived length n. Then $G \wr C_2$ is a soluble group of derived length $n + 1$.*

Proof. Let $H = \langle h \rangle$, of order 2, and let $W = G \wr H$, the regular wreath product. Let G^* be the base group of W: thus

$$G^* = G_1 \times G_2 = \{g_1 : g \in G\} \times \{g_2 : g \in G\},$$

where the maps $g \mapsto g_1$ and $g \mapsto g_2$ are isomorphisms of G onto G_1 and G_2, respectively. Then W is the semidirect product HG^*, where the action of H on G^* is given, for all $g \in G$, by the equations

$$g_1^h = g_2 \text{ and } g_2^h = g_1.$$

For any soluble group J, let $\delta(J)$ denote the derived length of J. Since G is soluble and $\delta(G) = n$, it follows that G^* is soluble (7.49) and $\delta(G^*) = n$ (373(iii)). Since $|W/G^*| = 2$, W/G^* is abelian. Therefore W is soluble (7.47) and $W' \leqslant G^*$ (3.52).

For every $g \in G$,

$$g_1^{-1}g_2 = g_1^{-1}g_1^h = [g_1, h] \in W'.$$

Let π denote the projection of G^* onto G_2 (8.16) and let $\pi' = \pi|_{W'}$. Then π' is a homomorphism $W' \to G_2$, and since, for every $g \in G$,

$$(g_1^{-1}g_2)\pi' = g_2,$$

π' is surjective. Since $\delta(G_2) = n$, it follows by the fundamental theorem on homomorphisms that

$$\delta(W') \geqslant n \quad (373(i)).$$

However, since $W' \leqslant G^*$ and $\delta(G^*) = n$,

$$\delta(W') \leqslant n \quad (373(i)).$$

Thus $$\delta(W') = n,$$

and therefore $$\delta(W) = n + 1.$$

9.23 Corollary. *For every positive integer n, there are soluble groups of derived length n.*
Proof. Let A be any non-trivial abelian group. Define groups $G_1, G_2, G_3, \ldots$ recursively as follows. Let $G_1 = A$ and, for each integer $n > 1$, let

$$G_n = G_{n-1} \wr C_2.$$

Then, by repeated application of 9.22, for every positive integer n, G_n is a soluble group of derived length n.

We shall show next that certain automorphism groups of direct products are *natural* wreath products.

9.24 Lemma. *Let n be a positive integer and let G be a non-trivial finite indecomposable group such that $Z(G) = 1$. For each $i = 1, \ldots, n$, let $g \mapsto g_i$ (defined for all $g \in G$) be an isomorphism of G onto a group G_i, and let*

$G^* = \mathrm{Dr} \prod\limits_{i=1}^{n} G_i$. *Then*

$$\mathrm{Aut}\, G^* \cong (\mathrm{Aut}\, G) \wr \Sigma_n,$$

where the wreath product is formed by means of the natural action of Σ_n on the set $\{1, 2, \ldots, n\}$.

Proof. Let $A = \mathrm{Aut}\, G$. For each $i = 1, \ldots, n$, let $\alpha \mapsto \alpha_i$ (defined for all $\alpha \in A$) be an isomorphism of A onto a group A_i, and let $A^* = \mathrm{Dr} \prod\limits_{i=1}^{n} A_i$. There is an obvious embedding of A^* in $\mathrm{Aut}\, G^*$ by which we identify A^* with the appropriate subgroup of $\mathrm{Aut}\, G^*$: namely, for each $\alpha \in \mathrm{Aut}\, G$ and each $i = 1, \ldots, n$, we identify α_i with the unique automorphism of G^* which maps g_i to $(g^\alpha)_i$ for all $g \in G$ and fixes every element of G_j for all $j \neq i$. It is clear that then

$$A^* = \{\beta \in \mathrm{Aut}\, G^* : G_i^\beta = G_i \text{ for every } i = 1, \ldots, n\}.$$

Let $\sigma \in \Sigma_n$. Then it is easy to verify that there is a unique automorphism σ^* of G^* such that, for all $g \in G$ and every $i = 1, \ldots, n$,

$$g_i^{\sigma^*} = g_{i\sigma}.$$

Moreover, the map $\sigma \mapsto \sigma^*$, defined for all $\sigma \in \Sigma_n$, is an injective homomorphism of Σ_n into $\mathrm{Aut}\, G^*$. (Here we need the hypothesis that $G \neq 1$.) Let

$$H = \{\sigma^* : \sigma \in \Sigma_n\}.$$

Thus $\qquad\qquad\qquad \Sigma_n \cong H \leqslant \mathrm{Aut}\, G^*.$

Since G_i is A^*-invariant for each $i = 1, \ldots, n$, we see that

$$H \cap A^* = 1.$$

Let $\sigma \in \Sigma_n, \alpha \in A, g \in G$ and $i, j \in \{1, 2, \ldots, n\}$. Then

$$
\begin{aligned}
g_j^{(\sigma^*)^{-1} \alpha_i \sigma^*} &= g_{j\sigma^{-1}}^{\alpha_i \sigma^*} \\
&= \begin{cases} g_{j\sigma^{-1}}^{\sigma^*} & \text{if } j\sigma^{-1} \neq i \\ (g^\alpha)_{j\sigma^{-1}}^{\sigma^*} & \text{if } j\sigma^{-1} = i \end{cases} \\
&= \begin{cases} g_j & \text{if } j \neq i\sigma \\ (g^\alpha)_j & \text{if } j = i\sigma \end{cases} \\
&= g_j^{\alpha_{i\sigma}}.
\end{aligned}
$$

Hence, for all $\sigma \in \Sigma_n, \alpha \in A$ and every $i = 1, \ldots, n$,

$$(\sigma^*)^{-1} \alpha_i \sigma^* = \alpha_{i\sigma} \in A^*. \qquad\qquad\qquad\text{(i)}$$

Since $A^* = \langle \alpha_i : \alpha \in A, i = 1, \ldots, n \rangle$, this shows that $H \leqslant N_{\mathrm{Aut}\ G^*}(A^*)$. Hence

$$A^* \trianglelefteq HA^* \leqslant \mathrm{Aut}\ G^*.$$

Let $W = A \wr \Sigma_n$, the *natural* wreath product. Then we may identify A^* with the base group of W, and then the action of Σ_n on A^* defining W is determined, for all $\sigma \in \Sigma_n$, $\alpha \in A$ and every $i = 1, \ldots, n$, by the equations

$$\alpha_i^\sigma = \alpha_{i\sigma}.$$

Then, in view of the equations (i), it is easy to check that the map

$$\sigma a^* \mapsto \sigma^* a^*,$$

defined for all $\sigma \in \Sigma_n$ and all $a^* \in A^*$, is an isomorphism of W onto HA^*.

We complete the proof by showing that $HA^* = \mathrm{Aut}\ G^*$. Let $\gamma \in \mathrm{Aut}\ G^*$. Then

$$\mathrm{Dr} \prod_{i=1}^{n} G_i = G^* = (G^*)^\gamma = \mathrm{Dr} \prod_{i=1}^{n} G_i^\gamma \quad (407).$$

But, for every $i = 1, \ldots, n$, $G_i^\gamma \cong G_i \cong G$, a finite indecomposable group with trivial centre. Then $Z(G^*) = 1$ (**406**) and, by the Krull–Remak–Schmidt theorem (8.18),

$$\{G_1, \ldots, G_n\} = \{G_1^\gamma, \ldots, G_n^\gamma\}.$$

Hence there is a permutation $\sigma \in \Sigma_n$ such that

$$G_i^\gamma = G_{i\sigma}$$

for every $i = 1, \ldots, n$. Let $\beta = (\sigma^*)^{-1}\gamma \in \mathrm{Aut}\ G^*$. Then

$$G_i^\beta = G_i$$

for every $i = 1, \ldots, n$. Hence $\beta \in A^*$, and so

$$\gamma = \sigma^*\beta \in HA^*.$$

Thus $\mathrm{Aut}\ G^* = HA^* \cong W.$

9.25 Theorem (Fitting [a27], 1938). *Let G be a non-trivial finite group with $Z(G) = 1$. By the Krull–Remak–Schmidt theorem (8.18), G is expressible as the direct product of finitely many non-trivial indecomposable normal subgroups and, apart from ordering of factors, this decomposition of G is unique: let this decomposition be*

$$G = G_{11} \times G_{12} \times \ldots \times G_{1n_1} \times G_{21} \times \ldots \times G_{2n_2} \times \ldots \times G_{s1} \times \ldots \times G_{sn_s},$$

where $s, n_1, n_2, \ldots, n_s$ are positive integers, the groups G_{ij} are non-trivial and indecomposable ($i = 1, \ldots, s, j = 1, \ldots, n_i$), and $G_{ij} \cong G_{kl}$ if and only if $i = k$. Then

$$\text{Aut } G \cong ((\text{Aut } G_{11}) \wr \Sigma_{n_1}) \times ((\text{Aut } G_{21}) \wr \Sigma_{n_2}) \times \dots \times ((\text{Aut } G_{s1}) \wr \Sigma_{n_s}),$$

where the wreath products are natural.

Proof. For each $i = 1, \dots, s$, let

$$G_i = G_{i1} \times \dots \times G_{in_i}.$$

Then
$$G = \text{Dr} \prod_{i=1}^{s} G_i.$$

Let $\gamma \in \text{Aut } G$. Then, by **407** and the Krull–Remak–Schmidt theorem, we see that for each $i = 1, \dots, s$ and each $j = 1, \dots, n_i$ there are integers k and l such that

$$G_{ij}^\gamma = G_{kl}.$$

Moreover, since $G_{ij}^\gamma \cong G_{ij}$, we must have $k = i$. It follows that, for each $i = 1, \dots, s$,

$$G_i^\gamma = G_i.$$

Thus $G_1, \dots, G_s$ are characteristic subgroups of G.

It follows easily that

$$\text{Aut } G \cong \text{Dr} \prod_{i=1}^{s} \text{Aut } G_i$$

(cf. **94**, **436**). Since the direct factors $G_{i1}, \dots, G_{in_i}$ of G_i are isomorphic non-trivial indecomposable groups with trivial centres (**406**), it follows from 9.24 that, for each $i = 1, \dots, s$,

$$\text{Aut } G_i \cong (\text{Aut } G_{i1}) \wr \Sigma_{n_i},$$

the natural wreath product. This gives the result.

527 Suppose that group H acts faithfully on the finite set X. Let G be any non-trivial group, W the corresponding wreath product $G \wr H$ and G^* the base group of W. Then

(i) $C_W(G^*) \leqslant G^*$.

(ii) If the action of H on X is also transitive then

$$Z(W) = \{f \in G^* : f(x) = f(y) \in Z(G) \text{ for all } x, y \in X\} \cong Z(G).$$

528 Suppose that group H acts transitively on the finite set X. Let G be a finite non-abelian simple group, W the corresponding wreath product $G \wr H$, and G^* the base group of W. Then

(i) G^* is a minimal normal subgroup of W.

(ii) If the action of H on X is also faithful then G^* is the unique minimal normal subgroup of W. (Hint. For (i) apply 8.9, and for (ii) apply **527**(i).)

529 (i) The Sylow p-subgroups of Σ_{p^2-1} are elementary abelian of order p^{p-1}.

(ii) The Sylow p-subgroups of Σ_{p^2} have order p^{p+1} and are isomorphic to $C_p \wr C_p$.

530 Let $A = \langle a \rangle$ and $H = \langle h \rangle$, groups of order p. Let $G = A \wr H$ and let A^* be the base group of G: thus

$$A^* = \langle a_1 \rangle \times \langle a_2 \rangle \times \dots \times \langle a_p \rangle,$$

$$a_i^h = a_{i+1} \quad \text{for each } i = 1, \ldots, p-1, \quad \text{and} \quad a_p^h = a_1.$$

Let $b_1 = a_1$ and, for each integer $i > 1$, let $b_i = [b_{i-1}, h]$. By induction on i, show that $b_i \in \Gamma_i(G)$ for every i (see 7.53), and also that for each $i = 2, 3, \ldots, p$,

$$b_i a_i^{-1} \in \mathrm{Dr} \prod_{j=1}^{i-1} \langle a_j \rangle.$$

Deduce that $\Gamma_p(G) \neq 1$.

Hence show that the p-group G has class p. (Hint. Apply **377**.)

531 $\mathrm{Aut}\,(\Sigma_3 \times \Sigma_3) \cong \Sigma_3 \wr C_2$.

532 Let $G = A_5 \wr C_2$. Show that G has subgroups J and K such that $\langle J, K \rangle = G$, K is subnormal in G and $j(J : J \cap K) > j(G : K)$ (see **339**).

533 Suppose that H acts on the infinite set X. Let G be any group and let $G^* = \mathrm{Dr}\, G^X$, the restricted direct power of G with index set X (see **445**). Verify that H acts on G^* when for each $h \in H$ and $f \in G^*$, f^h is defined as in 9.18.

534 Let N denote the set of all positive integers and let $H = A_N$ (see **291**). Consider the natural action of H on N. Let $G = C_2$, $G^* = \mathrm{Dr}\, G^N$, and let W be the semidirect product of G^* by H with action as in **533** (with $X = N$): thus W is the *restricted natural wreath product* of G by H (see 9.20(4)).

For each $f \in G^*$, let $s(f)$ denote the support of f (see **445**). Show that for all $f, f' \in G^*$ and all $h \in H$,

$$|s(f^h)| = |s(f)|$$

and

$$|s(ff')| = |s(f)| + |s(f')| - 2|s(f) \cap s(f')|.$$

Let $K = \{ f \in G^* : |s(f)| \text{ is even} \}$. Prove that $K \lhd W$, and that

$$1 < K < G^* < W$$

is a chief series of W. Show also that W does not have a composition series. (cf. **355**. Hint. Let $G = \langle g \rangle$. For each $n \in N$, let $g_n \in G^*$ be defined by

$$g_n(x) = \begin{cases} 1 & \text{whenever } x \neq n \\ g & \text{if } x = n. \end{cases}$$

If $f \in G^*$ and $s(f) = \{ j_1, j_2, \ldots, j_m \}$ with $j_1 < j_2 < \ldots < j_m$ then $f = g_{j_1} g_{j_2} \cdots g_{j_m}$.)

535 Let H be any infinite group and consider the action of H on itself by right multiplication (4.23). Let G be any non-trivial group, let $G^* = \mathrm{Dr}\, G^H$, and let W be the semidirect product of G^* by H with action as in **533** (with $X = H$): thus W is the *restricted regular* wreath product of G by H (see 9.20(4)).

(i) Then $Z(W) = 1$ (cf. **527**).

(ii) Let $G = C_p$ and let H be any infinite p-group (see **265**; for instance, we may choose $H = C_{p^\infty}$: see **144**). Then W is an infinite p-group and $Z(W) = 1$ (cf. 4.28).

We end this chapter with an application to group extensions.

9.26 Definition. Suppose that $K \trianglelefteq G$ and $G/K \cong H$. Then we shall call G *an extension of K by H*.

It follows (see **114**) that there is a homomorphism φ of G onto H with $\mathrm{Ker}\, \varphi = K$. In general there may be several such homomorphisms φ. Extension theory deals with the pairs (G, φ), which are then called the extensions of K by H, and seeks to classify these pairs. For further information on extension theory, see Gruenberg [b16] chapters 5 and 9, Kurosh [b27] vol. 2, chapter 12, MacLane [b31] chapter 4, Rotman [b34] chapter 7, Scott [b36] chapter 9. In this book, we shall not enter into a further discussion of general extension theory but merely prove a few special results.

We remark in passing that if H is finite then every extension of K by H can be embedded in the regular wreath product $K \wr H$; see Huppert [b21] p. 99, theorem 1.15.9 or Schenkman [b35] p. 100, theorem 3.5.k. However, we shall not use this fact here.

9.27 Theorem. *Let G be an extension of K by H. Then there is a homomorphism $\psi : G \rightarrow H \times \mathrm{Aut}\, K$ such that $\mathrm{Ker}\,\psi = Z(K)$. Moreover, if $\bar{G} = G\psi$ and π denotes the projection of $H \times \mathrm{Aut}\, K$ onto H (8.16), then*

$$\bar{G}\pi = H \quad \text{and} \quad \bar{G} \cap \mathrm{Aut}\, K = \mathrm{Inn}\, K.$$

In particular, if $Z(K) = 1$ then every extension of K by H can be embedded in $H \times \mathrm{Aut}\, K$.

Proof. There is a homomorphism φ of G onto H with $\mathrm{Ker}\,\varphi = K$ (**114**). Since $K \trianglelefteq G$, G acts on K by conjugation. Let $\sigma : G \rightarrow \mathrm{Aut}\, K$ be the corresponding action. We may suppose without loss of generality that H and $\mathrm{Aut}\, K$ have only the identity element in common and then that H and $\mathrm{Aut}\, K$ are identified with normal subgroups of $H \times \mathrm{Aut}\, K$ (see 8.1).

We define a map

$$\psi : G \rightarrow H \times \mathrm{Aut}\, K$$

by
$$\psi : g \mapsto (g\varphi)(g\sigma)$$

for every $g \in G$. Since φ and σ are homomorphisms and since every element of H commutes with every element of $\mathrm{Aut}\, K$ in $H \times \mathrm{Aut}\, K$, it is clear that ψ is a homomorphism. Moreover,

$$\begin{aligned}
\mathrm{Ker}\,\psi &= \mathrm{Ker}\,\varphi \cap \mathrm{Ker}\,\sigma \\
&= K \cap C_G(K) \quad \text{(by } \mathbf{476}) \\
&= Z(K).
\end{aligned}$$

Now let $\bar{G} = G\psi \leqslant H \times \mathrm{Aut}\, K$, and let π be the projection of $H \times \mathrm{Aut}\, K$ onto H. Then

$$(g\psi)\pi = g\varphi$$

for every $g \in G$. Since φ maps G onto H, it follows that π maps $\bar{G}$ onto H; that is,

$$\bar{G}\pi = H.$$

Also

$$\bar{G} \cap \mathrm{Aut}\, K = \{(g\varphi)(g\sigma) : g \in G \text{ and } g\varphi = 1\}$$

$$= \{g\sigma : g \in \text{Ker } \varphi\}$$
$$= K\sigma \text{ (since Ker } \varphi = K)$$
$$= \text{Inn } K.$$

If $Z(K) = 1$ then $G \cong \text{Im } \psi \leqslant H \times \text{Aut } K$, so that G can be embedded in $H \times \text{Aut } K$.

Recall that for any group K, Inn $K \trianglelefteq \text{Aut } K$ (**92**).

9.28 Corollary. *Let $\mathfrak{X}$ and $\mathfrak{Y}$ be classes of groups with the following three properties:*

(i) *Every quotient group of every $\mathfrak{X}$-group is an $\mathfrak{X}$-group.*

(ii) *Every subgroup of every $\mathfrak{Y}$-group is a $\mathfrak{Y}$-group.*

(iii) *The trivial group is the only group which is both an $\mathfrak{X}$-group, and a $\mathfrak{Y}$-group.*

Let H be $\mathfrak{X}$-group and let K be a group such that $Z(K) = 1$ and Aut $K/\text{Inn } K$ is a $\mathfrak{Y}$-group. Then every extension of K by H is isomorphic to $H \times K$.

Proof. Let G be an extension of K by H. Let $\psi : G \to H \times \text{Aut } K$ be the homomorphism defined in 9.27, $\bar{G} = G\psi$, and let π and ρ be the projections of $H \times \text{Aut } K$ onto H and Aut K, respectively. Since $Z(K) = 1$, 9.27 shows that $G \cong \bar{G}$. By 8.19(i),

$$\bar{G}\pi/(H \cap \bar{G}) \cong \bar{G}\rho/((\text{Aut } K) \cap \bar{G});$$

that is, by 9.27,

$$H/(H \cap \bar{G}) \cong \bar{G}\rho/\text{Inn } K \leqslant \text{Aut } K/\text{Inn } K.$$

Since H is an $\mathfrak{X}$-group, property (i) implies that $H/(H \cap \bar{G})$ is an $\mathfrak{X}$-group; since Aut $K/\text{Inn } K$ is a $\mathfrak{Y}$-group, property (ii) implies that $\bar{G}\rho/\text{Inn } K$ is a $\mathfrak{Y}$-group. Hence, by property (iii), $|H/(H \cap \bar{G})| = 1$ so that

$$\bar{G}\pi = H = H \cap \bar{G} \quad \text{and} \quad \bar{G}\rho = \text{Inn } K.$$

Hence, by 8.19(ii),

$$\bar{G} = H \times \text{Inn } K.$$

Since $Z(K) = 1$, Inn $K \cong K$ (**117**). Hence

$$G \cong \bar{G} \cong H \times K.$$

Remarks. (1) If for $\mathfrak{X}$ we choose the class of all groups and for $\mathfrak{Y}$ the class consisting of the trivial group alone, then $\mathfrak{X}$ and $\mathfrak{Y}$ obviously satisfy (i), (ii), (iii). Thus we deduce from 9.28 that if K is a group such that $Z(K) = 1$ and Aut $K = \text{Inn } K$ then, for any group H, every extension of K by H is isomorphic to $H \times K$. (This is the case of a *complete* group K: see **509**.)

(2) .Let H be a simple group. If for $\mathfrak{X}$ we choose the class consisting of all groups isomorphic to H, together with the trivial group, and for $\mathfrak{Y}$ the class of all groups in which H cannot be embedded then again $\mathfrak{X}$ and

$\mathfrak{Y}$ satisfy (i), (ii), (iii). We deduce from 9.28 that if K is a group such that $Z(K) = 1$ and H cannot be embedded in Aut $K/$Inn K then every extension of K by H is isomorphic to $H \times K$.

We shall make use of this remark in proving the last result of this chapter. Before stating this, we mention a famous conjecture.

9.29. *Schreier's conjecture* is that for every finite simple group G, Aut $G/$Inn G is soluble. No counter-example to this conjecture is known.

9.30 Theorem. *Let G be a non-trivial finite group, of composition length n. Suppose that in a composition series of G there are n_1 factors isomorphic to H_1, n_2 factors isomorphic to $H_2, \ldots, n_s$ factors isomorphic to H_s, where $s, n_1, \ldots, n_s$ are positive integers such that $\sum_{i=1}^{s} n_i = n$, and $H_1, \ldots, H_s$ are mutually non-isomorphic simple groups. Suppose further that, for each $i = 1, \ldots, s$,*
    (i) *H_i is non-abelian and satisfies Schreier's conjecture, and*
    (ii) *$n_i \leqslant 4$.*
Then G is completely reducible (8.8).
Proof. We argue by induction on n. If $n = 1$ then G is simple and the result is trivial. Suppose that $n > 1$ and let

$$1 = G_0 \lhd G_1 \lhd \ldots \lhd G_n = G$$

be a composition series of G. Let $K = G_{n-1} \lhd G$. Then

$$1 = G_0 \lhd G_1 \lhd \ldots \lhd G_{n-1} = K$$

is a composition series of K, of length $n - 1$. Now it is clear that the induction hypothesis applies to show that K is completely reducible. Hence, by (i), K is a direct product of non-abelian simple groups, and so $Z(K) = 1$.

We may suppose that $G/K \cong H_1$, so that G is an extension of K by H_1. Then

$$K \cong H_1 \times \ldots \times H_1 \times H_2 \times \ldots \times H_2 \times \ldots \times H_s \times \ldots \times H_s,$$

where, on the right, there are $n_1 - 1$ copies of H_1, n_2 copies of $H_2, \ldots, n_s$ copies of H_s (and where the copies of H_1 are omitted if $n_1 = 1$). Hence, by 9.25,

$$\text{Aut } K \cong ((\text{Aut } H_1) \wr \Sigma_{n_1 - 1}) \times ((\text{Aut } H_2) \wr \Sigma_{n_2}) \times \ldots \times ((\text{Aut } H_s) \wr \Sigma_{n_s}),$$

where the wreath products are natural (and the first factor on the right is omitted if $n_1 = 1$). Then, by means of **111** and **526**(ii), we see that

$$\text{Aut } K/\text{Inn } K \cong ((\text{Aut } H_1/\text{Inn } H_1) \wr \Sigma_{n_1 - 1}) \times ((\text{Aut } H_2/\text{Inn } H_2) \wr \Sigma_{n_2}) \times \ldots$$
$$\times ((\text{Aut } H_s/\text{Inn } H_s) \wr \Sigma_{n_s}).$$

When m is an integer not exceeding 4, Σ_m is soluble (**364**): and, by (ii), Aut H_i/Inn H_i is soluble, for each $i = 1, \ldots, s$. Hence, by 7.47 and 7.49, Aut K/Inn K is soluble. Since H_1 is a non-abelian simple group, it follows that H_1 cannot be embedded in Aut K/Inn K. Therefore, by remark (2) after 9.20,

$$G \cong H_1 \times K.$$

Thus G is completely reducible. This completes the induction argument.

536 Suppose that $Z(K) = 1$ and that Aut K/Inn K is soluble. Then every extension of K by any perfect group H (**168**) is isomorphic to $H \times K$.

537 (i) Suppose that K is a finite group with $Z(K) = 1$. Let ϖ be the set of all prime divisors of $|$ Aut K/Inn $K|$ and let ϖ' be the set of all primes which do not belong to ϖ. Then every extension of K by any finite ϖ'-group H is isomorphic to $H \times K$.

(ii) Let $K = D_{2p}$, the dihedral group of order $2p$, where p is any odd prime. Let H be any finite group such that $(|H|, (p-1)/2) = 1$. Then every extension of K by H is isomorphic to $H \times K$. (Hint. See **485** and **488**.)

(iii) Let $K = \Sigma_3 \times \Sigma_3$, and let H be any finite group of odd order. Then every extension of K by H is isomorphic to $H \times K$.

538 (i) Show by an example that the result of **537**(i) does not remain true in general if the condition that $Z(K) = 1$ is omitted. (Hint. Consider $K = C_p$.)

(ii) Show by an example that the result of **537**(ii) does not remain true in general if $p > 3$ and the condition that $(|H|, (p-1)/2) = 1$ is omitted.

(iii) Show by an example that the result of **537**(iii) does not remain true in general if the condition that H has odd order is omitted.

539 Show by an example that the result of 9.30 does not remain true in general if the condition (ii) is relaxed to $n_i \leqslant 5$. (Hint. Let $H = A_5$, and let K be any finite non-abelian simple group satisfying Schreier's conjecture and not isomorphic to H. Consider the natural wreath product $K \wr H$.)

10

TRANSFER AND SPLITTING THEOREMS

We shall establish some fundamental sufficient conditions for a finite group to split over a normal subgroup; and also define and apply some important homomorphisms, called *transfer* maps, of a group G into abelian sections of G. We follow an elegant approach due to H. Wielandt which is based on a consideration of group actions on suitable sets.

10.1 Definition. Let $H \leqslant G$. A subset T of G which contains just one element from each right coset of H in G is called a *right transversal to H in G*. Then $|T| = |G : H|$ and $HT = G$. Similarly, a *left transversal to H in G* is a subset S of G which contains just one element from each left coset of H in G: then $|S| = |G : H|$ and $SH = G$.

10.2. Let $H \leqslant G$ and let T be a right transversal to H in G. Let $g \in G$. Then the set $Tg = \{tg : t \in T\}$ is again a right transversal to H in G: for if we consider any right coset Hx of H in G (where $x \in G$) then, by hypothesis, $|T \cap Hxg^{-1}| = 1$ and so $|Tg \cap Hx| = |(T \cap Hxg^{-1})g| = 1$. Now it is clear that G *acts* (on the right) by right multiplication on the set $\mathcal{T}$ of all right transversals to H in G.

Also, let $h \in H$. Then the set $hT = \{ht : t \in T\}$ is a right transversal to H in G: for if Hx is any right coset of H in G (with $x \in G$) then $|hT \cap Hx| = |T \cap h^{-1}Hx| = |T \cap Hx| = 1$. Hence H *acts on the left* on the set $\mathcal{T}$ by left multiplication. We shall study these right and left actions on $\mathcal{T}$ and certain other actions determined by them.

540 Let $H \leqslant G$ and $T \subseteq G$. Then T is a right transversal to H in G if and only if every element of G is uniquely expressible in the form ht with $h \in H$ and $t \in T$.

541 Let $H \leqslant G$. For each non-empty subset X of G, let $X^* = \{x^{-1} : x \in X\}$. Then X is a right transversal to H in G if and only if X^* is a left transversal to H in G.

542 Let $H \leqslant G$, a finite group, and let $\mathcal{T}$ be the set of all right transversals to H in G. Let $|H| = m$ and $|G : H| = n$. Then $|\mathcal{T}| = m^n$.

*543 Let $H \leqslant G$.

    (i) If $H \trianglelefteq G$, then every right transversal to H in G is also a left transversal to H in G and every left transversal to H in G is also a right transversal to H in G. In this case we speak simply of *transversals* to H in G.

    (ii) If $H \trianglelefteq G$, then there is a transversal to H in G which is a subgroup of G if and only if G splits over H.

(iii) If every right transversal to H in G is also a left transversal to H in G, then $H \trianglelefteq G$.

(Remark. It is in fact always true that if G is a finite group then there is *some* right transversal to H in G which is also a left transversal to H in G. This depends on reasoning which lies outside the domain of group theory, for instance on graph theory. See Zassenhaus [b41] pp. 11–13 or Wilson [b39] p. 126, ex. 27d.)

10.3 Definitions. Let $J \trianglelefteq H \leqslant G$, and suppose that $|G : H| = n < \infty$ and H/J is abelian. Let $\mathscr{T}$ be the set of all right transversals to H in G. To each ordered pair T, U of elements of $\mathscr{T}$, we shall associate an element of the group H/J which we denote by T/U and define as follows. Now $|T| = n$: say $T = \{t_1, \ldots, t_n\}$. For each $i = 1, \ldots, n$ there is, by hypothesis, a unique element $u_i \in U$ such that $Ht_i = Hu_i$, and then $U = \{u_1, \ldots, u_n\}$. Then $t_i u_i^{-1} \in H$ for $i = 1, \ldots, n$, and we define

$$T/U = \prod_{i=1}^{n} Jt_i u_i^{-1}.$$

Because the group H/J is abelian, the order in which we multiply together the n elements $Jt_1 u_1^{-1}, \ldots, Jt_n u_n^{-1}$ does not affect the product which we obtain, and so T/U is a well-defined element of H/J.

Now let $T, U, V \in \mathscr{T}$, say with

$$T = \{t_1, \ldots, t_n\}, \qquad U = \{u_1, \ldots, u_n\}, \qquad V = \{v_1, \ldots, v_n\},$$

where, for $i = 1, \ldots, n$,

$$Ht_i = Hu_i = Hv_i.$$

Then
$$T/T = \prod_{i=1}^{n} Jt_i t_i^{-1} = J, \qquad \text{(i)}$$

the identity element of H/J;

$$U/T = \prod_{i=1}^{n} Ju_i t_i^{-1} = \prod_{i=1}^{n} (Jt_i u_i^{-1})^{-1} = (T/U)^{-1}, \qquad \text{(ii)}$$

since H/J is abelian; and

$$T/V = \prod_{i=1}^{n} Jt_i v_i^{-1} = \prod_{i=1}^{n} (Jt_i u_i^{-1})(Ju_i v_i^{-1}) = T/U \cdot U/V, \qquad \text{(iii)}$$

again since H/J is abelian.

We define next a relation $\sim$ on $\mathscr{T}$ by setting $T \sim U$ if and only if $T, U \in \mathscr{T}$ and $T/U = J$. Then equations (i), (ii) and (iii) above show that $\sim$ is an equivalence relation on $\mathscr{T}$. Let Ω denote the set of equivalence classes of this equivalence relation $\sim$. Thus Ω is a set of sets of right transversals to H in G.

10.4 Definition. Let $J \trianglelefteq H \leqslant G$ with $|G : H| < \infty$ and H/J abelian. Let

T be a right transversal to H in G and let $g \in G$. By 10.2, Tg is a right transversal to H in G, and so there is an element $Tg/T \in H/J$, defined as in 10.3. We define a map

$$\tau : G \to H/J$$

by $$\tau : g \mapsto Tg/T,$$

and call this *the transfer of G into H/J*. This was first investigated in [a90] by I. Schur (1875–1941). We shall show that τ is independent of the choice of right transversal used to define it, and that τ is a homomorphism.

10.5. *Let* $J \trianglelefteq H \leqslant G$, *with* $|G : H| = n < \infty$ *and* H/J *abelian. Let* T *and* U *be right transversals to* H *in* G, $g \in G$ *and* $h \in H$. *Then*

(a) $Tg/Ug = T/U = hT/hU$.

(b) *The transfer* τ *of* G *into* H/J *is independent of the choice of right transversal of* H *in* G *used to define it, and is a homomorphism.*

Proof. (a) Let $T = \{t_1, \ldots, t_n\}$ and $U = \{u_1, \ldots, u_n\}$, with $Ht_i = Hu_i$ for $i = 1, \ldots, n$. Then, for $i = 1, \ldots, n$,

$$Ht_i g = Hu_i g$$

and $$Hht_i = Ht_i = Hu_i = Hhu_i.$$

Hence

$$Tg/Ug = \prod_{i=1}^{n} J(t_i g)(u_i g)^{-1} = \prod_{i=1}^{n} Jt_i u_i^{-1} = T/U$$

and $$hT/hU = \prod_{i=1}^{n} J(ht_i)(hu_i)^{-1} = \prod_{i=1}^{n} (Jh)(Jt_i u_i^{-1})(Jh)^{-1}$$

$$= \prod_{i=1}^{n} Jt_i u_i^{-1} \text{ (since } H/J \text{ is abelian)}$$

$$= T/U.$$

(b) By 10.3 (iii), we have

$$Tg/T = Tg/Ug \, . \, Ug/U \, . \, U/T$$
$$= T/U \, . \, Ug/U \, . \, (T/U)^{-1} \text{ (by (a) and 10.3(ii))}$$
$$= Ug/U \text{ (since } H/J \text{ is abelian).}$$

This shows that τ is independent of the choice of right transversal of H in G used to define it.

Now let $x, y \in G$. Then

$$(xy)\tau = Txy/T$$
$$= Txy/Ty \, . \, Ty/T \text{ (by 10.3(iii))}$$
$$= Tx/T \, . \, Ty/T \text{ (by (a))}$$
$$= (x\tau)(y\tau).$$

Thus τ is a homomorphism.

10.6. In 10.5, let $\mathscr{T}$ be the set of all right transversals to H in G and let $\sim$ be the equivalence relation on $\mathscr{T}$ defined in 10.3. Then 10.5(a) shows that if $T \sim U$ then also $Tg \sim Ug$ and $hT \sim hU$ for every $g \in G$ and $h \in H$. Thus the right action of G on $\mathscr{T}$ and the left action of H on $\mathscr{T}$ defined in 10.2 respect the equivalence relation $\sim$.

It follows that these actions induce in a natural way a right action of G on the set Ω of equivalence classes of $\sim$, and a left action of H on Ω, when we define, for any $\omega \in \Omega, g \in G, h \in H$, with T an element of $\mathscr{T}$ in the equivalence class ω,

$$\omega g = \text{the equivalence class containing } Tg, \text{ and}$$

$$h\omega = \text{the equivalence class containing } hT.$$

The remarks above show that these are well defined, and thus obviously define right and left actions on Ω.

10.7. Let the notation and hypotheses be as in the preceding paragraphs. Let $T = \{t_1, \ldots, t_n\} \in \mathscr{T}$ and $h \in H$. Then, since $Hht_i = Ht_i$ for $i = 1, \ldots, n$,

$$hT/T = \prod_{i=1}^{n} J(ht_i)t_i^{-1} = Jh^n.$$

In particular, $jT \sim T$ for every $j \in J$. Hence, for every $j \in J$ and every $\omega \in \Omega$,

$$j\omega = \omega.$$

We shall now make a simple application of the transfer.

10.8 Theorem. *Let* $J \trianglelefteq H \leqslant G$, *with* $|G : H| = n < \infty$, $|H/J| = m < \infty$ *and* H/J *abelian. Suppose that* $(n, m) = 1$. *Then* $H \cap G' \cap Z(G) \leqslant J$.
Proof. Let τ be the transfer of G into H/J and let $h \in H \cap G' \cap Z(G)$. Then, if T is a right transversal to H in G,

$$
\begin{aligned}
h\tau &= Th/T \\
&= hT/T \quad \text{(since } h \in Z(G)) \\
&= Jh^n \quad \text{(by 10.7).}
\end{aligned}
$$

Since τ is a homomorphism (10.5), the fundamental theorem shows that

$$G/\mathrm{Ker}\,\tau \cong \mathrm{Im}\,\tau \leqslant H/J.$$

Hence $G/\mathrm{Ker}\,\tau$ is abelian and so, by 3.52, $G' \leqslant \mathrm{Ker}\,\tau$. Therefore $h \in \mathrm{Ker}\,\tau$. Now

$$J = h\tau = Jh^n,$$

and so $\qquad\qquad h^n \in J.$

Since $(n, m) = 1$ it follows **(105)** that

$$h \in J.$$

544 Let G be a finite group and H an abelian subgroup of G with $|H| = m$ and $|G : H| = n$. Let $\mathscr{T}$ be the set of all right transversals to H in G, $\sim$ the equivalence relation on $\mathscr{T}$ defined in 10.3 and Ω the set of equivalence classes of $\sim$, where $J = 1$. Then each equivalence class of $\sim$ contains m^{n-1} elements of $\mathscr{T}$ and $|\Omega| = m$.

545 Let $K \trianglelefteq G$, with $|K| = n < \infty$ and G/K abelian. Suppose that G splits over K and let H be a complement to K in G. Let τ be the transfer of G into H; ψ the map $G \to H$ defined by $\psi : hk \mapsto h$, for all $h \in H$ and $k \in K$ (see 9.9); and ν the map $h \mapsto h^n$ of H into itself. Then $\tau = \psi\nu$.

546 Let N be an integer, $N > 1$, $G = \Sigma_N$, $H = \langle (12) \rangle$ and τ the transfer of G into H. Then τ is the trivial homomorphism if and only if $N \geqslant 4$. (Hint. Use **545**.)

547 Let N be an integer, $N > 1$, F a finite field, and $G = \mathrm{GL}_N(F)$.
   (i) Let H be the set of all diagonal matrices in G for which all the diagonal entries other than the first are equal to 1, the identity element of F. Then $F^\times \cong H \leqslant G$.
   (ii) The transfer of G into H is trivial.
(Hint. Use **545**; cf. **491**.)

548 Let n be an integer, $n > 2$, and $G = \Sigma_n$. Then
   (i) $G = \langle (12), (13), \dots, (1n) \rangle$ (see **21** and **2.30**(i)).
   (ii) Consider the natural action of G on the set $\{1, 2, \dots, n\}$ and let $H = \mathrm{Stab}_G(1)$ and $J = A_n \cap H$. Then $J \trianglelefteq H$ with $|H/J| = 2$. The transfer of G into H/J is nontrivial if and only if n is odd. (Hint. Observe that the set $\{1, (12), (13), \dots, (1n)\}$ is a right transversal to H in G.)

549 Let G be a finite group.
   (i) If G has an abelian Sylow p-subgroup then p does not divide $|G' \cap Z(G)|$. Hence if all Sylow subgroups of G are abelian then $G' \cap Z(G) = 1$.
   (ii) If $G/Z(G)$ is a ϖ-group then G' is a ϖ-group.
   (iii) Show by an example that the converse of (ii) is false.
   (iv) If G is a non-abelian p-group then $G' \cap Z(G) \neq 1$.
(Remark. If G is a not necessarily finite group such that $G/Z(G)$ is a finite ϖ-group then G' is a finite ϖ-group. This is a result of Schur: see Huppert [b21] p. 417, theorem 4.2.3. J. Wiegold [a99] proved by a neat elementary argument that if $|G/Z(G)| = p^n$ then $|G'|$ divides $p^{n(n-1)/2}$.)

550 Let G be a finite soluble group with an abelian Sylow p-subgroup P. Let p' denote the set of all primes distinct from p, and suppose that $O_{p'}(G) = 1$. Then $P \trianglelefteq G$. (Hints. Argue by induction on $|G|$. Hence show that for every $K \triangleleft G, (P \cap K) \triangleleft G$. Then consider $C_G(O_p(G))$ and use **157**, **252**, **381** and **549**(i).)

Now we shall study the left action of H on Ω introduced in 10.6. For this purpose we need two preliminary lemmas.

10.9 Lemma. *Let $J \trianglelefteq H$, and suppose that H acts on the left on the set X. Suppose also that the action by restriction of J on X is trivial, that is, $jx = x$ for every $j \in J$ and every $x \in X$. Then we obtain a left action of H/J on X when we define (for all $h \in H$, $x \in X$)*

$$(Jh)x = hx.$$

Proof. Once we know that the defining equation

$$(Jh)x = hx$$

makes sense, in that it does not depend on the choice of element h in the coset Jh, then we obviously obtain in this way a left action of H/J on X. Let $h, h' \in H$ with $Jh = Jh'$, and let $x \in X$. Since $J \trianglelefteq H, Jh' = h'J$, and so there is an element $j \in J$ such that

$$h = h'j.$$

Then $$hx = h'(jx) = h'x,$$

since the action of J on X is trivial.

We shall also need the analogue for a left action of 4.3. We include it here explicitly, since one point of care is needed in defining the appropriate permutation representation because of our invariable convention of placing permutations on the right of the symbols on which they operate (see the remarks in chapter 2 before **17**).

10.10 Lemma. *Suppose that G acts on the left on the set X. Then, to each $g \in G$ there corresponds a map $\lambda_g : X \to X$, defined by $\lambda_g : x \mapsto gx$, and this is a permutation of X. Moreover, the map $\lambda^* : G \to \Sigma_X$ defined by $\lambda^* : g \mapsto \lambda_{g^{-1}}$ is a homomorphism (cf. **186**).*
Proof. For $g_1, g_2 \in G$ and $x \in X$,

$$x\lambda_{g_1 g_2} = (g_1 g_2)x = g_1(g_2 x) = g_1(x\lambda_{g_2}) = x\lambda_{g_2}\lambda_{g_1}$$

and so

$$\lambda_{g_1 g_2} = \lambda_{g_2}\lambda_{g_1}.$$

Hence also

$$\lambda_{(g_1 g_2)^{-1}} = \lambda_{g_2^{-1}g_1^{-1}} = \lambda_{g_1^{-1}}\lambda_{g_2^{-1}}.$$

Certainly $\lambda_1 = 1 \in \Sigma_X$ and therefore, for every $g \in G$,

$$\lambda_g \lambda_{g^{-1}} = 1 = \lambda_{g^{-1}}\lambda_g.$$

Hence $\lambda_g \in \Sigma_X$; and then the equation above shows that the map $\lambda^* : g \mapsto \lambda_{g^{-1}}$ is a homomorphism of G into Σ_X.

10.11. *Let the notation be as in 10.6, and assume further that $|H/J| = m < \infty$ and that n and m are co-prime integers. Consider the left action of H on Ω defined in 10.6. By 10.7 and 10.9, this induces naturally a left action of H/J on Ω. Then*
  (i) *the left action of H on Ω is transitive and $\mathrm{Stab}_H(\omega) = J$ for every $\omega \in \Omega$, and*
  (ii) *the left action of H/J on Ω is regular. In particular, $|\Omega| = m$ (cf. **544**).*
Proof. (i) To show that the left action of H on Ω is transitive, it is enough to show that, for any $T, U \in \mathcal{T}$, there is some element $h \in H$ such that $hT \sim U$.

By 10.7, for any $h \in H$,

$$hT/T = Jh^n,$$

and, by 10.3(iii), $$hT/U = hT/T . T/U.$$

Since $(n, m) = 1$, there are integers a and b such that

$$an + bm = -1.$$

Moreover, since $T/U \in H/J$ and $|H/J| = m$,

$$(T/U)^m = J.$$

Hence $$(T/U)^{an+1} = J.$$

Therefore, if we set $(T/U)^a = Jh$ with $h \in H$,

then $$hT/U = Jh^n . T/U = (T/U)^{an+1} = J,$$

and so $$hT \sim U.$$

Now let $\omega \in \Omega$. By 10.7,

$$J \leqslant \mathrm{Stab}_H(\omega).$$

Let $h \in \mathrm{Stab}_H(\omega)$ and let T be an element of $\mathcal{T}$ in the equivalence class ω: then $hT \sim T$. Hence, by 10.7, $Jh^n = J$, that is, $h^n \in J$. Since $(n, m) = 1$ it follows (**105**) that

$$h \in J.$$

Hence $$\mathrm{Stab}_H(\omega) = J.$$

(ii) By (i), the left action of H/J on Ω is transitive. Moreover, for any $\omega \in \Omega$,

$$\begin{aligned}\mathrm{Stab}_{H/J}(\omega) &= \{ Jh : h \in H \text{ and } h\omega = \omega \} \\ &= J/J \text{ (by (i))}.\end{aligned}$$

Therefore the action is regular.

We shall apply the following lemma.

10.12 Lemma. *Let X be a non-empty set and let A be an abelian subgroup of Σ_X. If the natural action of A on X is transitive then $C_{\Sigma_X}(A) = A$ (cf. **518**).*
Proof. Let $C = C_{\Sigma_X}(A)$. Since A is abelian, $A \leqslant C$. Let $\sigma \in C, x \in X$ and $x\sigma = y \in X$. Since the action of A is transitive, $y = x\alpha$ for some $\alpha \in A$. Then, for every $\beta \in A$,

$$\begin{aligned}(x\beta)\sigma &= (x\sigma)\beta \text{ (since } \sigma \in C) \\ &= (x\alpha)\beta \\ &= (x\beta)\alpha \text{ (since } A \text{ is abelian)}.\end{aligned}$$

Since the action of A on X is transitive, this shows that

$$w\sigma = w\alpha$$

for every $w \in X$. Therefore, since the action is faithful,

$$\sigma = \alpha \in A.$$

Hence $C \leqslant A$, and so $C = A$.

With this lemma, we can show that, under suitable conditions, the right action of G on Ω in 10.6 and the left action of H/J on Ω in 10.11 are related in a nice way.

10.13 Theorem (H. Wielandt). *Let $J \trianglelefteq H \leqslant G$, with $|G:H| = n < \infty$, $|H/J| = m < \infty, (n,m) = 1$ and H/J abelian. Let Ω be the set defined in 10.3, and let G act on the right on Ω as in 10.6 and H/J act on the left on Ω as in 10.11. Further, let τ be the transfer of G into H/J. Then, for every $g \in G$, there is a unique element $g^* \in H/J$ such that*

$$\omega g = g^* \omega \text{ for every } \omega \in \Omega.$$

Moreover, the map $g \mapsto g^*$ is a homomorphism of G into H/J, and $g\tau = (g^*)^n$.

Proof. For each $g \in G$, let ρ_g denote the permutation $\omega \mapsto \omega g$ of Ω. For each $h \in H$, let λ_{Jh} denote the permutation $\omega \mapsto (Jh)\omega$ of Ω, and let λ^* denote the homomorphism $Jh \mapsto \lambda_{(Jh)^{-1}}$ of H/J into Σ_Ω: see 10.10. Let $A = \operatorname{Im} \lambda^* \leqslant \Sigma_\Omega$. Since H/J is abelian, A is abelian. Since the left action of H/J on Ω is transitive (10.11), the natural action of A on Ω is transitive.

Let $g \in G$. Then, by the associativity of multiplication in G,

$$h(\omega g) = (h\omega)g$$

for all $h \in H$ and $\omega \in \Omega$, hence

$$(Jh)(\omega g) = ((Jh)\omega)g,$$

that is,

$$\rho_g \lambda_{Jh} = \lambda_{Jh} \rho_g.$$

Since this is true for all $h \in H$, and since $A = \{\lambda_{Jh} : h \in H\}$,

$$\rho_g \in C_{\Sigma_\Omega}(A) = A,$$

by 10.12. Hence there is an element $g^* \in H/J$ such that

$$\rho_g = (g^*)^{-1}\lambda^* = \lambda_{g^*},$$

and since the left action of H/J on Ω is faithful (10.11), g^* is uniquely determined by g. Then, for every $\omega \in \Omega$,

$$\omega g = g^* \omega.$$

Now let $g_1, g_2 \in G$. Then, for all $\omega \in \Omega$,

$$(g_1 g_2)^* \omega = \omega(g_1 g_2) = (\omega g_1)g_2 = g_1^* \omega g_2 = g_2^*(g_1^* \omega)$$
$$= (g_2^* g_1^*)\omega = (g_1^* g_2^*)\omega,$$

since $g_1^*, g_2^* \in H/J$, which is abelian. Again since the left action of H/J is faithful, it follows that

$$(g_1 g_2)^* = g_1^* g_2^*.$$

Let T be a right transversal to H in $G, g \in G$ and, say, $g^* = Jh$, where $h \in H$. Then, since $\omega g = g^* \omega = h\omega$ for every $\omega \in \Omega$,

$$Tg \sim hT,$$

in the notation of 10.3. Hence

$$
\begin{aligned}
g\tau &= Tg/T \\
&= Tg/hT . hT/T \text{ (by 10.3(iii))} \\
&= J . Jh^n \text{ (by 10.7)} \\
&= Jh^n \\
&= (g^*)^n.
\end{aligned}
$$

551 Let n be an integer, $n > 1$. Then the cyclic subgroup $\langle (12 \ldots n) \rangle$ is a maximal abelian subgroup of Σ_n (see **235**).

*552 Let J, H, G, n, m, Ω and τ be as in 10.13; and let $\theta : G \rightarrow H/J$ be the homomorphism $g \mapsto g^*$ defined in 10.13, and $K = \text{Ker } \theta$. Prove that
  (i) $\text{Ker } \tau = K$;
  (ii) for the action of G on Ω defined in 10.6, $\text{Stab}_G(\omega) = K$ for every $\omega \in \Omega$; and
  (iii) if the action of G on Ω is transitive then the action is equivalent to the action of G by right multiplication on the set of all cosets of K in G, and $G/K \cong H/J$.
Is the action of G on Ω necessarily transitive?

10.14. Let $J \trianglelefteq H \leqslant G$, with $|G : H| = n < \infty$ and H/J abelian, and let τ be the transfer of G into H/J. Then, for any $g \in G, g\tau = Tg/T$, where, by 10.5, we may choose for T any right transversal to H in G. For use in the proof of the next theorem, we observe that, for a particular g, we can choose T in an especially convenient way for calculating $g\tau$.

Let $g \in G$ and let X denote the set of all right cosets of H in G. Then the action of G on X by right multiplication (4.13) restricts to an action of $\langle g \rangle$ on X. Suppose that in this action of $\langle g \rangle$ on X there are just s orbits $X_1, \ldots, X_s$ (where $1 \leqslant s \leqslant n$). For each $i = 1, \ldots, s$, let $|X_i| = n_i$, so that $n_1 + \ldots + n_s = n$, and let $Hx_i \in X_i$, where $x_i \in G$. Then

$$X_i = \{Hx_i, Hx_i g, Hx_i g^2, \ldots, Hx_i g^{n_i - 1}\}$$

and

$$Hx_i g^{n_i} = Hx_i. \tag{i}$$

Now we may choose

$$T = \{x_i g^r : r = 0, 1, \ldots, n_i - 1; i = 1, \ldots, s\}:$$

this *is* a right transversal to H in G.

Then

$$Tg = \{x_i g^r : r = 1, 2, \ldots, n_i; i = 1, \ldots, s\}.$$

Since $x_i g^r \in Tg \cap T$ whenever $0 < r < n_i$, and by (i), we get

$$g\tau = Tg/T = \prod_{i=1}^{s} Jx_i g^{n_i} x_i^{-1},$$

where $x_i g^{n_i} x_i^{-1} \in H$ for $i = 1, \ldots, s$.

10.15 Definition. We have introduced in chapter 9 the notion of a *complement* to a normal subgroup in a group G. It is convenient now to speak of complements to arbitrary subgroups or, more generally, of sections of G (324).

Let $J \trianglelefteq H \leqslant G$ and $K \leqslant G$. We say now that K is a *complement to* H/J *in* G if $HK = G$ and $H \cap K = J$. If also $K \trianglelefteq G$, we say that K is a *normal complement to* H/J *in* G.

Note that if K is a complement to H/J in G then $HK = KH = G$ (95). Note also that if K is a normal complement to H/J in G then, by 3.40, $G/K = HK/K \cong H/(H \cap K) = H/J$.

We shall now connect the idea of a normal complement to a section in a group with the ideas already developed in this chapter in the following main result. It is closely related to results in papers of Frobenius [a30] and Schur [a90].

10.16. Theorem (H. Wielandt). *Let $J \trianglelefteq H \leqslant G$, with $|G:H| = n < \infty$, $|H/J| = m < \infty, (n, m) = 1$ and H/J abelian. Let $\sim$ be the equivalence relation on the set $\mathcal{F}$ of all right transversals to H in G defined in 10.3, and let τ denote the transfer $G \to H/J$. Then the following statements are equivalent:*
  (i) *There is a normal complement to H/J in G.*
  (ii) *Whenever $h_1, h_2 \in H$ and h_1, h_2 are conjugate in $G, Jh_1 = Jh_2$.*
  (iii) *For every $h \in H, h\tau = Jh^n$.*
  (iv) *For every $h \in H$ and every $T \in \mathcal{F}, hT \sim Th$.*
Proof. (i) $\Rightarrow$ (ii) Suppose that $K \trianglelefteq G$, with $HK = G$ and $H \cap K = J$. Let $h \in H$ and $g \in G$, with $h^g \in H$. We may express g in the form $g = h_1 k$, with $h_1 \in H$ and $k \in K$. Let

$$h_2 = h^{h_1} \in H.$$

Then also $$h_2^k = h^g \in H.$$

Hence, since $K \trianglelefteq G$,

$$h_2^k h_2^{-1} = k^{-1} k^{h_2^{-1}} \in H \cap K = J,$$

by hypothesis. Therefore $h^g = jh_2$ for some $j \in J$, and so

$$Jh^g = Jh_2 = (Jh)^{Jh_1} = Jh,$$

since H/J is abelian.

(ii) $\Rightarrow$ (iii) Suppose that whenever h_1, h_2 are elements of H conjugate in $G, Jh_1 = Jh_2$. Let $h \in H$. By 10.14 (with $g = h$), there are positive integers

$s, n_1 \ldots, n_s$ and elements $x_1, \ldots, x_s \in G$ such that

$$n_1 + \ldots + n_s = n,$$

$$x_i h^{n_i} x_i^{-1} \in H$$

for $i = 1, \ldots, s$, and $$h\tau = \prod_{i=1}^{s} J x_i h^{n_i} x_i^{-1}.$$

Then $x_i h^{n_i} x_i^{-1}$ and h^{n_i} are elements of H which are conjugate in G. Therefore, by hypothesis,

$$J x_i h^{n_i} x_i^{-1} = J h^{n_i},$$

for $i = 1, \ldots, s$. Hence

$$h\tau = J h^{n_1} . J h^{n_2} \ldots J h^{n_s} = J h^n.$$

(iii) $\Rightarrow$ (iv) Suppose that $h\tau = Jh^n$ for every $h \in H$. Let $h \in H$ and $T \in \mathcal{T}$. Then, by hypothesis,

$$Th/T = Jh^n.$$

Also, by 10.7, $$hT/T = Jh^n.$$

Hence, by 10.3,

$$hT/Th = hT/T . (Th/T)^{-1} = J,$$

and so $$hT \sim Th.$$

(iv) $\Rightarrow$ (i) Suppose that $hT \sim Th$ for every $h \in H$ and $T \in \mathcal{T}$. Let Ω be the set of equivalence classes of $\sim$, and consider the action of G on Ω defined in 10.6. Let $\omega \in \Omega$ and $K = \mathrm{Stab}_G(\omega)$. By hypothesis, $h\omega = \omega h$ for every $h \in H$. Hence, by 10.11, the (right) action of H on Ω by restriction of the action of G is transitive and $\mathrm{Stab}_H(\omega) = J$; that is,

$$H \cap K = J.$$

Let $g \in G$. Then, by the transitivity of the action of H, $\omega g = \omega h$ for some $h \in H$. Then

$$g h^{-1} \in K.$$

Hence $$G = KH = HK.$$

Thus K is a complement to H/J in G. Finally, by **552**, K is the kernel of the transfer of G into H/J, and therefore $K \trianglelefteq G$. Thus K is a normal complement to H/J in G.

10.17 Definition. A subgroup H of a finite group G is said to be a *Hall subgroup* of G if $(|G : H|, |H|) = 1$.

Any Sylow subgroup of G is a Hall subgroup of G. In chapter 11, we shall prove P. Hall's fundamental generalization for finite soluble groups of Sylow's theorem. This generalization deals with the existence and properties of Hall subgroups.

As an immediate consequence of 10.16, we note

10.18 Corollary. *Suppose that A is an abelian Hall subgroup of the finite group G. Then there is a normal complement to A in G if and only if no two distinct elements of A are conjugate in G.*
Proof. In 10.16, choose $H = A, J = 1$.

10.19 Definitions. Let G be a finite group. A complement to a Sylow p-subgroup of G is called a *p-complement* of G. Note that a subgroup H of G is a p-complement of G if and only if $|G : H|$ is a power of p and p does not divide $|H|$. In particular, a p-complement of G is a Hall subgroup of G.

If G has a normal p-complement then G is said to be *p-nilpotent*. We shall see that a finite group is nilpotent if and only if it is p-nilpotent for every prime p: see **563**.

A group need not possess a p-complement. For instance, 5.25 shows that the alternating group A_5 of degree 5 does not possess either a 2-complement or a 3-complement; although it does possess 5-complements (namely, the subgroups isomorphic to A_4). See also **561**.

553 Let $J \trianglelefteq H \leqslant G$, with $|G : H| = n < \infty$ and H/J abelian. Let τ be the transfer of G into H/J. Then, for every $g \in Z(G), g\tau = Jg^n$. (Hint. Use 10.14.)

554 Let $J \trianglelefteq H \leqslant G$, and suppose that there is a complement K to H/J in G. Then every conjugate of K in G is a complement to H/J in G. (Hint. Note that every conjugate of K in G is of the form K^h with $h \in H$.)

555 Let $J \trianglelefteq H \leqslant G$. There is a complement to H/J in G if and only if there is an action of G on some set X which restricts to a transitive action of H on X and such that, for some $x \in X$, $\text{Stab}_H(x) = J$.

556 (i) Let $H \leqslant G$, with $|G : H| < \infty$. Suppose that H/H' is finite and that $(|G : H|, |H/H'|) = 1$. Then $(H \cap [H, G]) \trianglelefteq H$, and there is a normal complement to $H/(H \cap [H, G])$ in G.
(ii) Let G be a finite group such that $O^p(G) = G$. If P is a Sylow p-subgroup of G then $P \leqslant [P, G]$.

*557 (Schur [a90]) Let G be a finite group and suppose that H is a Hall subgroup of G such that $H \leqslant Z(G)$. Then there is a subgroup K of G such that $G = H \times K$. (Remark. This result will be generalized in 10.30 and 10.31.)

558 Let P be a Sylow p-subgroup of the finite group G and let A be a maximal abelian normal subgroup of P (see **251**). Then $C_G(A) = A \times B$ for some subgroup B of G such that p does not divide $|B|$. (Hints. Use **235**, **236**, **251** and **252** to show that A is a Sylow p-subgroup of $C_G(A)$. Then use **557**.)

559 (Frobenius [a31]) Let G be a group of order mn, where m and n are co-prime positive integers. Let $X = \{x \in G : x^m = 1\}$ and $Y = \{y \in G : y^n = 1\}$. Suppose that

$|X| \leqslant m$ and $|Y| \leqslant n$. Then

(i) $X \cap Y = 1, G = XY, |X| = m, |Y| = n$, and every element of X commutes with every element of Y. (Hint. Use **7**.)

(ii) Let $H = C_G(X) \leqslant G$ (by **4.35**), and let $X_0 = X \cap H$ and $|X_0| = m_0$. Then $X_0 \leqslant G, |H| = m_0 n$, and m_0 divides m.

(iii) $H = X_0 \times Y_0$ for some subgroup Y_0 of H. (Hint. Use **557**.)

(iv) $Y_0 = Y$ and hence $Y \leqslant G$.

(v) $X \leqslant G$ and $G = X \times Y$.

(Warning. It is not obvious that X and Y are subgroups of G until (i), (ii), (iii) and (iv) have been proved.)

*560 Let G be a finite group.

(i) If G has a p-complement L then $L^G = O^p(G)$ (where L^G is the normal closure of L in G : see **180**).

(ii) If G is p-nilpotent then G has just one p-complement and this is $O^p(G)$.

561 There is no prime divisor p of $|A_6|$ for which A_6 has a p-complement. (Hints. Suppose to the contrary that A_6 has a p-complement for some p dividing $|A_6|$ and derive a contradiction. If $p = 2$, consider a subgroup of A_6 isomorphic to A_5 and use **99**(i) and **5.25**. If $p = 3$, show that a 3-complement of A_6 would have a normal Sylow 5-subgroup, and calculate the total number of Sylow 5-subgroups of A_6. If $p = 5$, use **4.14** and **5.28**.)

562 If G is a p-nilpotent finite group then every subgroup and every quotient group of G is p-nilpotent.

*563 Let G be a finite group. Then the following statements are equivalent:

(i) G is p-nilpotent.

(ii) Every chief factor of G of order divisible by p is central.

Hence G is nilpotent if and only if G is p-nilpotent for every prime p.

(Hints. For (i) $\Rightarrow$ (ii), it is enough to show that if L is a minimal normal subgroup of G such that p divides $|L|$, then $L \leqslant Z(G)$. Use **3.53** and **5.8**. For (ii) $\Rightarrow$ (i), use induction on $|G|$: hence if L is a minimal normal subgroup of $G, G/L$ is p-nilpotent. If p divides $|L|$, use **557**.)

564 Let G be a finite group with $|G| = 3r$, where r is an odd positive integer not divisible by 3. Then G is 3-nilpotent. (Hint. Use **302** and **10.18**. Remark. Note that the result of **205** follows at once from **10.18**. In **10.24** we shall prove a more general result.)

We shall apply **10.18** to establish a useful criterion for a finite group to be p-nilpotent. For the proof we also need

10.20 Lemma (Burnside [b3], p. 155). *Let P be a Sylow p-subgroup of the finite group G. Then any two elements of $Z(P)$ which are conjugate in G are in fact conjugate in $N_G(P)$.*

Proof. Let $x \in Z(P), g \in G$ and $x^g \in Z(P)$. Then $P \leqslant C_G(x) \cap C_G(x^g) = C_G(x) \cap C_G(x)^g$, by **229**. Hence (by **252**) P and $P^{g^{-1}}$ are Sylow p-subgroups of $C_G(x)$, and are therefore conjugate in $C_G(x)$: for some $y \in C_G(x)$,

$$P^{g^{-1}} = P^y.$$

Then $yg \in N_G(P)$ and $x^g = (x^y)^g = x^{yg}$.

10.21 Theorem (Burnside [a9]). *Let P be a Sylow p-subgroup of the finite group G. If $P \leqslant Z(N_G(P))$ then G is p-nilpotent.*

Proof. Let $H = N_G(P)$. Since $P \leqslant Z(H)$, P is in particular abelian. Therefore we may apply 10.18 with $A = P$. Let $x_1, x_2 \in P$. If x_1 and x_2 are conjugate in G then, by 10.20 (and since P is abelian), x_1 and x_2 are conjugate in H. But then, since $P \leqslant Z(H), x_1 = x_2$. Hence, by 10.18, G is p-nilpotent.

Remark. It would not be enough in 10.21 merely to suppose P abelian. In fact, every known finite non-abelian simple group G has cyclic Sylow p-subgroups for some prime divisor p of $|G|$; and although (by **262** or 10.24) such a p must be odd, there are many examples of finite non-abelian simple groups with abelian Sylow 2-subgroups.

We use this theorem to establish another fact about orders of finite simple groups.

10.22 Corollary. *Let G be a finite simple group of even order greater than 2. Then $|G|$ is divisible either by 8 or by 12.*

Proof. Suppose that 8 does not divide $|G|$, and let T be a Sylow 2-subgroup of G. Since $|G|$ is even and $|G| \neq 2, 1 < T < G$, by 4.29; and by our supposition, $|T| \leqslant 4$. Hence, by **77**, T is isomorphic to C_2 or $C_2 \times C_2$ or C_4. (In fact, by **262** or 10.24, $T \cong C_2 \times C_2$, but we do not need to appeal to these results here.) In particular, T is abelian, so that, by 4.36,

$$T \leqslant C_G(T) \trianglelefteq N_G(T).$$

Since T is a Sylow 2-subgroup of $G, N_G(T)/C_G(T)$ must have odd order. If $C_G(T) = N_G(T)$ then $T \leqslant Z(N_G(T))$ and so, by 10.21, G would be 2-nilpotent, in contradiction to the simplicity of G. Therefore

$$C_G(T) < N_G(T).$$

By 4.36, $N_G(T)/C_G(T)$ can be embedded in Aut T. By **40, 46, 47, 48** and 2.36, Aut T is isomorphic to $\mathbb{Z}_2^\times$ or $GL_2(\mathbb{Z}_2)$ or $\mathbb{Z}_4^\times$, hence $|$Aut $T|$ is 1 or 6 or 2 (see 2.16 or **44**). But we have shown that $|N_G(T)/C_G(T)|$ is odd and greater than 1, and so the only possibility is that $|N_G(T)/C_G(T)| = 3$ and $T \cong C_2 \times C_2$. Therefore $|G|$ is divisible by 12, as claimed.

10.23. The alternating group A_5 of degree 5 is an example of a finite non-abelian simple group of even order not divisible by 8. Any such simple group G must have its Sylow 2-subgroups isomorphic to $C_2 \times C_2$ (by the proof of 10.22). There is in fact a complete classification theorem for finite simple groups with Sylow 2-subgroups isomorphic to $C_2 \times C_2$: any such group is isomorphic to $PSL_2(F)$ (see 3.61), where F is a finite field with $|F| \geqslant 5$ and $|F|$ congruent to either 3 or 5 mod 8. (In this connexion, see Gorenstein [b13] chapter 15, where this classification theorem is proved under the additional assumption that the Sylow 2-subgroups are self-centralizing. For the general result, see Gorenstein [a43].)

In his book ([b3] p. 330, footnote), Burnside remarks that 'An examination of the orders of the known non-cyclical simple groups brings out the remarkable fact that all of them are divisible by 12'. It was long thought that the orders of finite non-abelian simple groups would all prove to be divisible by 12; but in 1960, M. Suzuki [a92] announced the existence of an infinite family of finite non-abelian simple groups with orders which are not divisible by 3; see also [a93]. The order of the smallest Suzuki group is $2^6.5.7.13$. No other examples have been discovered of finite non-abelian simple groups with orders not divisible by 3, and it is thought likely that the Suzuki groups are the only such groups.

A refinement of the proof of 10.22, made by analysing the possible Sylow 2-subgroups of order 8, shows that a finite simple group of even order greater than 2 must have order divisible by 12, 16 or 56: see **640**.

By means of Burnside's theorem 10.21, we can establish a strong structure theorem for finite groups in which all Sylow subgroups are cyclic. First we prove

10.24 Corollary. *Let G be a finite group and p the smallest prime divisor of $|G|$. If the Sylow p-subgroups of G are cyclic then G is p-nilpotent.* (This generalizes the results of **262** and **564**.)

Proof. Let P be a Sylow p-subgroup of G and suppose that P is cyclic. Then $P \leqslant C_G(P) \trianglelefteq N_G(P)$ and, by 4.36, $N_G(P)/C_G(P)$ can be embedded in Aut P. Since $P \leqslant C_G(P)$, $|N_G(P)/C_G(P)|$ is not divisible by p. Let $|P| = p^m$, where m is a positive integer. Then, since P is cyclic, $|\text{Aut } P| = p^m - p^{m-1}$ (**243**). Hence $|N_G(P)/C_G(P)|$ must divide $p - 1$. Since p is the smallest prime divisor of $|G|$, it follows that $|N_G(P)/C_G(P)| = 1$, hence that $C_G(P) = N_G(P)$. Thus $P \leqslant Z(N_G(P))$ and so, by Burnside's theorem, G is p-nilpotent.

We also use

10.25 Lemma (H. J. Zassenhaus [b41]). *Let G be a soluble group. If, in the derived series of G (7.51), the factors G'/G'' and G''/G''' are both cyclic then $G'' = G''' = 1$.*

Proof. Since G is soluble, there is a positive integer r such that $G^{(r)} = 1$ (7.52). Hence, if $G'' = G'''$, it follows that $G'' = G''' = 1$. Therefore (by replacing G by G/G''') we may assume that $G''' = 1$, and try to prove that when G'/G'' and G'' are both cyclic then $G'' = 1$.

Since $G'' \trianglelefteq G$, 4.36 shows that $C_G(G'') \trianglelefteq G$ and $G/C_G(G'')$ can be embedded in Aut G''. Since G'' is cyclic, Aut G'' is abelian, by 4.38, and therefore, by 3.52, $G' \leqslant C_G(G'')$. Hence $G'' \leqslant Z(G')$. Since G'/G'' is cyclic, it follows, by 3.30, that $G'/Z(G')$ is cyclic and therefore (**125**) that G' is abelian. Hence $G'' = 1$.

10.26 Theorem (Hölder, Burnside, Zassenhaus [a108], [b41]). *Let G be a finite group such that all Sylow subgroups of G are cyclic. Then G is*

soluble. Moreover, G/G' and G' are both cyclic (so that G is metacyclic (**152**)), *G splits over G', and G' is a Hall subgroup of G.*

Remarks. If G is a finite cyclic group then all subgroups of G are cyclic (**3.32**), in particular all Sylow subgroups of G are cyclic. If G is a finite abelian group and all its Sylow subgroups are cyclic then G is cyclic (**410**). But there are also finite non-abelian groups with all their Sylow subgroups cyclic: for instance, the dihedral group D_{2n} for every odd integer $n \geqslant 3$ (**259**).

Proof of the theorem. (i) We observe first that every subgroup and every quotient group of G has the same property as G. Let $H \leqslant G$. By Sylow's theorem, every Sylow subgroup of H is a subgroup of some Sylow subgroup of G. Since subgroups of cyclic groups are cyclic (**3.32**), all Sylow subgroups of H are cyclic. Let $K \trianglelefteq G$. Every Sylow subgroup of G/K is of the form PK/K, where P is a Sylow subgroup of G (**252**). Since $PK/K \cong P/(P \cap K)$ (**3.40**) and P is cyclic, PK/K is cyclic. Thus all Sylow subgroups of G/K are cyclic.

(ii) Now we prove by induction on $|G|$ that G is soluble. This is trivial if $|G| = 1$, so we assume that $|G| > 1$. Let p be the smallest prime divisor of $|G|$. By **10.24**, G is p-nilpotent. Let K be the normal p-complement of G. Then $K \triangleleft G$ and, by (i), all Sylow subgroups of K are cyclic. Hence, by the induction hypothesis, K is soluble. Since G/K is isomorphic to a Sylow p-subgroup P of G and P is cyclic, it follows (**7.47**) that G is soluble.

(iii) Now $G/G', G'/G''$ and G''/G''' are abelian groups and all their Sylow subgroups are cyclic (by (i)). Hence (**410**) these abelian groups are cyclic. Therefore, by (ii) and **10.25**, $G'' = 1$, and G/G' and G' are cyclic.

(iv) Let $|G/G'| = n$ and $|G'| = m$. By (iii), there are elements $x, y \in G$ such that

$$G' = \langle x \rangle \quad \text{and} \quad G/G' = \langle yG' \rangle.$$

Then (**108**) $G = \langle x, y \rangle$ and $x^y \in G'$, so that

$$x^y = x^r$$

for some integer r. Hence

$$[x, y] = x^{r-1}.$$

Let $L = \langle x^{r-1} \rangle$. Then L is characteristic in G' (**138**) and therefore $L \trianglelefteq G$ (**3.15**). Now

$$G/L = \langle xL, yL \rangle$$

and

$$[xL, yL] = [x, y]L = L,$$

since $[x, y] \in L$. Thus G/L is generated by 2 elements which commute. Hence G/L is abelian (**69**), and so $G' \leqslant L$ (**3.52**). Since also $L \leqslant G'$,

$$\langle x^{r-1} \rangle = L = G' = \langle x \rangle.$$

Therefore, since $o(x) = m$,

$$(r - 1, m) = 1 \; (5).$$

Because $|G/G'| = n$, $y^n \in G'$ (105), so that

$$y^n = x^s$$

for some integer s. Then

$$x^s = y^n = (y^n)^y = (x^s)^y = x^{rs},$$

and therefore $x^{(r-1)s} = 1$. Hence $(r - 1)s$ is divisible by $o(x) = m$. Since $(r - 1, m) = 1$, it follows that s is divisible by m.

Hence $\qquad\qquad\qquad y^n = 1.$

Since $G/G' = \langle yG' \rangle$, of order n, this implies that

$$o(y) = n.$$

Now $G = \langle y \rangle G'$ and, by 3.40,

$$G/G' \cong \langle y \rangle / \langle y \rangle \cap G'.$$

Since $|G/G'| = n = |\langle y \rangle|$, it follows that

$$\langle y \rangle \cap G' = 1.$$

Thus $\langle y \rangle$ is a complement to G' in G.

Suppose that $(n, m) > 1$. Then there is a prime q which divides both n and m. Let

$$n_1 = n/q \text{ and } m_1 = m/q.$$

Then $\qquad\qquad o(x^{m_1}) = q = o(y^{n_1}).$

Just as $L \trianglelefteq G$, so $\langle x^{m_1} \rangle \trianglelefteq G$. Then (by 3.38 and 3.40)

$$J = \langle x^{m_1} \rangle \langle y^{n_1} \rangle \leqslant G \text{ and } |J| = q^2.$$

Therefore, by Sylow's theorem, $J \leqslant Q$, some Sylow q-subgroup of G. By hypothesis, Q is cyclic and therefore Q has a unique subgroup of order q (3.32). But this contradicts the fact that $\langle x^{m_1} \rangle$ and $\langle y^{n_1} \rangle$ are distinct subgroups of J of order q. We conclude therefore that $(n, m) = 1$.

---

565 Let P be a Sylow p-subgroup of the finite group G. Let X and Y be non-empty subsets of P such that $P \leqslant N_G(X) \cap N_G(Y)$ (see 4.32). If X and Y are conjugate in G then they are conjugate in $N_G(P)$. (This generalizes 10.20.)

566 Let P be a Sylow p-subgroup of the finite group G. Suppose that $P \cap P^g = 1$ whenever $g \in G$ and $P \neq P^g$. Then any two elements of P which are conjugate in G are conjugate in $N_G(P)$.

567 Suppose that $|G| = pr$, where r is a positive integer such that p does not divide r. If G has r distinct subgroups of order p then G has a normal subgroup of order r.

568 If $|G| = p^2q^2$, where p, q are distinct primes, then G has either a normal Sylow p-subgroup or a normal Sylow q-subgroup; and so G is not simple. (Cf. 5.19. Hint. Use 4.30, Sylow's theorem and Burnside's theorem.)

569 If $|G| = p^3q$, where p, q are distinct primes, then G is not simple. (Hints. Suppose that G is a simple group of order p^3q. Show that $p < q$ and use Sylow's theorem and Burnside's theorem to show that the number of Sylow q-subgroups of G is p^2. Deduce that $p = 2, q = 3$. Remark. In general, G need not have either a normal Sylow p-subgroup or a normal Sylow q-subgroup; consider for example $G = \Sigma_4$.)

570 Let G be a simple group of order p^2qr, where p, q, r are distinct primes, and let P be a Sylow p-subgroup of G. Then
   (i) p is the smallest prime divisor of $|G|$; so we may assume without loss of generality that $p < q < r$;
   (ii) $P \cong C_p \times C_p$, and $|N_G(P)/C_G(P)|$ is either q or r;
   (iii) $p = 2$ and $q = 3$;
   (iv) $G \cong A_5$.
(Hint. See the proof of 10.22, and use **294**.)

571 Prove that every group of odd order less than 1000 is soluble.
(Hints. Assume that the result is false. Show that this implies that there is a non-abelian simple group G of order n for some odd integer $n < 1000$. Let $n = \prod_{i=1}^{s} p_i^{m_i}$, where $s, m_1, \ldots, m_s$ are positive integers and $p_1, \ldots, p_s$ distinct odd primes. Apply 4.29, 5.17, 5.19, 5.20, **568**, **569**, **570** and arithmetic to show that $\sum_{i=1}^{s} m_i = 5$. Deduce that 3 divides n, hence that n must be one of the five numbers $3^4 \times 5, 3^4 \times 7, 3^4 \times 11$, $3^3 \times 5^2, 3^3 \times 5 \times 7$. Eliminate the first four of these possibilities by means of **279** and Sylow's theorem. If $n = 3^3 \times 5 \times 7$, use Sylow's theorem to show that G has a subgroup of index 7, and then apply 4.14 to derive a final contradiction. Remark. This result is merely a special case of the Feit–Thompson theorem : see 1.12 and **383**.)

572 Let $|G| = p^mq^n$, where p, q are distinct primes and m, n non-negative integers, and suppose that the Sylow p-subgroups and Sylow q-subgroups of G are abelian. Then G is soluble. (Hints. By induction on $|G|$, it is enough to show that G cannot be simple and non-abelian. Suppose that it is. Let Q be a Sylow q-subgroup of G and use **100** and **264**(ii) to show that there is no subgroup L of G such that $Q < L < G$. Then use Burnside's theorem. Remark. By another theorem of Burnside – see 11.27 – the condition of abelian Sylow subgroups is in fact superfluous here.)

573 Let T be a Sylow 2-subgroup of the finite group G.
   (i) If $T \cong C_2 \times C_2$ and $|G|$ is not divisible by 3 then G is 2-nilpotent.
   (ii) If $T \cong C_2 \times C_2 \times C_2$ and $|G|$ is not divisible by either 3 or 7 then G is 2-nilpotent.

574 Suppose that the finite group G has an abelian Sylow p-subgroup P, and let $H = N_G(P)$. Then G is p-nilpotent if and only if $H = P \times Q$ for some subgroup Q of H.

575 Let $n = \prod_{i=1}^{s} p_i^{m_i}$, where $s, m_1, \ldots, m_s$ are positive integers and $p_1, \ldots, p_s$ distinct primes. Then the following two statements are equivalent :
   (i) Every group G of order n is cyclic.
   (ii) For all $i, j \in \{1, 2, \ldots, s\}$, $m_i = 1$ and $p_i \not\equiv 1 \bmod p_j$.

(This is the result stated in 1.4. Hints. To show that if (ii) does not hold then (i) does not hold, if $m_i > 1$ for some i consider abelian groups; if $m_i = 1$ for all i, but $p_i \equiv 1 \bmod p_j$ for some i,j, apply 9.16. To show that (ii) $\Rightarrow$ (i), let G be a group of order n and apply 10.26. Note that if $G' \neq 1$ then, by **125**, $C_G(G') < G$. Then apply 4.36, **46** and **138**.)

We are going to prove that whenever H is a normal Hall subgroup of a finite group G, G splits over H: cf. the special case of this result in **557**. Recall that when $H \trianglelefteq G$, there is no distinction between left and right transversals to H in G and we refer simply to transversals (**543**). We observe that then the left action of H on the set $\mathcal{T}$ of all transversals to H in G defined in 10.2 extends to a left action of G on $\mathcal{T}$.

10.27. *Let* $H \trianglelefteq G$. *Then* G *acts on the left by left multiplication on the set* $\mathcal{T}$ *of all transversals to* H *in* G.
Proof. Let $g \in G$ and $T \in \mathcal{T}$. Then, for any $x \in G$, using 3.19 we have

$$|gT \cap Hx| = |T \cap g^{-1}Hx| = |T \cap Hg^{-1}x| = 1,$$

since $T \in \mathcal{T}$. Hence $gT \in \mathcal{T}$. The result is now clear.

10.28. *Let* $J \trianglelefteq H \trianglelefteq G$, *with* $|G/H| = n < \infty$ *and* H/J *abelian. Let* $\mathcal{T}$ *be the set of all transversals to* H *in* G. *Whenever* $g \in G$ *and* $T, U \in \mathcal{T}$,

$$gT/gU = Jg(T/U)Jg^{-1}.$$

Proof. Let $T = \{t_1, \ldots, t_n\} \in \mathcal{T}$ and $U = \{u_1, \ldots, u_n\} \in \mathcal{T}$ with $Ht_i = Hu_i$ for $i = 1, \ldots, n$. Then (by 3.19) $Hgt_i = gHt_i = gHu_i = Hgu_i$ for $i = 1, \ldots, n$. Hence

$$gT/gU = \prod_{i=1}^{n} J(gt_i)(gu_i)^{-1} = \prod_{i=1}^{n} Jgt_i u_i^{-1} g^{-1} = Jg(T/U)Jg^{-1}.$$

10.29 Theorem. *Let* A *be an abelian normal Hall subgroup of the finite group* G. *Then* G *splits over* A. *Moreover, the complements to* A *in* G *form a single conjugacy class of subgroups of* G.
Proof. In 10.3, let $H = A$ and $J = 1$. Let $\mathcal{T}$ denote the set of all transversals to A in G, $\sim$ the equivalence relation on $\mathcal{T}$ defined in 10.3, and Ω the set of equivalence classes of $\sim$. By 10.27, G acts on the left on $\mathcal{T}$ by left multiplication. Moreover, this left action induces naturally a left action of G on Ω: for if $g \in G$ and $T, U \in \mathcal{T}$ with $T \sim U$, then 10.28 shows that $gT \sim gU$. The restriction to A of this left action of G on Ω is of course the left action of A on Ω already introduced in 10.6: by 10.11 (with $J = 1$), it is regular.

Now let $\omega \in \Omega$ and $K = \operatorname{Stab}_G(\omega)$ (where this stabilizer refers to the left action of G on Ω). Let $g \in G$. Then, by transitivity of the left action of A on Ω,

$$g\omega = a\omega$$

for some $a \in A$, and then $a^{-1}g \in K$.

Hence $AK = G$.

Moreover, $A \cap K = \text{Stab}_A(\omega) = 1,$

since the left action of A on Ω is regular. Thus K is a complement to A in G.

Now let L be any complement to A in G. Then clearly $L \in \mathcal{T}$ (cf. **543**). Let λ be the element of Ω which contains L. Then, for every $l \in L$,

$$lL = L,$$

and so $l\lambda = \lambda.$

Hence $L \leqslant \text{Stab}_G(\lambda).$

Because the left action of A on Ω is transitive,

$$\lambda = b\omega$$

for some $b \in A$, and then, by the left action analogue of **187**,

$$\text{Stab}_G(\lambda) = b\,\text{Stab}_G(\omega)b^{-1}$$
$$= K^{b^{-1}}.$$

Since $|L| = |G/A| = |K^{b^{-1}}|,$

$$L = K^{b^{-1}}.$$

Since also any conjugate in G of a complement to A in G is again a complement to A in G (9.11), the theorem is proved.

We now show that in 10.29, the splitting conclusion holds without the condition that A is abelian. This important result is called the Schur–Zassenhaus theorem.

10.30 Theorem (Schur, Zassenhaus [b41]). *Let K be a normal Hall subgroup of the finite group G. Then G splits over K.*
Proof. Let $|G/K| = n$ and $|K| = m$. We note that it is enough to show that G has a subgroup H of order n: for then, since $(n, m) = 1, H \cap K = 1,$ hence also, by 3.40, $|HK| = nm$ and so $HK = G$.

We prove by induction on m that G has a subgroup of order n. This is trivial if $m = 1$, so we assume that $m > 1$. Let p be a prime divisor of m, P a Sylow p-subgroup of K, and $N = N_G(P)$. Then, by Frattini's lemma (5.13), $G = NK$. By 3.40, $N \cap K \trianglelefteq N$ and $N/N \cap K \cong G/K$, of order n. If $N < G$ then $N \cap K < K$, and so $|N \cap K|$ is a proper divisor of m. Then, by the inductive assumption, N has a subgroup H of order n. Then also H is a subgroup of G of order n.

Therefore we may assume that $N = G$, that is, that $P \trianglelefteq G$. Since p divides $m, P \neq 1$. Let $Z = Z(P)$. Then, by 4.28 and **121**, $1 < Z \trianglelefteq G$. By 3.30,

$K/Z \trianglelefteq G/Z$ and $G/Z \big/ K/Z \cong G/K$, of order n. Moreover, $|K/Z|$ is a proper divisor of m, and so, by the inductive assumption, G/Z has a subgroup L/Z of order n, where $L \leqslant G$.

Certainly p does not divide n, and so Z is a normal Hall subgroup of L. Since Z is also abelian, it follows from 10.29 that L splits over Z, hence that L has a subgroup H of order n. Then H is also a subgroup of G of order n. Thus the induction argument goes through.

We now consider whether in 10.30, as in 10.29, complements are conjugate. We prove

10.31 Theorem (Zassenhaus [b41]). *Let K be a normal Hall subgroup of the finite group G, and suppose that either K or G/K is soluble. Then the complements to K in G form a single conjugacy class of subgroups of G.*

Proof. Let H and H^* be complements to K in G. It is enough to show that H and H^* are conjugate in G. We argue by induction on $|G|$. Let $|G/K| = n$ and $|K| = m$. The result is trivial if either $n = 1$ or $m = 1$, so we may assume that $n > 1$ and $m > 1$.

(i) Suppose first that K is soluble. By 3.51 and 3.15, $K' \trianglelefteq G$. By 3.30, K/K' is a normal Hall subgroup of G/K'. Also HK'/K' and H^*K'/K' are complements to K/K' in G/K' (by 3.40). Since K/K' is abelian (3.52), it follows from 10.29 that HK'/K' and H^*K'/K' are conjugate in G/K'. Hence there is an element $g \in G$ such that

$$H^*K' = (HK')^g = H^g K'.$$

Let $|K'| = m'$. Then m' divides m, and since $|H^*K'/K'| = n, K'$ is a normal Hall subgroup of H^*K'. Moreover, H^* and H^g are complements to K' in H^*K', and K' is soluble (7.46). Since K is soluble and $K \neq 1$, $K' < K$ (7.52). Therefore $|H^*K'| = nm' < nm = |G|$. Hence, by the inductive assumption, H^* and H^g are conjugate in H^*K'. Therefore H and H^* are conjugate in G. This completes the induction argument in this case.

(ii) Now suppose that G/K is soluble. Since $K < G$, there is a chief factor J/K of G, and since G/K is soluble, J/K is an elementary abelian p-group for some p dividing n (7.56). Then $K < J \leqslant G = HK$, so that, by Dedekind's rule (7.3), $J = (H \cap J)K$. Similarly, $J = (H^* \cap J)K$. Since p does not divide m, J is p-nilpotent, with normal p-complement K, and $H \cap J$ and $H^* \cap J$ are Sylow p-subgroups of J. Hence, by Sylow's theorem, for some $x \in J$,

$$H^* \cap J = (H \cap J)^x.$$

Now $H \cap J \trianglelefteq H$, and so $H^* \cap J = (H \cap J)^x \trianglelefteq H^x$. Also $H^* \cap J \trianglelefteq H^*$. Let $L = H^* \cap J$ and $N = N_G(L)$. Thus N contains both H^x and H^*. In particular, $NK = G$. By 3.40, $N \cap K \trianglelefteq N$ and $N/(N \cap K) \cong NK/K = G/K$. Therefore $N/(N \cap K)$ is soluble and $N \cap K$ is a Hall subgroup of N.

Moreover, $|H^*| = n = |G/K| = |N/N \cap K|$. Now $(N \cap K)L \trianglelefteq N$ (3.39), so that, by 3.30, $(N \cap K)L/L \trianglelefteq N/L$ and

$$N/L \Big/ (N \cap K)L/L \cong N/(N \cap K)L \cong N/(N \cap K) \Big/ (N \cap K)L/(N \cap K),$$

which is soluble (7.46). Since also, by 3.40,

$$(N \cap K)L/L \cong (N \cap K)/(N \cap K \cap L) \cong N \cap K,$$

$(N \cap K)L/L$ is a Hall subgroup of N/L. Moreover, $|(N \cap K)L| = |N \cap K||L|$, and so

$$|H^*/L| = |N/(N \cap K)| \Big/ |L| = |N/(N \cap K)L|.$$

Hence H^*/L and H^x/L are complements to $(N \cap K)L/L$ in N/L. Since $K < J, L \neq 1$ and so $|N/L| < |G|$. Hence, by the inductive assumption, H^*/L and H^x/L are conjugate in N/L. Therefore H and H^x are conjugate in G (**230**). This completes the induction argument in this case.

Remark. In this theorem, since $(|G/K|, |K|) = 1$, either $|G/K|$ or $|K|$ is odd. It follows by the Feit–Thompson Theorem (1.12; see **383**) that either G/K or K is necessarily soluble. Therefore the hypothesis in the theorem that either K or G/K is soluble is superfluous. But no proof is known of the conjugacy of complements in the general case which does not an appeal to this very deep result 1.12.

576 Let $J \trianglelefteq H \trianglelefteq G$, with $|G/H| = n < \infty, |H/J| = m < \infty, (n, m) = 1$ and H/J abelian. Let $\sim$ be the equivalence relation on the set $\mathscr{T}$ of all transversals to H in G defined in 10.3. Then the following statements are equivalent:
  (i) There is a normal complement to H/J in G.
  (ii) $J \trianglelefteq G$ and $G/J = (H/J) \times (K/J)$ for some subgroup K of G.
  (iii) For every $g \in G$ and every $T \in \mathscr{T}, gT \sim Tg$.
(cf. 10.16. Hints. For (ii) $\Rightarrow$ (iii), let $g \in G$ and show that there are positive integers $s, n_1, \ldots, n_s$ and elements $k_1, \ldots, k_s \in K$ such that $n_1 + \ldots + n_s = n, Hg^{-n_i}k_i g^{n_i} = Hk_i$ for $i = 1, \ldots, s$, and $U = \{g^{-j}k_i g^j : j = 0, 1, \ldots, n_i - 1; i = 1, \ldots, s\}$ is a transversal to H in G. For this $U, U^g \sim U$ and hence $Ug \sim gU$. Deduce that for every $T \in \mathscr{T}$, $T^g \sim T$ and hence $Tg \sim gT$.)

577 Let $J \trianglelefteq H \trianglelefteq G$, with G/H finite and H/J abelian. Let τ be the transfer $G \to H/J$ and let $\operatorname{Im} \tau = I/J$, where $J \leqslant I \leqslant H$. Then $[G, I] \leqslant J$.

578 (Zassenhaus [b41]) Let G be a finite group with an abelian normal Hall subgroup A, and let τ be the transfer of G into A. Then
  (i) $\operatorname{Im} \tau = Z(G) \cap A$.
  (ii) $\operatorname{Ker} \tau \cap \operatorname{Im} \tau = 1$.
  (iii) $G = \operatorname{Ker} \tau \times \operatorname{Im} \tau$.
  (iv) $A \cap \operatorname{Ker} \tau \leqslant [A, G]$.
  (v) $A \cap \operatorname{Ker} \tau = [A, G] = G' \cap A$.
  (vi) $A = (G' \cap A) \times (Z(G) \cap A)$.
  (vii) If $A < G$ and G is indecomposable (**81**) then $Z(G) \cap A = 1$ and $A \leqslant G'$.
(Hints. For (i) and (ii), use **553** and **577**; for (iv), use 10.14; for (v), use **162**; for (vi), use 9.12.)

579 Give an example of a finite group G with a normal subgroup K such that G

splits over K but the complements to K in G do not form a single conjugacy class of subgroups of G (cf. 10.29, 10.30, 10.31).

580 Let K be a normal Hall subgroup of the finite group G.

(i) Then K is characteristic in G.

(ii) Let H be a complement to K in G (which exists, by 10.30) and suppose that either K or G/K is soluble. Then H is both intravariant and pronormal in G : see **267** and **268**.

581 Let $K \trianglelefteq G$, a finite group. Suppose that Aut K splits over Inn K. If $(|G/K|, |Z(K)|) = 1$ then G splits over K. (Hint. Consider the action of G on K by conjugation, and apply 9.12, 10.29, and **501**.)

582 Let G be a finite group and p a prime divisor of $|G|$. Suppose that K is a normal subgroup of G such that p does not divide $|G/K|$. Then there is a subgroup H of G such that $G = HK$ and p does not divide $|H|$. (Hint. Use 5.13 and 10.30.)

Before making another major application of 10.16, we introduce the notion of *fusion*. This has figured prominently in the modern analysis of finite simple groups.

10.32 Definitions. (i) Let $H \leqslant G$. Two elements (or subsets) of H are said to be *fused in H by G* if they are conjugate in G but not in H.

For instance, the permutations (123) and (132) are fused in $\langle (123) \rangle$ by Σ_3.

Thus 10.16(ii) may be reformulated as follows: 'whenever h_1, h_2 are elements of H fused in H by G, $Jh_1 = Jh_2$'. (This is an equivalent statement, for if $h_1, h_2 \in H$ and h_1, h_2 are conjugate *in H* then automatically $Jh_1 = Jh_2$, because H/J is abelian.) Again, Burnside's lemma 10.20 may be stated as 'if P is a Sylow p-subgroup of a finite group G, then any two elements of $Z(P)$ which are fused in P by G are also fused in P by $N_G(P)$'.

(ii) Let $H \leqslant G$. Then the *focal subgroup of H in G* is defined to be

$$\text{Foc}_G(H) = \langle [h, g] : h \in H, g \in G, [h, g] \in H \rangle.$$

Equivalently,

$$\text{Foc}_G(H) = \langle h_1 h_2^{-1} : h_1, h_2 \text{ elements of } H \text{ conjugate in } G \rangle.$$

Then clearly

$$H' \leqslant \text{Foc}_G(H) \leqslant G' \cap H.$$

If $H \trianglelefteq G$ then $[h, g] \in H$ for every $h \in H$ and every $g \in G$ (**162**), and then $\text{Foc}_G(H) = [H, G]$. However, in general we may have $[H, G] \not\leqslant H$ and then $\text{Foc}_G(H) < [H, G]$.

If there is no fusion in H by G, that is, if elements of H conjugate in G are already conjugate in H, then $\text{Foc}_G(H) = H'$. In general, the quotient $\text{Foc}_G(H)/H'$ may be thought of as measuring in some way the amount of fusion which takes place in H by G.

10.33 Theorem. *Let* $H \leqslant G$. *Suppose that* $|G : H| = n < \infty$,

$|H/H'| = l < \infty$, and $(n,l) = 1$. Then $\mathrm{Foc}_G(H) = G' \cap H$, and there is a normal complement to $H/(G' \cap H)$ in G.

Proof. Let $J = \mathrm{Foc}_G(H)$. Then

$$H' \leqslant J \leqslant H,$$

and so (by 3.30 and 3.52) $J \trianglelefteq H$ and H/J is abelian. Moreover, if $|H/J| = m$, then m divides l and so $(n,m) = 1$. Thus we may apply 10.16. Let h_1 and h_2 be elements of H which are conjugate in G. Then, by definition of J, $h_1 h_2^{-1} \in J$ and so $Jh_1 = Jh_2$. Hence, by 10.16, there is a normal complement K to H/J in G.

Then
$$G/K \cong H/J.$$

Hence G/K is abelian, so that (3.52)

$$G' \leqslant K.$$

Now
$$J \leqslant G' \cap H \leqslant H \cap K = J.$$

Therefore
$$J = G' \cap H.$$

The most important special case of this result is called the focal subgroup theorem.

10.34 Corollary (Focal subgroup theorem: D. G. Higman [a56], 1953). *Let P be a Sylow p-subgroup of the finite group G. Then $\mathrm{Foc}_G(P) = G' \cap P$.*

The following facts make clear the significance of this result.

10.35. (i) *For any set ϖ of prime numbers, every finite group G has an $\mathfrak{X}$-residual when $\mathfrak{X}$ is the class of finite abelian ϖ-groups: this 'abelian ϖ-residual' of G is $G/(G'O^\varpi(G))$ (see 3.44, 3.45).*

(ii) *For any finite group G and any Sylow p-subgroup P of G, the abelian p-residual of G is isomorphic to $P/(G' \cap P)$.*

Proof. Let G be a finite group.

(i) Let $K \trianglelefteq G$. Then G/K is an abelian ϖ-group if and only if $G' \leqslant K$ and $O^\varpi(G) \leqslant K$ (by 3.52 and 3.44), hence if and only if $G'O^\varpi(G) \leqslant K$. Moreover (by 3.39), $G'O^\varpi(G) \trianglelefteq G$.

(ii) Let P be a Sylow p-subgroup of G. Then, by 3.40, $(P/G' \cap P) \cong PG'/G'$, a Sylow p-subgroup of G/G' (**252**).

Let
$$R = G'O^p(G) \trianglelefteq G.$$

Then
$$R/G' = O^p(G/G'),$$

by 3.30 and (i). Hence, since G/G' is abelian, R/G' is the (unique) p-complement of G/G' and $G/G'\big/R/G'$ is isomorphic to the (unique) Sylow p-subgroup of G/G' (by **560** and **563**). Thus, by 3.30,

$$G/R \cong P/(G' \cap P).$$

We now see that the focal subgroup theorem relates the abelian p-residual of a finite group G to fusion by G of elements of a Sylow p-subgroup of G.

We shall use the focal subgroup theorem to obtain some information about finite groups with abelian Sylow p-subgroups. The result stated is a special case of a theorem of O. Grün [a45] to which we shall return later.

10.36 Theorem. *Let G be a finite group with an abelian Sylow p-subgroup P, and let $H = N_G(P)$. Then $G/O^p(G) \cong H/O^p(H)$.*
Proof. Since P is abelian and $PO^p(G) = G$ (by **252**), $G/O^p(G)$ is abelian, and hence is the abelian p-residual of G. Also P is the (unique) Sylow p-subgroup of H, and so, by the same remark, $H/O^p(H)$ is the abelian p-residual of H. Hence, by 10.35, what we have to prove is that

$$P/(G' \cap P) \cong P/(H' \cap P).$$

Obviously $H' \cap P \leqslant G' \cap P$; it remains to show that $G' \cap P \leqslant H' \cap P$, or, equivalently, by the focal subgroup theorem, that

$$\mathrm{Foc}_G(P) \leqslant \mathrm{Foc}_H(P).$$

Let x_1 and x_2 be elements of P which are conjugate in G. Then, since P is abelian, Burnside's lemma 10.20 shows that x_1 and x_2 are conjugate in H. Hence

$$\mathrm{Foc}_G(P) \leqslant \mathrm{Foc}_H(P),$$

and the result is proved.

Next, we shall apply several of the preceding results to prove another splitting theorem, in this case due to W. Gaschütz. We shall need the following lemma.

10.37 Lemma. *Suppose that G is a finite group with an abelian normal Sylow p-subgroup P, and suppose also that $O^p(G) = G$. Let Q be a p-complement of G (which exists, by 10.29). Then $N_G(Q) = Q$.*
Proof. Let $R = N_G(Q)$ and $P_0 = P \cap R$. Since $P \trianglelefteq G, P_0 \trianglelefteq R$. Moreover, by definition, $Q \trianglelefteq R$, and since P_0 is a p-group and p does not divide $|Q|, P_0 \cap Q = 1$. Hence, by 3.53,

$$[P_0, Q] = 1.$$

Therefore, since also P is abelian,

$$C_G(P_0) \geqslant PQ = G.$$

Thus $$P_0 \leqslant Z(G).$$

Since, by hypothesis, $O^p(G) = G$, p does not divide $|G/G'|$ (**297**), and therefore $P \leqslant G'$ (see **252**). Hence

$$P_0 \leqslant P \cap G' \cap Z(G) = 1,$$

by 10.8 (with $H = P, J = 1$). Since

$$Q \leqslant R \leqslant G = PQ,$$

it follows, by Dedekind's rule (7.3), that

$$R = P_0 Q = Q,$$

as asserted.

10.38 Theorem (W. Gaschütz [a34], 1952). *Let G be a finite group. If the Sylow p-subgroups of $O^p(G)$ are abelian then G splits over $O^p(G)$.*

Proof. We argue by induction on $|G|$. The result is trivial if $|G| = 1$, so we assume that $|G| > 1$. Let $K = O^p(G)$, and let P be a Sylow p-subgroup of K. Let $L = N_K(P)$. By hypothesis, P is abelian, and from the definition of $K, O^p(K) = K$ (see **156**). Hence, by 10.36, $O^p(L) = L$.

Let $H = N_G(P)$. Then $H \cap K = L$, and, since $K \trianglelefteq G, L \trianglelefteq H$. By Frattini's lemma (5.13), $HK = G$. Thus K is a normal complement to H/L in G. In particular,

$$H/L \cong G/K,$$

a p-group. Since we have shown that $O^p(L) = L$, it follows that

$$L = O^p(H).$$

Since $L \leqslant K$ and the Sylow p-subgroups of K are abelian, the Sylow p-subgroups of L are abelian. Hence if $H < G$, it follows by the induction hypothesis that H splits over L. Then, if P_1 is a complement to L in H, $P_1 K = P_1 LK = HK = G$ and $P_1 \cap K = P_1 \cap H \cap K = P_1 \cap L = 1$, so that P_1 is also a complement to K in G.

Therefore we may assume that $H = G$, that is, that $P \trianglelefteq G$. Then P is an abelian normal Sylow p-subgroup of K, and so, by 10.29, K splits over P and the complements to P in K form a single conjugacy class of subgroups of K. Let Q be a complement to P in K, that is, a p-complement of K. Then a repetition of the proof of Frattini's lemma (5.13; see also **267**) shows that

$$N_G(Q)K = G.$$

Since also $O^p(K) = K$, 10.37 shows that

$$N_G(Q) \cap K = N_K(Q) = Q.$$

Thus K is a normal complement to $N_G(Q)/Q$ in G. In particular,

$$N_G(Q)/Q \cong G/K,$$

a p-group. Then, if P_2 is a Sylow p-subgroup of $N_G(Q)$,

$$P_2 Q = N_G(Q).$$

Hence $$P_2K = P_2QK = N_G(Q)K = G$$

and $$P_2 \cap K = P_2 \cap N_G(Q) \cap K = P_2 \cap Q = 1,$$

since p does not divide $|Q|$. Thus P_2 is a complement to K in G and the induction argument is complete.

583 (i) Let P be a Sylow p-subgroup of the finite group G, where p is a prime divisor of $|G|$. Prove that if no two distinct elements of P are fused in P by G, then $P \not\leqslant G'$ and $O^p(G) < G$.

(ii) Give an example of a finite soluble group G with a Sylow p-subgroup P for some prime divisor p of $|G|$ such that $P \leqslant G'$.

584 (i) Prove that if H is an abelian subgroup of G and there is a normal complement to H in G, then $\mathrm{Foc}_G(H) = 1$.

(ii) Let $H \leqslant G$, and suppose that $\mathrm{Foc}_G(H) = 1$. Prove that H is abelian and that no two distinct elements of H are fused in H by G. Show by an example that there need not be a normal complement to H in G (cf. 10.18).

585 Give an example of a group G with a subgroup H such that

$$\mathrm{Foc}_G(H) < H \cap [H, G].$$

586 (i) Suppose that $H \leqslant G$, with $|G:H| < \infty$, $|H/H'| < \infty$ and $(|G:H|, |H/H'|) = 1$. Prove that $H \cap [H,G] = G' \cap H$ (cf. **556** and 10.33).

(ii) Give an example of a group G with a subgroup H such that $H \cap [H,G] < G' \cap H$.

587 Let $H \leqslant G$. Suppose that $|G:H| = n < \infty$, $|H/H'| = l < \infty$ and $(n, l) = 1$. Let τ be the transfer of G into $H/(G' \cap H)$. Then $\mathrm{Im}\, \tau = H/(G' \cap H)$.

588 Let P be a Sylow p-subgroup of the finite group G. Then the following statements are equivalent:

(i) There is a normal complement to P/P' in G.

(ii) $G' \cap P = P'$.

589 Give an example of a finite group G with a Sylow p-subgroup P, for some prime p, such that $G/O^p(G) \not\cong H/O^p(H)$, where $H = N_G(P)$ (cf. 10.36).

590 Let G be a finite group with a cyclic Sylow p-subgroup. Then either G is p-nilpotent or $O^p(G) = G$. (Hint. Use 10.38, **502** and **492**.)

We shall end this chapter by proving some important generalizations of transfer theorems already obtained. It will be convenient to introduce a new notation for conjugacy of elements: this notation is due to H. Wielandt. Recall that conjugacy is an equivalence relation on the set of all elements of a group (**49**).

10.39 Definition. Let $x, y \in G$. We write $x \underset{G}{=} y$ if and only if x and y are conjugate in G.

We shall prove a generalization of part of 10.16, in which the hypothesis that H/J is abelian is weakened to the assumption that H/J is nilpotent.

10.40 Theorem (H. Wielandt). *Let* $J \trianglelefteq H \leqslant G$, *with* $|G:H| = n < \infty$, $|H/J| = m < \infty$, $(n, m) = 1$ *and* H/J *nilpotent. Then the following statements are equivalent:*

(i) *There is a normal complement to H/J in G.*

(ii) *Whenever $h_1, h_2 \in H$ and $h_1 \underset{G}{=} h_2$, $Jh_1 \underset{H/J}{=} Jh_2$.*

Proof. (i) $\Rightarrow$ (ii) The argument is exactly the same as for 10.16, (i) $\Rightarrow$ (ii), with the last equation omitted.

(ii) $\Rightarrow$ (i) We argue by induction on m. The assertion in (i) is trivial if $m = 1$, and so we assume that $m > 1$. Let $Z(H/J) = J_1/J$. Then, by 3.30 and 7.46, $J_1 \trianglelefteq H$ and H/J_1 is nilpotent. Let $|H/J_1| = m_1$. Then m_1 divides m, so that $(n, m_1) = 1$; and, by 7.54, $m_1 < m$. Let $h_1, h_2 \in H$ with $h_1 \underset{G}{=} h_2$. By hypothesis, $Jh_1 \underset{H/J}{=} Jh_2$, and so evidently $J_1 h_1 \underset{H/J_1}{=} J_1 h_2$. Now it follows, by the induction hypothesis, that there is a normal complement K_1 to H/J_1 in G. Then

$$H/J_1 \cong G/K_1.$$

Now $J \trianglelefteq J_1 \leqslant K_1$, and $|K_1 : J_1| = |G : H| = n$. Let $m_2 = |J_1/J|$: then m_2 is a divisor of m, and so $(n, m_2) = 1$. Moreover, J_1/J is abelian. Let $x_1, x_2 \in J_1$, with $x_1 \underset{K_1}{=} x_2$. Since $x_1, x_2 \in H$, it follows, by hypothesis, that $Jx_1 \underset{H/J}{=} Jx_2$. Hence, since $Jx_1, Jx_2 \in J_1/J = Z(H/J)$, it follows that $Jx_1 = Jx_2$. Now, by 10.16, there is a normal complement K to J_1/J in K_1. But now

$$HK = HJ_1 K = HK_1 = G$$

and

$$H \cap K = H \cap K_1 \cap K = J_1 \cap K = J.$$

Thus K is a complement to H/J in G.

Finally, we show that $K \trianglelefteq G$. We have

$$J \leqslant K \trianglelefteq K_1 \trianglelefteq G.$$

Let $h \in H$. Then, since also $J \trianglelefteq H$,

$$J = J^h \leqslant K^h \trianglelefteq K_1^h = K_1.$$

By 3.40,

$$|K^h/(K^h \cap K)| = |K^h K/K|.$$

The left side of this equation divides

$$|K^h : J^h| = |K : J| = |G : H| = n,$$

whereas the right side divides

$$|K_1/K| = |J_1/J| = m_2.$$

Since $(n, m_2) = 1$, it follows that $K^h \leqslant K$. Similarly $K \leqslant K^h$. Thus

$$K^h = K.$$

Since this is true for every $h \in H$ and since $G = HK$, this shows that

$$K \trianglelefteq G.$$

This completes the induction argument.

As a special case we note the following generalization of 10.18.

10.41 Corollary. *Let H be a nilpotent Hall subgroup of the finite group G. Then there is a normal complement to H in G if and only if no two elements of H are fused in H by G.*

In [a94], M. Suzuki establishes necessary and sufficient conditions for a (not necessarily nilpotent) Hall subgroup of a finite group G to have a normal complement in G. The proof makes use of character theory.

As another consequence of 10.40, we prove

10.42 Corollary. *Let $J \trianglelefteq H \leqslant V \leqslant G$, with $|G : H| = n < \infty, |H/J| = m < \infty, (n, m) = 1$ and H/J nilpotent. Assume that whenever $h_1, h_2 \in H$ and $h_1 \underset{G}{=} h_2, h_1 \underset{V}{=} h_2$. Then the following statements are equivalent:*

(i) *There is a normal complement to H/J in G.*

(ii) *There is a normal complement to H/J in V.*

Proof. (i) $\Rightarrow$ (ii) Suppose that K is a normal complement to H/J in G: thus $K \trianglelefteq G = HK$ and $H \cap K = J$. Then $V \cap K \trianglelefteq V$ and, by Dedekind's rule (7.3), $V = H(V \cap K)$. Moreover $H \cap (V \cap K) = H \cap K = J$. Thus $V \cap K$ is a normal complement to H/J in V.

(ii) $\Rightarrow$ (i) Let $h_1, h_2 \in H$ with $h_1 \underset{G}{=} h_2$. Then, by hypothesis, $h_1 \underset{V}{=} h_2$. Hence, if there is a normal complement to H/J in V, then, by 10.40 (applied to V in place of G), $Jh_1 \underset{H/J}{=} Jh_2$; and then, also by 10.40 (applied to G), there is a normal complement to H/J in G.

We now use 10.42 to prove an important generalization of 10.36. First we need a definition.

10.43 Definition. *Let G be a finite group and P a Sylow p-subgroup of G. Suppose that for every Sylow p-subgroup P^* of G which contains $Z(P)$, $Z(P^*) = Z(P)$. Then G is said to be p-normal.*

It is easy to see that the definition of p-normality depends only on p and not on the choice of Sylow p-subgroup P of G (see **593**(i) and (iii)). Also it is obvious that G is p-normal in particular if P is abelian.

We shall use

10.44 Lemma. *Let G be a p-normal finite group, and let P be a Sylow p-subgroup of G. If $x_1, x_2 \in P$ and $x_1 \underset{G}{=} x_2$ then $x_1 \underset{W}{=} x_2$, where $W = N_G(Z(P))$.*

Proof. Let $x \in P, g \in G$ and $x^g \in P$. Then

$$Z(P) \leqslant C_G(x) \cap C_G(x^g) = C_G(x) \cap C_G(x)^g$$

(229). By Sylow's theorem, there is a Sylow p-subgroup P_1 of $C_G(x)$ containing $Z(P)$; and then there is also a Sylow p-subgroup P^* of G containing P_1. Since $Z(P)^{g^{-1}} \leqslant C_G(x)$,

$$Z(P)^{g^{-1}} \leqslant P_1^y$$

for some $y \in C_G(x)$. Now

$$Z(P) \leqslant P \cap P^* \cap (P^*)^{yg}.$$

Hence, since G is p-normal and by **229**,

$$Z(P) = Z(P^*) = Z(P^*)^{yg}.$$

Therefore

$$yg \in N_G(Z(P^*)) = N_G(Z(P)) = W.$$

Thus

$$yg = w$$

for some $w \in W$. Since $y \in C_G(x)$,

$$x^g = x^{yg} = x^w,$$

and so x and x^g are conjugate in W.

Now we can prove the main theorem about p-normal groups. It is an improvement by P. Hall of a result of O. Grün. It generalizes 10.36.

10.45 Theorem (O. Grün [a45], P. Hall). *Let G be a p-normal finite group and P a Sylow p-subgroup of G. Then, for any subgroup V of G which contains $N_G(Z(P))$, $G/O^p(G) \cong V/O^p(V)$.*
Proof. Let $W = N_G(Z(P))$. If $x_1, x_2 \in P$ and $x_1 \underset{G}{=} x_2$, then, by 10.44, $x_1 \underset{W}{=} x_2$ and therefore, since $W \leqslant V, x_1 \underset{V}{=} x_2$. Certainly $O^p(V) \leqslant O^p(G)$ **(157)**. Since $P \leqslant V, P$ is a Sylow p-subgroup of V and therefore $PO^p(V) = V$ **(252)**. Let $J = P \cap O^p(V) \trianglelefteq P$. Then $O^p(V)$ is a normal complement to P/J in V. Now, since P/J is nilpotent (7.44), we can apply 10.42 with $H = P$ and J and V as above. Hence there is a normal complement K to P/J in G. Then $K \trianglelefteq G$ and $G/K \cong P/J$, a p-group. Hence

$$O^p(G) \leqslant K.$$

Hence $P \cap O^p(G) \leqslant P \cap K = J = P \cap O^p(V) \leqslant P \cap O^p(G),$

and so $P \cap O^p(G) = P \cap K.$

Since $O^p(G)$ and K are normal subgroups of $G, P \cap O^p(G)$ and $P \cap K$ are Sylow p-subgroups of $O^p(G)$ and K respectively **(252)**. Hence, since $K/O^p(G)$ is a p-group,

$$K = O^p(G).$$

Therefore, since also $O^p(V)$ is a normal complement to P/J in V,

$$G/O^p(G) = G/K \cong P/J \cong V/O^p(V).$$

Now we shall use 10.42 to prove a fundamental theorem of Frobenius giving necessary and sufficient conditions for a finite group to be p-nilpotent. First we prove a lemma. Recall that whenever $H \leqslant G$, $C_G(H) \trianglelefteq N_G(H)$ (4.36).

10.46 Lemma. *Let G be a finite group and P a Sylow p-subgroup of G. Suppose that, for every subgroup Q of P, $N_G(Q)/C_G(Q)$ is a p-group. Then, for any Sylow p-subgroup P^* of G and any $x \in P \cap P^*$, there is an element $y \in C_G(x)$ such that $P^* = P^y$.*

Proof. Let P^* be a Sylow p-subgroup of G and $Q = P \cap P^*$. We argue by induction on $|P : Q|$. If $|P : Q| = 1$ then $P^* = P$ and the assertion is obvious. Now assume that $|P : Q| > 1$. Then $Q < P$ and so, by 5.6, $Q < N_P(Q)$. By Sylow's theorem, there is a Sylow p-subgroup Q_1 of $N_G(Q)$ with $N_P(Q) \leqslant Q_1$, and there is a Sylow p-subgroup P_1 of G with $Q_1 \leqslant P_1$. Let $x \in P \cap P^* = Q$. Then also $x \in P \cap P_1$. Moreover, since $Q < N_P(Q) \leqslant P \cap P_1$, $|P : P \cap P_1| < |P : Q|$. Hence, by the induction hypothesis, there is an element $y_1 \in C_G(x)$ such that $P_1 = P^{y_1}$.

Also $Q < P^*$ and therefore (5.6) $Q < N_{P^*}(Q)$. Since $N_{P^*}(Q)$ is a p-subgroup of $N_G(Q)$ there is, by Sylow's theorem, an element $w \in N_G(Q)$ such that $N_{P^*}(Q) \leqslant Q_1^w$. By hypothesis, $N_G(Q)/C_G(Q)$ is a p-group, and therefore $N_G(Q) = Q_1 C_G(Q)$ **(252)**. Hence we may choose $w \in C_G(Q) \leqslant C_G(x)$, since $x \in Q$.

Now $$Q < N_{P^*}(Q) \leqslant P^* \cap P_1^w = P^* \cap P^{y_1 w}.$$

Let $u = (y_1 w)^{-1} \in C_G(x)$. Then

$$x = x^u \in Q^u < N_{P^*}(Q)^u \leqslant P \cap (P^*)^u.$$

Hence $|P : P \cap (P^*)^u| < |P : Q|$ and so, by the induction hypothesis, there is an element $y_2 \in C_G(x)$ such that $(P^*)^u = P^{y_2}$. Then $P^* = P^{y_2 y_1 w}$ and $y_2 y_1 w \in C_G(x)$. This completes the induction argument.

10.47 Theorem (Frobenius [a32]). *Let G be a finite group and P a Sylow p-subgroup of G. Then G is p-nilpotent if and only if $N_G(Q)/C_G(Q)$ is a p-group for every subgroup Q of P.*

Proof. Suppose that $N_G(Q)/C_G(Q)$ is a p-group for every $Q \leqslant P$. Then we want to show that there is a normal complement to P in G. Let $V = N_G(P)$. Then $P \trianglelefteq V$ and $(|V/P|, |P|) = 1$, so that, by the Schur–Zassenhaus theorem (10.30), there is a complement W to P in V. Then **(560)**

$$W \leqslant O^p(V) \leqslant C_G(P),$$

since, by hypothesis, $N_G(P)/C_G(P)$ is a p-group. Hence

$$P \leqslant C_G(W) \leqslant N_G(W),$$

and so $$W \trianglelefteq PW = V.$$

Thus W is a normal complement to P in V.

Now we apply 10.42 with $J = 1, H = P$ and $V = N_G(P)$. Note that P is nilpotent (7.44). It only remains to verify the fusion condition. Let $x \in P, g \in G$ and $x^g \in P$. Then $x \in P \cap P^{g^{-1}}$, and therefore, by 10.46,

$$P^{g^{-1}} = P^y$$

for some $y \in C_G(x)$. Then $yg \in V$ and

$$x^g = x^{yg}.$$

Hence x and x^g are conjugate in V. Hence, by 10.42, there is a normal complement to P in G.

Suppose, conversely, that G has a normal p-complement K and let $Q \leqslant P$. Then

$$Q \trianglelefteq N_G(Q), \qquad K \cap N_G(Q) \trianglelefteq N_G(Q)$$

and

$$Q \cap K \cap N_G(Q) = Q \cap K = 1,$$

since p does not divide $|K|$. Hence, by 3.53,

$$[Q, K \cap N_G(Q)] = 1,$$

and so

$$K \cap N_G(Q) \leqslant C_G(Q) \trianglelefteq N_G(Q).$$

Now, by 3.40, $N_G(Q)/(K \cap N_G(Q)) \cong KN_G(Q)/K \leqslant G/K$, a p-group. Hence

$$O^p(N_G(Q)) \leqslant K \cap N_G(Q) \leqslant C_G(Q),$$

and so $N_G(Q)/C_G(Q)$ is a p-group.

10.48. For p odd, there is a powerful and important improvement of Frobenius's theorem 10.47 due to J. G. Thompson: see Huppert [b21] p. 438, theorem 4.6.2 or Schenkman [b35] p. 273, theorem 9.3.a. Further refinements of Thompson's result have been made. For instance, G. Glauberman has proved that if P is a non-trivial Sylow p-subgroup of the finite group G, with p odd, there is an explicitly defined non-trivial abelian characteristic subgroup A of P such that if $N_G(A)$ is p-nilpotent then G is p-nilpotent: see Gorenstein [b13] p. 280, theorem 8.3.1.

For a different approach to transfer theorems, based on delicate fusion results discovered by J. L. Alperin, see Gorenstein [b13], chapters 7 and 8, and the lectures of G. Glauberman presented in Powell and Higman [b33] chapter 1.

591 (G. Zappa [a105]; see also [a106]) Let H be a nilpotent Hall subgroup of the finite group G. Then the following statements are equivalent:
(i) There is a normal complement to H in G.
(ii) There is a right transversal T to H in G such that, for every $h \in H, hT = Th$.

592 The group Σ_4 is not 2-normal.

593 Let $J \leqslant H \leqslant G$. Then J is said to be *weakly closed in H with respect to G* if H contains no conjugate of J in G other than J itself.

(i) Show that if J is weakly closed in H with respect to G, and $g \in G$, then J^g is weakly closed in H^g with respect to G.

(ii) Show that if J is weakly closed in H with respect to G, then $J \trianglelefteq N_G(H)$. Show by an example that the converse is not true in general.

(iii) Suppose that G is finite and that P is a Sylow p-subgroup of G. Prove that G is p-normal if and only if $Z(P)$ is weakly closed in P with respect to G.

(iv) Suppose that G is finite and that H is a Sylow p-subgroup of G. Prove that J is weakly closed in H with respect to G if and only if J is pronormal in G. (See **268**. Hint. Use 7.14 and **334**.)

594 Let P be a Sylow p-subgroup of the finite group G.

(a) If G is p-nilpotent then every normal subgroup of P is weakly closed in P with respect to G (cf. **593**(ii)).

(b) The following two statements are equivalent:

(i) G is p-nilpotent.

(ii) G is p-normal and $N_G(Z(P))$ is p-nilpotent.

595 Let P be a Sylow p-subgroup of the finite group G. Suppose that $1 < P < G$, that $N_G(P) = P$ and that, whenever P_1, P_2 are distinct Sylow p-subgroups of G, $P_1 \cap P_2 = 1$. Then G is p-nilpotent. Moreover, if K is the normal p-complement of G then $K = \{1\} \cup (G \setminus \bigcup_{g \in G} P^g)$.

(Remarks. In the terminology of **248**, the hypothesis is that G is a Frobenius group, with P as a Frobenius complement in G. The result is a special case of the theorem of Frobenius mentioned in **248**. Hint. Show that if $g \in G$ and $Z(P) \leqslant P^g$ then $P^g = P$, hence in particular that G is p-normal and $N_G(Z(P)) = P$.)

596 Let G be a p-normal finite group and P a Sylow p-subgroup of G. Then, for any subgroup V of G which contains $N_G(Z(P))$, $P \cap G' = P \cap V'$. (Hint. Use 10.34 and 10.44.)

597 Suppose that the finite group G has a non-trivial Sylow p-subgroup P. Suppose further that for every non-trivial abelian subgroup Q of P, $N_G(Q) = P$. Then G is p-nilpotent. (Hint. Use **595**.)

598 Suppose that the finite group G has a non-trivial Sylow p-subgroup P. Then G is p-nilpotent if and only if for every non-trivial subgroup Q of P, $N_G(Q)$ is p-nilpotent.

599 (N. Itô [a62].) Let G be a finite group in which every proper subgroup is p-nilpotent. If G is not itself p-nilpotent then $O^p(G) = G$, G has a normal Sylow p-subgroup P, and G/P is a cyclic q-group for some prime $q \neq p$; moreover, every proper subgroup of G is nilpotent. (Hints. Argue by induction on $|G|$. Assume that G is not p-nilpotent and deduce from **598** that there is a non-trivial subgroup Q of P which is normal in G. By induction, reduce to the case $Q = P$. Then use 10.47 to show that there is a subgroup P_1 of P and an element $x \in N_G(P_1) \setminus C_G(P_1)$ such that $o(x)$ is a power of a prime $q \neq p$. Show that $\langle x \rangle P = G$.)

600 (O. J. Schmidt [a84], K. Iwasawa [a64].) Let G be a finite group with every proper subgroup nilpotent. Then either G is nilpotent, or there are primes p, q and positive integers m, n such that $|G| = p^m q^n$, G has a normal Sylow p-subgroup P and G/P is cyclic. In any case G is soluble. (Hint. Apply **563** and **599**. Remark. For another proof of this result, due to W. Gaschütz and avoiding the use of transfer, see Huppert [b21] pp. 280–3, theorems 3.5.1, 3.5.2.)

10.49. We end the chapter by mentioning what is perhaps the most important of the several substantial classification theorems which have been established during the past ten years: J. G. Thompson's classification of the insoluble N-groups ([a98]). An *N-group* is a finite group in which the normalizer of every non-trivial soluble subgroup is soluble. Thompson's result gives the simple N-groups and, in particular, provides a list of the so-called *minimal simple groups*, that is, the finite simple groups all of whose proper subgroups are soluble. For further information about classification of simple groups, see the references mentioned at the end of 3.61.

11

FINITE NILPOTENT AND SOLUBLE GROUPS

Throughout this chapter, G denotes a finite group. We begin by associating to G a new characteristic subgroup $\Phi(G)$, and then prove a result promised in chapter 7 which relates the normal structure and the arithmetical structure of a finite nilpotent group in a pleasing way.

11.1 Definitions. (i) Recall from **140** the definition of a maximal subgroup. A proper subgroup M of G is said to be a *maximal subgroup* of G if there is no subgroup L such that $M < L < G$.

(ii) If $G \neq 1$ then (because G is finite) G certainly contains at least one maximal subgroup. (Indeed, every proper subgroup of G lies in a maximal subgroup of G: see **140**(ii)). We define $\Phi(G)$ to be the intersection of all the maximal subgroups of G. If $G = 1$ we define $\Phi(G) = 1$.

$\Phi(G)$ is called the *Frattini subgroup* of G (after G. Frattini, 1852–1925, who first investigated its properties).

Let $\mathcal{M}$ denote the set of all maximal subgroups of G and let $\alpha \in \text{Aut } G$. From 3.29, if $M \in \mathcal{M}$ then $M^{\alpha} \in \mathcal{M}$. Moreover, since α is bijective, $\{M^{\alpha} : M \in \mathcal{M}\} = \mathcal{M}$. It follows that $\Phi(G)$ is a characteristic subgroup of G.

11.2 Lemma. *Let $K \lhd G$. Then K is a maximal subgroup of G if and only if G/K has prime order.*
Proof. Since $K \lhd G, |G/K| > 1$. Now, by 3.30, K is a maximal subgroup of G if and only if G/K has no non-trivial proper subgroup; that is, (by **29**) if and only if $|G/K| = p$ for some prime p.

11.3 Theorem. *The following seven statements are equivalent:*
  (i) *G is nilpotent.*
  (ii) *Every subgroup of G is subnormal in G.*
  (iii) *Whenever $H < G, H < N_G(H)$.*
  (iv) *Every maximal subgroup of G is normal in G.*
  (v) *$G' \leqslant \Phi(G)$.*
  (vi) *Every Sylow subgroup of G is normal in G.*
  (vii) *G is a direct product of groups of prime power orders.*
Proof. We may assume that $G \neq 1$, since otherwise all the statements are trivially true.

  (i) $\Rightarrow$ (ii) This is a special case of 7.59.

(ii) $\Rightarrow$ (iii) Let $H < G$. By (ii), there is a series from H to G. Hence there is a proper series from H to G, say

$$H = H_0 \lhd H_1 \lhd \ldots \lhd H_n = G.$$

Since $H < G, n > 0$. Then $H < H_1 \leqslant N_G(H)$.

(iii) $\Rightarrow$ (iv) Let M be a maximal subgroup of G. By (iii), $M < N_G(M) \leqslant G$. Now the maximality of M implies that $N_G(M) = G$. Thus $M \lhd G$.

(iv) $\Rightarrow$ (v) Let M be a maximal subgroup of G. By (iv), $M \lhd G$. Then, by 11.2, G/M is cyclic, of prime order. Therefore, by 3.52, $G' \leqslant M$. This is true for every maximal subgroup M of G and so, by definition of $\Phi(G)$, $G' \leqslant \Phi(G)$.

(v) $\Rightarrow$ (iv) Let M be a maximal subgroup of G. Then (v) implies that $G' \leqslant M$. Hence M/G' is a subgroup of the abelian group G/G' and so $M/G' \lhd G/G'$. Hence, by 3.30, $M \lhd G$.

(iv) $\Rightarrow$ (vi) Let P be a Sylow p-subgroup of G. Suppose that $N_G(P) < G$. Then there is a maximal subgroup M of G which contains $N_G(P)$. But then, by 5.14, $N_G(M) = M$, in contradiction to (iv). Hence (iv) implies that $N_G(P) = G$, that is, $P \trianglelefteq G$.

(vi) $\Rightarrow$ (vii) Let the distinct prime divisors of $|G|$ be $p_1, \ldots, p_s$, where s is a positive integer. By (vi), G has a normal Sylow p_i-subgroup P_i for each $i = 1, \ldots, s$. Then, by 8.6,

$$G = \text{Dr} \prod_{i=1}^{s} P_i.$$

(vii) $\Rightarrow$ (i) This follows by 7.44 and repeated application of 7.49(i).

*601 Suppose that A is an abelian normal subgroup of G such that G splits over A. Let H be a complement to A in G. Then A is a minimal normal subgroup of G if and only if H is a maximal subgroup of G (cf. 359).

602 Suppose that A is an abelian minimal normal subgroup of G. Then either $A \leqslant \Phi(G)$ or G splits over A (cf. 359).

603 (W. Gaschütz [a35].) (i) Suppose that M is a maximal subgroup of G. Then either $M \geqslant Z(G)$ or $M \geqslant G'$. (Hint. Show that if $M \not\geqslant Z(G)$ then $M \lhd G$.)

(ii) $G' \cap Z(G) \leqslant \Phi(G)$.

604 G has a maximal subgroup of order 2 if and only if $|G| = 2p$ for some prime p. (Hint. If G has a maximal subgroup of order 2 which is not normal in G, apply 6.11 and 6.13.)

605 (i) A normal maximal subgroup of G is necessarily a maximal normal subgroup of G (see 363).

(ii) If G is soluble then a maximal normal subgroup of G is necessarily a normal maximal subgroup of G.

(iii) Show by an example that if G is insoluble then a maximal normal subgroup of G need not be a maximal subgroup of G.

606 For every prime p, $O_p(G)$ is the unique Sylow p-subgroup of $F(G)$, the Fitting subgroup of G. (Hint. See 7.44 and 11.3.)

607 If G is nilpotent then, for every set ϖ of primes,

$$G = O_\varpi(G) \times O^\varpi(G).$$

608 Suppose that G has a maximal subgroup M such that $M_G = 1$ and $|G : M| = 4$. Then $M \ntrianglelefteq G$ and G is isomorphic to either A_4 or Σ_4. (Hints. Apply 4.14 and **289**. Note that by 11.3, G cannot be a 2-group. Remark. Both A_4 and Σ_4 have maximal subgroups of index 4 with trivial core: see **288** and **289**.)

609 Suppose that G is non-trivial and supersoluble (see **389**). Show that if p is the largest prime divisor of $|G|$ then G has a normal Sylow p-subgroup. By induction on the number of distinct prime divisors of $|G|$, deduce that if q is the smallest prime divisor of $|G|$ then G is q-nilpotent. (Hint. Apply 398(iii) and **603**.)

610 (B. Huppert [a61].) If every proper subgroup of G is supersoluble then G is soluble. (cf. **600**. Hint. Use **609** and **599**.)

611 Suppose that G acts transitively on the finite set X. For each subset Y of X and each $g \in G$, let $Yg = \{yg : y \in Y\} \subseteq X$ (see **187**); and for each $x \in X$ and each subgroup H of G, let $xH = \{xh : h \in H\} \subseteq X$.

A subset Y of X is said to be a *block* (or *set of imprimitivity*) for the action if, for each $g \in G$, either $Yg = Y$ or $Yg \cap Y = \emptyset$. In particular, $\emptyset$, X and all 1-element subsets of X are obviously blocks: these are called the *trivial blocks*.

The action is said to be *primitive* if the only blocks are the trivial blocks; otherwise the action is said to be *imprimitive*.

Prove the following statements.

(i) If Y is a block for the action then, for every $g \in G$, Yg is also a block. Moreover, if $Y \neq \emptyset$ then $|Y|$ divides $|X|$.

Let $x \in X$ and let $L = \mathrm{Stab}_G(x)$.

(ii) For any subgroup H of G containing L, xH is a block.

(iii) Any block containing x is of the form xH, where $L \leqslant H \leqslant G$. (Hint. If Y is a block containing x, let $H = \{h \in G : Yh = Y\}$.)

(iv) Now suppose that $|X| > 1$. Then the action is primitive if and only if L is a maximal subgroup of G.

612 If G acts transitively on a set X such that $|X| = p$, a prime number, then the action is primitive. (Hint. Apply **611**(i).)

613 Suppose that G acts on the finite set X and let $K \trianglelefteq G$. Let ρ be the permutation representation of G corresponding to the action. If the action is primitive, then either $K \leqslant \mathrm{Ker}\,\rho$ or the action of K on X (by restriction of the action of G) is transitive. (See **611**. Hint. Show that any K-orbit is a block for the action of G.)

614 Suppose that G acts on the finite set X, where $|X| \geqslant 2$. The action is said to be *2-transitive* (or *doubly transitive*) if, whenever (x, x') and (y, y') are ordered pairs of distinct elements of X, there is an element $g \in G$ such that $xg = y$ and $x'g = y'$.

Prove the following statements.

(i) Let $x \in X$ and let $L = \mathrm{Stab}_G(x)$. Then the action is 2-transitive if and only if the action is transitive and, furthermore, the action of L on $X \setminus \{x\}$, defined by restriction of the action of G, is transitive.

(ii) If the action is 2-transitive and if $|X| = n$, then $|G|$ is divisible by $n(n-1)$. (Hint. Use (i) and 4.11.)

(iii) If the action is 2-transitive then it is primitive (see **611**).

615 Let n be an integer, $n \geqslant 2$, and let $X = \{1, 2, \ldots, n\}$.

(i) The natural action of Σ_n on X is 2-transitive, and if $n \geqslant 4$, the natural action of A_n on X is 2-transitive (see **614**).

(ii) Σ_n has a maximal subgroup of index n, and if $n \geqslant 3$, A_n has a maximal subgroup of index n. (cf. 5.29, **292**. Hint. Apply **611** and **614**.)

616 Let $H < G$ and let X be the set of right cosets of H in G. Then the action of G on X by right multiplication is 2-transitive if and only if there is an element $g \in G$ such that $G = H \cup HgH$ (see 4.13, **614**). Moreover, if the action is 2-transitive then $|H|$ is divisible by $|G : H| - 1$.

We shall now establish a few properties of $\Phi(G)$.

11.4 Lemma. *Let $K \trianglelefteq G$. Then $K \leqslant \Phi(G)$ if and only if there is no proper subgroup H of G such that $HK = G$.*
Proof. Suppose first that $K \leqslant \Phi(G)$. Let $H < G$. Then there is a maximal subgroup M of G such that $H \leqslant M < G$. Since $K \leqslant \Phi(G)$, $K \leqslant M$. Hence $HK \leqslant M < G$. Thus there is no proper subgroup H of G such that $HK = G$.

Now suppose that $K \nleqslant \Phi(G)$. Then $G \neq 1$ and, by definition of $\Phi(G)$, there is a maximal subgroup M of G such that $K \nleqslant M$. Then $M < MK \leqslant G$ (3.38). The maximality of M implies that $MK = G$. Thus in this case M is a proper subgroup of G such that $MK = G$.

It is now easy to establish the fundamental property of $\Phi(G)$ discovered in 1885 by Frattini.

11.5 Theorem (Frattini [a29]). *$\Phi(G)$ is nilpotent.*
Proof. Let P be any Sylow subgroup of $\Phi(G)$. Since $\Phi(G) \trianglelefteq G$, Frattini's lemma (5.13) shows that

$$G = N_G(P)\Phi(G).$$

Hence, by 11.4, $N_G(P) = G$. Thus $P \trianglelefteq G$, and so $P \trianglelefteq \Phi(G)$. Now it follows from 11.3 that $\Phi(G)$ is nilpotent.

In view of the definition of $F(G)$, the Fitting subgroup of G (7.64), the following corollary is immediate.

11.6 Corollary. $\Phi(G) \leqslant F(G)$.

11.7 Lemma. *Let $H \leqslant G$ and $K \trianglelefteq G$.*
  (i) *If $K \leqslant \Phi(H)$ then $K \leqslant \Phi(G)$.*
  (ii) *$\Phi(K) \leqslant \Phi(G)$* (cf. **395**).
Proof. (i) Suppose that $K \nleqslant \Phi(G)$. Then, by 11.4, there is a proper subgroup J of G such that

$$JK = G.$$

Assume that $K \leqslant \Phi(H)$. Then

$$K \leqslant H \leqslant G = JK,$$

so that, by Dedekind's rule (7.3),

$$H = (H \cap J)K.$$

By 11.4, the assumption that $K \leqslant \Phi(H)$ implies that

$$H \cap J = H.$$

Then

$$K \leqslant H \leqslant J,$$

and therefore

$$G = JK = J < G,$$

a contradiction. Hence, if $K \nleqslant \Phi(G)$, it follows that $K \nleqslant \Phi(H)$.

(ii) Since $\Phi(K)$ is characteristic in K and $K \unlhd G$, 3.15 shows that $\Phi(K) \unlhd G$. Application now of (i), with $\Phi(K)$ in place of K and K in place of H, gives the result.

Remark. It is not true in general that if $H \leqslant G$ then $\Phi(H) \leqslant \Phi(G)$: see **629**.

11.8 Lemma. *Let* $K \lhd G$. *Then*

(i) $\Phi(G)K/K \leqslant \Phi(G/K)$.

(ii) *If* $K \leqslant \Phi(G)$ *then* $\Phi(G)/K = \Phi(G/K)$.

Proof. It is clear from 3.30 that every maximal subgroup of G/K is of the form M/K, where M is a maximal subgroup of G containing K. Moreover, for every such $M, M/K$ is a maximal subgroup of G/K. The statements (i) and (ii) follow.

Remarks. (1) It is not true in general that if $K \lhd G$ then $\Phi(G)K/K = \Phi(G/K)$: see **630**.

(2) Note that, by (ii), $G/\Phi(G)$ always has trivial Frattini subgroup.

11.9 Lemma. *Let G be a p-group. Then $\Phi(G) = 1$ if and only if G is elementary abelian.*

Proof. We may assume that $G \neq 1$. Suppose first that G is elementary abelian. Since $G \neq 1, \Phi(G) < G$ (by definition of $\Phi(G)$). Moreover, by 7.41, G is characteristically simple. Therefore, since $\Phi(G)$ is a proper characteristic subgroup of $G, \Phi(G) = 1$.

Now suppose, conversely, that $\Phi(G) = 1$. By 7.44, G is nilpotent. Therefore, by 11.3, $G' \leqslant \Phi(G)$. Hence $G' = 1$, so that G is abelian. Let M be a maximal subgroup of G and let $g \in G$. Then $M \lhd G$ and so, by 11.2, G/M has prime order. Since G is a p-group, $|G/M| = p$. Therefore **(105)** $g^p \in M$. This is true for every maximal subgroup M of G, and so

$$g^p \in \Phi(G) = 1.$$

Hence

$$g^p = 1$$

for every $g \in G$. Thus G is elementary abelian.

11.10 Corollary. *Let G be a p-group. Then $\Phi(G)$ is the unique smallest* *normal subgroup K of G such that G/K is elementary abelian. (In other words, $G/\Phi(G)$ is the $\mathfrak{X}$-residual of G, when $\mathfrak{X}$ is the class of elementary abelian p-groups.)*

Proof. Certainly $G/\Phi(G)$ is a p-group, and, by 11.8, $G/\Phi(G)$ has trivial Frattini subgroup. Hence, by 11.9, $G/\Phi(G)$ is elementary abelian.

Suppose that $K \trianglelefteq G$, with G/K elementary abelian. Then, by 11.8 and 11.9,

$$\Phi(G)K/K \leqslant \Phi(G/K) = K/K.$$

Hence $$\Phi(G) \leqslant K.$$

This completes the proof.

Remark. It follows that if G is a p-group then $G/\Phi(G)$ may be viewed in a natural way as a finite-dimensional vector space over $\mathbf{Z}_p$: see 7.40.

617 Let $x \in G$. Then x is said to be a *non-generator* of G if, whenever X is a set of generators of G with $x \in X$, $X \setminus \{x\}$ is also a set of generators of G.

Prove that $\Phi(G)$ is the set of all non-generators of G. (Hint. To show that each element of $\Phi(G)$ is a non-generator of G, apply 11.4.)

618 Let $K \trianglelefteq G$. A subgroup H of G is said to be a *supplement to K in G* if $HK = G$. In particular, G itself is a supplement to K in G; and any complement to K in G is a supplement to K in G.

Let H be a *minimal* supplement to K in G; that is, let H be a supplement to K in G such that no proper subgroup of H is a supplement to K in G. Prove that then $H \cap K \leqslant \Phi(H)$. (Hint. Apply 11.4.)

619 (i) Suppose that G is nilpotent. If G/G' is cyclic then G is cyclic. (Hint. Use 11.3 and 11.4. Remark. The condition that G is nilpotent is needed, as the group Σ_3 shows.)
(ii) Let G be a non-trivial p-group. If G has derived length n then $|G| \geqslant p^{2n-1}$. (Hint. Use (i) and induction on n. Remark. This bound can be improved substantially. P. Hall [a48] proved that a non-trivial p-group of derived length n has order $\geqslant p^{2^{n-1}+n-1}$: see Huppert [b21] p. 307, theorem 3.7.11.)

620 If $G/\Phi(G)$ is a ϖ-group then G is a ϖ-group. (Hint. Suppose the result does not hold and apply the Schur–Zassenhaus theorem 10.30, and 11.4.)

621 (W. Gaschütz [a35].) If $K/\Phi(G)$ is a nilpotent normal subgroup of $G/\Phi(G)$ then K is nilpotent. Thus $F(G/\Phi(G)) = F(G)/\Phi(G)$. In particular, if $G/\Phi(G)$ is nilpotent then G is nilpotent. (cf. 11.6. Hint. Show that for any Sylow subgroup P of K, $P\Phi(G) \trianglelefteq G$, and adapt the proof of 11.5.)

622 Let $\mathfrak{X}$ be a class of groups with the following two properties:
  (i) Every quotient group of every $\mathfrak{X}$-group is an $\mathfrak{X}$-group.
  (ii) Whenever $H/\Phi(H) \in \mathfrak{X}$, $H \in \mathfrak{X}$.
(For instance, $\mathfrak{X}$ may be the class of finite ϖ-groups or the class of finite nilpotent groups: see **620, 621**.)

Let $K \trianglelefteq G$ and suppose that $G/K \in \mathfrak{X}$. Prove that there is an $\mathfrak{X}$-subgroup of G which is a supplement to K in G (see **618**). Show by an example that there need not be a complement to K in G.

623 Let G be a non-trivial soluble group. Then $\Phi(G) < F(G)$. (cf. 11.6. Hint. Apply **396**(i) and **621**.)

624 (i) $F(G)/\Phi(G)$ is abelian. (cf. 11.6. Hint. Apply 11.3 and 11.7.)

(ii) If G is soluble then $C_G(F(G)/\Phi(G)) = F(G)$; hence $G/F(G)$ can be embedded in $\mathrm{Aut}(F(G)/\Phi(G))$. (See **390**. Hint. Apply **621**, 7.67 and (i).)

625 (i) If $F(G)$ is a p-group then $F(G)/\Phi(G)$ is elementary abelian. (Hint. Apply 11.7 and 11.10.)

(ii) Suppose that G is soluble and that $F(G)$ is a 2-generator p-group, where p is either 2 or 3. Let $\varpi = \{2, 3\}$. Then G is a ϖ-group. Moreover, if $p = 2$ then the Sylow 3-subgroups of G have order at most 3, and if $p = 3$ then the Sylow 2-subgroups of G have order at most 2^4. (Hint. Apply (i), 7.40, **47** and **624**.)

626 (W. Gaschütz [a 35].) (i) If A is an abelian normal subgroup of G such that $A \cap \Phi(G) = 1$ then G splits over A. (Hint. Apply **618** and 11.7(i).)

(ii) Suppose that $\Phi(G) = 1$. If $F(G) \ne 1$ then every non-trivial normal subgroup A of G contained in $F(G)$ is a direct product of abelian minimal normal subgroups of G and G splits over A. In particular, if G is a non-trivial soluble group such that $\Phi(G) = 1$ then $F(G) = S(G)$, the socle of G (see **397**), and G splits over $F(G)$. (Hints. Argue by induction on $|A|$. Let B be a minimal normal subgroup of G contained in A and apply **624** and (i).)

627 (W. Gaschütz [a35]) $\Phi(G) = 1$ if and only if G splits over $S_1(G)$, where $S_1(G)$ is defined as in **412**. (Hints. See **626**. If $\Phi(G) \ne 1$, consider a minimal normal subgroup H of G with $H \le \Phi(G)$. Show that $H \le S_1(G)$, and apply **414**. Then use 11.4 to show that G cannot split over $S_1(G)$.)

628 Suppose that H and K are normal subgroups of G such that $G = H \times K$.

(i) For any maximal subgroup L of K, $H \times L$ is a maximal subgroup of G; and every maximal subgroup of G containing H is of the form $H \times L$, where L is a maximal subgroup of K. (Hint. Apply **402**.)

(ii) If $K \ne 1$ then the intersection of all maximal subgroups of G containing H is $H \times \Phi(K)$.

(iii) $\Phi(G) = \Phi(H) \times \Phi(K)$. (Hint. Apply (ii) and 11.7.)

(iv) Suppose that M is a subgroup of G which contains neither H nor K. Then M is a maximal subgroup of G if and only if M is a subdirect product of H and K and $H/H \cap M$ is simple (see **441** and 8.19). (Hint. If M is a subdirect product of H and K, and if $M \le L \le G$. show by means of 8.19 and **439** that $|L| = |H \cap L||K|$.)

(v) Suppose that either H or K is soluble. Then any maximal subgroup M of G which contains neither H nor K is necessarily normal in G. Moreover, G has such a maximal subgroup M if and only if $(|H/H'|, |K/K'|) > 1$. (Hints. Apply (iv), 7.55, 8.19 and **297**.)

(vi) If H and K are isomorphic non-abelian simple groups then G has maximal subgroups which are isomorphic to H (and are not normal in G). (Hint. Apply (iv) and **438**.)

629 Let $G = \Sigma_4$. Then $|F(G)| = 4$, $F(G)$ is the unique minimal normal subgroup of G, and $\Phi(G) = 1$. If H is a Sylow 2-subgroup of G then $\Phi(H) \ne 1$. (cf. 11.7, **623**, **624**, **625**, **626**. Hint. See **289**.)

630 Let $K = C_p$, $M = \mathrm{Aut}\, C_p$ and $G = \mathrm{Hol}\, C_p$, where p is any prime. Then

(i) M is a maximal subgroup of G, and $M_G = 1$.

(ii) $F(G) = K$ and $\Phi(G) = 1$.

(iii) For $p = 5$, $\Phi(G/K)$ is non-trivial (cf. 11.8).

(Hints. See **499**, 7.65, 9.15(iii).)

631 Let M be a maximal subgroup of G. Suppose that M has a normal Sylow p-subgroup P which is not normal in G. Then $N_G(P) = M$ and P is a Sylow p-subgroup of G. (Hint. Suppose that P is not a Sylow p-subgroup of G and apply Sylow's theorem and 5.6.)

632 Suppose that there is an insoluble finite group with an abelian maximal subgroup and let G be such a group of least possible order. Let M be an abelian maximal subgroup of G. Establish the following consequences.

(i) $M_G = 1$. (Hint. If not, apply 7.47.)

(ii) G' is the unique minimal normal subgroup of G.

(iii) M is a Hall subgroup of G. (Hint. Apply **631**.)

(iv) G is p-nilpotent for every prime p dividing $|M|$. (Hint. Apply Burnside's theorem 10.21.)

(v) M is a complement to G' in G.

(vi) Every complement to G' in G is conjugate to M. (Hint. 10.31 is applicable.)

(vii) Every subgroup of G isomorphic to M is conjugate to M.

(viii) Let q be a prime divisor of $|G'|$ and let Q be a Sylow q-subgroup of G'. Then $N_G(Q)$ has a subgroup M^* isomorphic to M. (Hint. Apply Frattini's lemma 5.13 and the Schur–Zassenhaus theorem 10.30.)

(ix) M^* is a maximal subgroup of G and $Q \trianglelefteq G$.

(x) If follows that G is soluble, a contradiction.

We conclude that a finite group is soluble if it has an abelian maximal subgroup. (See also **633**, **634**. Remarks. This result was proved in a more general version by B. Huppert [a61], and independently by I. N. Herstein [a55]. An insoluble finite group *can* have a nilpotent maximal subgroup: for instance, it is known that the Sylow 2-subgroups of the simple group $PSL_2(\mathbf{Z}_{17})$ are maximal subgroups. However, J. G. Thompson [a96] has proved the important result that a finite group is soluble if it has a nilpotent maximal subgroup of odd order: see Gorenstein [b13] p. 340, theorem 10.3.2, or Huppert [b21] p. 445, theorem 4.7.4, or Schenkman [b35] p.277, theorem 9.3.b.)

633 Suppose that G has an abelian maximal subgroup (so that, by **632**, G is soluble). Then G has derived length at most 3. (Remark. We shall see in **634** that G can have derived length 3.)

634 Let $G = GL_2(\mathbf{Z}_3)$, $G_1 = SL_2(\mathbf{Z}_3)$ and $K = Z(G)$. Note that $|G/G_1| = 2$ and, by **123**, $|K| = 2$ and $K < G_1$.

(i) Show that $G_1/K \cong A_4$ and that G_1 has a normal Sylow 2-subgroup, J say. (Hint. See **193**, **288** and **289**.)

(ii) Show that J is non-abelian. Deduce that $G_1' = J$. (Hints. Note that, by Sylow's theorem, J contains every element of G_1 of order a power of 2. To prove that $G_1' = J$, use **164** and **288**.)

(iii) Deduce that $G' = G_1$. Hence G_1 has derived length 3 and G has derived length 4.

(iv) Show that G_1 has a cyclic subgroup M of order 6 and prove that M is a maximal subgroup of G_1. (cf. **633**. Hint. If M were not maximal in G_1 then $|G_1/G_1'|$ would be even.)

(v) Show that G_1 has a unique involution. Deduce from 9.33 that $J \cong Q_8$, the quaternion group.

We shall prove next a fundamental result of Burnside on sets of generators of finite p-groups.

11.11 Definition. Let X be a set of generators of G (2.29). We say that X is a *minimal set of generators* of G if, for every proper subset Y of X, $\langle Y \rangle$ is a proper subgroup of G.

It is clear from the definition that any set of generators of G contains a minimal set of generators of G.

This notion of a minimal set of generators of a group is one possible analogue of the concept of a base of a vector space. However, even for a finite abelian group there can be minimal sets of generators containing different numbers of elements. For instance, let G be a cyclic group of order 6, say $G = \langle x \rangle$. Then $\{x\}$ is a minimal set of generators of G. But also $G = \langle x^2, x^3 \rangle$ and, because $\langle x^2 \rangle < G$ and $\langle x^3 \rangle < G, \{x^2, x^3\}$ is a minimal set of generators of G.

What we are going to prove is that if G is a p-group then any two minimal sets of generators of G do contain the same number of elements.

11.12 Burnside's basis theorem. *Let G be a non-trivial p-group, $\bar{G} = G/\Phi(G)$ and $|\bar{G}| = p^d$. For each $x \in G$, let $\bar{x} = x\Phi(G) \in \bar{G}$, and for each non-empty subset X of G, let $\bar{X} = \{\bar{x} : x \in X\} \subseteq \bar{G}$. If X is a minimal set of generators of G then $\bar{X}$ is a base of $\bar{G}$ (viewed in the natural way as a vector space over $\mathbf{Z}_p$) and $|X| = d$. Conversely, if X is a subset of G such that $|X| = d$ and $\bar{X}$ is a base of $\bar{G}$ then X is a minimal set of generators of G.*

Proof. Since $G \neq 1, d > 0$. By 11.10, $\bar{G}$ is elementary abelian and therefore (as in 7.40) can be viewed as a vector space over $\mathbf{Z}_p$. Since $|\bar{G}| = p^d$, the dimension of this vector space is d.

Let $\emptyset \subset X \subseteq G$. If $\langle X \rangle = G$ then $\langle \bar{X} \rangle = \bar{G}$ (**108**). Then, regarded as a set of vectors of the vector space $\bar{G}, \bar{X}$ certainly spans $\bar{G}$. Suppose, conversely, that $\bar{X}$ is a set of vectors spanning $\bar{G}$. Then, by definition of the vector space structure on $\bar{G}$ (see 7.40), $\bar{X}$ is a set of generators of the elementary abelian group $\bar{G}$. Let $\langle X \rangle = H \leqslant G$. Then

$$\bar{G} = \langle \bar{X} \rangle = H\Phi(G)/\Phi(G).$$

Hence $\qquad\qquad G = H\Phi(G).$

Now it follows from 11.4 that $H = G$.

Thus X is a set of generators of G if and only if $\bar{X}$ is a spanning set of vectors of the vector space $\bar{G}$. Now suppose that X is a minimal set of generators of G. Then there is no proper subset Y of X such that $\bar{Y}$ spans $\bar{G}$. Hence $\bar{X}$ is a minimal spanning set of vectors of $\bar{G}$, that is, a base of $\bar{G}$. Since there is no proper subset Y of X such that $\bar{Y} = \bar{X}$, it follows that

$$|X| = |\bar{X}| = d.$$

Suppose, conversely, that $|X| = d$ and $\bar{X}$ is a base of $\bar{G}$. Then, since $\bar{X}$ spans $\bar{G}$, it follows from above that X is a set of generators of G. We have shown that there is no set of generators of G with fewer than d elements. Therefore, since $|X| = d, X$ is a minimal set of generators of G.

11.13 Corollary (P. Hall [a48], 1933). *Let G be a p-group and $\bar{G} = G/\Phi(G)$, where, say, $|\bar{G}| = p^d$ and $|\Phi(G)| = p^m$.*

(i) *Let $\varphi : \text{Aut } G \to \text{Aut } \bar{G}$ be the homomorphism defined by $\varphi : \alpha \mapsto \bar{\alpha}$, where, for each $\alpha \in \text{Aut } G, x \in G$ and $\bar{x} = x\Phi(G) \in \bar{G}$,*

$$\bar{x}^{\bar{\alpha}} = \overline{x^\alpha}(= x^\alpha \Phi(G))$$

(see 136). Then $\text{Ker } \varphi$ is a normal p-subgroup of G of order at most p^{dm}, and $\text{Aut } G/\text{Ker } \varphi$ can be embedded in $\text{GL}_d(\mathbf{Z}_p)$.

(ii) $|\text{Aut } G|$ *divides* $p^{dm}(p^d - 1)(p^d - p)(p^d - p^2)\ldots(p^d - p^{d-1})$.

Proof. (i) Since $\Phi(G)$ is a characteristic subgroup of G, it is straightforward to verify that, for each $\alpha \in \text{Aut } G, \bar{\alpha}$ is well defined and is an automorphism of $\bar{G}$, and that the map φ is a homomorphism (136).

Let X be a minimal set of generators of G. By 11.12, $|X| = d$: say

$$X = \{x_1, \ldots, x_d\}.$$

Then $\bar{X} = \{\bar{x}_1, \ldots, \bar{x}_d\}$ is a base of $\bar{G}$. Also, by 11.12, for any choice of d elements (not necessarily distinct) of $\Phi(G)$, say $u_1, \ldots, u_d \in \Phi(G)$, the set

$$\{x_1 u_1, \ldots, x_d u_d\}$$

is a minimal set of generators of G.

Now let $\mathscr{M}$ be the set of all *ordered* subsets of G of the form

$$(x_1 u_1, \ldots, x_d u_d),$$

with $u_1, \ldots, u_d \in \Phi(G)$. Then clearly

$$|\mathscr{M}| = p^{dm}.$$

Note that if $y_1, \ldots, y_d \in G$, then the ordered subset $(y_1, \ldots, y_d)$ of G belongs to $\mathscr{M}$ if and only if

$$\bar{y}_i = \bar{x}_i$$

for every $i = 1, \ldots, d$. Let $K = \text{Ker } \varphi$. If $\alpha \in K$ and $(y_1, \ldots, y_d) \in \mathscr{M}$ then

$$\overline{y_i^\alpha} = \bar{y}_i^{\bar{\alpha}} = \bar{y}_i = \bar{x}_i$$

for every $i = 1, \ldots, d$, and therefore $(y_1^\alpha, \ldots, y_d^\alpha) \in \mathscr{M}$.

From this it is clear that K acts in a natural way on the set $\mathscr{M}$. If α belongs to the stabilizer in K of an ordered set $(y_1, \ldots, y_d) \in \mathscr{M}$, then

$$(y_1^\alpha, \ldots, y_d^\alpha) = (y_1, \ldots, y_d),$$

hence (since this is an equation between *ordered* sets)

$$y_i^\alpha = y_i$$

for every $i = 1, \ldots, d$. Then, since $\langle y_1, \ldots, y_d \rangle = G$, it follows that $\alpha = 1$ (72). Hence, by 4.11, the length of each orbit of the action of K on $\mathscr{M}$ is equal to $|K|$. Hence $|\mathscr{M}|$ is a multiple of $|K|$. Therefore, since $|\mathscr{M}| = p^{dm}$, K is a p-group and $|K| \leqslant p^{dm}$.

By the fundamental theorem on homomorphisms, $K \trianglelefteq \operatorname{Aut} G$ and $(\operatorname{Aut} G)/K$ can be embedded in $\operatorname{Aut} \bar{G}$. Since $\bar{G}$ is elementary abelian of order p^d (11.10), $\bar{G}$ is isomorphic to the additive group of a vector space V of dimension d over $\mathbf{Z}_p$ (7.40). Therefore, by 47, $\operatorname{Aut} \bar{G} \cong \operatorname{GL}_d(\mathbf{Z}_p)$.

(ii) This follows immediately from (i), together with 2.16 and 2.17.

Remark. W. Gaschütz [a37] has proved the important result that if G is a p-group with $|G| > p$ then $|\operatorname{Aut} G/\operatorname{Inn} G|$ is divisible by p: see Huppert [b21] p. 403, theorem 3.19.1.

635 Let G be a non-trivial p-group and view $\bar{G} = G/\Phi(G)$ as a vector space over $\mathbf{Z}_p$ in the natural way. For any base B of $\bar{G}$ there is a minimal set of generators, X say, of G such that $\bar{X} = B$ (where $\bar{X}$ is defined as in 11.12).

636 Let G be a 2-generator p-group.

(i) If q is any prime divisor of $|\operatorname{Aut} G|$ distinct from p then $p^2 \equiv 1 \bmod q$.

(ii) If $p > 2$ then the largest prime divisor of $|\operatorname{Aut} G|$ does not exceed p, while if $p = 2$ then $\operatorname{Aut} G$ either is itself a 2-group or has order $2^r \times 3$, where r is some positive integer. (cf. 625. Note that, by 141, $|\operatorname{Aut} G| \neq 3$.)

637 If a 3-generator 3-group has an automorphism of prime order $q \neq 3$ then q is either 2 or 13. (Remark. $C_3 \times C_3 \times C_3$ is a 3-generator 3-group with automorphisms of orders 2 and 13.)

638 Let P be a Sylow p-subgroup of G, and let d be a positive integer such that every subgroup of G is a d-generator group. If the integers $|G|$ and $(p^d - 1)(p^{d-1} - 1)\ldots(p^2 - 1)(p - 1)$ are co-prime then G is p-nilpotent. (Hint. Apply 4.36, 10.47 and 11.13. Remark. This is an improved version of a result of Frobenius [a32].)

639 Let P be a Sylow p-subgroup of G, where p is the smallest prime divisor of $|G|$. Prove that if P is metacyclic (152) then G is p-nilpotent, unless $p = 2$ and $|G|$ is divisible by 3. Show by an example that if $p = 2$ and $|G|$ is divisible by 3 then G need not be p-nilpotent. (cf. 10.24. Hint. Use 108, 152 and 638.)

640 Suppose that G is a simple group of even order greater than 2. Then $|G|$ is divisible by 12, 16, or 56. (cf. 10.22, 10.23. Hint. Use 638.)

641 (i) Prove that every group of order $3^2 \times 5 \times 17$ is abelian and that every group of order $3^3 \times 5 \times 17$ is nilpotent. (cf. 278, 367. Hint. Use 563 and 638.)

(ii) Show that there is a group of order $3^4 \times 5 \times 17$ which is not nilpotent. (Hint. Use 486 to show that there is a group of order $3^4 \times 5$ which is not nilpotent.)

We shall now prove an important result partially known to Galois. Other versions of parts of the result are due to O. Ore [a76] and R. Baer [a6].

11.14 Theorem. *Suppose that $O_p(G) \neq 1$. Let $L = O_p(G)$ and $|L| = p^n$. Suppose also that G has a maximal subgroup M such that $M_G = 1$. Then*

(i) *L is elementary abelian,*

(ii) *M is a complement to L in G; in particular, $|G : M| = p^n$,*

(iii) *L is minimal normal in G, and*

(iv) *$C_G(L) = L$; hence L is the unique minimal normal subgroup of G.*

Suppose further that there is a prime q such that $O_q(M) \neq 1$. Then, for any such q,

(v) $p^n \equiv 1 \bmod q$; *in particular, $q \neq p$, and*

(vi) *every complement to L in G is conjugate to M.*

Remark. If G is a non-trivial soluble group with a maximal subgroup M such that $M_G = 1$ then, by **381**, there is a prime p such that $O_p(G) \neq 1$ and so statements (i) to (iv) hold. In particular, there is only one prime p such that $O_p(G) \neq 1$. Furthermore, unless $|G| = p, M \neq 1$. Then, since M is soluble (7.46), there is a prime q such that $O_q(M) \neq 1$ and then statements (v) and (vi) also hold. There may be several such primes q: see **630**.

Proof of the theorem. (i) Since M is a maximal subgroup of G, $\Phi(G) \leqslant M$, and since $\Phi(G) \trianglelefteq G$, it follows that $\Phi(G) \leqslant M_G = 1$. Thus $\Phi(G) = 1$. Then, because $L \trianglelefteq G$, 11.7 shows that $\Phi(L) = 1$. As L is a p-group, it follows from 11.9 that L is elementary abelian.

(ii) Since $1 \neq L \trianglelefteq G$ and $M_G = 1$, $L \not\leqslant M$. Therefore $M < ML \leqslant G$ (3.38), and so the maximality of M implies that $ML = G$. Since $L \trianglelefteq G$, $M \cap L \trianglelefteq M$ and since, by (i), L is abelian, $M \cap L \trianglelefteq L$. Hence $M \cap L \trianglelefteq ML = G$, so that $M \cap L \leqslant M_G = 1$. Thus $ML = G$ and $M \cap L = 1$; that is, M is a complement to L in G. In particular, $|G : M| = |L| = p^n$.

(iii) Let $1 < K \trianglelefteq G$ with $K \leqslant L$. Then $M \leqslant MK \leqslant G$. Since $M_G = 1$, $K \not\leqslant M$ and so $MK \neq M$. Therefore, by the maximality of M, $MK = G$. Moreover, $M \cap K \leqslant M \cap L = 1$, by (ii). Hence M is a complement to K in G, and so $|K| = |G : M| = |L|$, by (ii). Since $K \leqslant L$, it follows that $K = L$. Thus L is minimal normal in G.

(iv) By (i) and (ii),

$$L \leqslant C_G(L) \leqslant G = ML.$$

Therefore, by Dedekind's rule (7.3),

$$C_G(L) = (M \cap C_G(L))L.$$

By 4.36, $C_G(L) \trianglelefteq G$. Therefore $M \cap C_G(L) \trianglelefteq M$. Moreover L centralizes and therefore normalizes $M \cap C_G(L)$.

Hence $\qquad\qquad M \cap C_G(L) \trianglelefteq ML = G.$

Since $M_G = 1$, it follows that $M \cap C_G(L) = 1$.

Thus $\qquad\qquad C_G(L) = L.$

Suppose that G has a minimal normal subgroup $N \neq L$. Then $L \cap N \trianglelefteq G$ and, by (iii), $N \not\leqslant L$, so that $L \cap N < N$. Since N is minimal normal in G, it follows that $L \cap N = 1$. Then, by 3.53,

$$[L, N] = 1.$$

Hence $N \leqslant C_G(L) = L$. This is a contradiction. Therefore L is the unique minimal normal subgroup of G.

Now suppose that $O_q(M) \neq 1$ for some prime q. Let $Q = O_q(M)$.

(v) Since $Q \trianglelefteq M$ and $L \trianglelefteq G$,

$$QL \trianglelefteq ML = G,$$

by 7.4 and (ii). Also, by (ii), $Q \cap L = 1$ and so, by 3.40,

$$|QL| = |Q| |L|.$$

Therefore, since $Q \neq 1$ and $L = O_p(G)$, it follows that $q \neq p$. Thus Q is a Sylow q-subgroup of QL.

Now $1 < Q \trianglelefteq M$ and, since $M_G = 1, Q \ntrianglelefteq G$. Therefore, the maximality of M implies that

$$N_G(Q) = M.$$

Hence

$$N_{QL}(Q) = (QL) \cap M = Q(L \cap M) = Q,$$

by Dedekind's rule (7.3) and (ii). Thus

$$|L| = |QL : Q| = |QL : N_{QL}(Q)| \equiv 1 \bmod q,$$

by Sylow's theorem; that is,

$$p^n \equiv 1 \bmod q.$$

(vi) Let M^* be any complement to L in G, and let $Q^* = O_q(M^*)$. The natural homomorphism v of G onto G/L maps Q to QL/L and Q^* to Q^*L/L. Since, by (ii), M is a complement to L in G, the restriction of v to M is an isomorphism of M onto G/L. Therefore, v must map $Q = O_q(M)$ to $O_q(G/L)$. Similarly, v must map Q^* to $O_q(G/L)$. Thus

$$QL/L = O_q(G/L) = Q^*L/L,$$

so that

$$QL = Q^*L.$$

Since, by (v), $q \neq p$, it follows that Q and Q^* are Sylow q-subgroups of QL. Therefore, by Sylow's theorem,

$$Q^* = Q^x$$

for some $x \in QL$. As in (v),

$$M = N_G(Q).$$

Since $Q^* \trianglelefteq M^*$, it follows that

$$M^* \leqslant N_G(Q^*) = N_G(Q^x) = M^x,$$

by **229**. Because M and M^* are complements to L in G,

$$|M^*| = |G/L| = |M| = |M^x|.$$

Hence

$$M^* = M^x.$$

11.15 Corollary (Galois). *Suppose that G is non-trivial and soluble, and let M be a maximal subgroup of G. Then $|G:M| = p^n$ for some prime p and positive integer n.*

Proof. Let $\bar{G} = G/M_G$ and $\bar{M} = M/M_G$. Then, by 3.30, $\bar{M}$ is a maximal subgroup of $\bar{G}$ such that $\bar{M}_{\bar{G}} = 1$. Since $\bar{G}$ is soluble (7.46) and non-trivial, there is a prime p such that $O_p(\bar{G}) \neq 1$ (**381**). Let $|O_p(\bar{G})| = p^n$. Then, by 11.14,

$$|\bar{G}:\bar{M}| = p^n.$$

Hence
$$|G:M| = p^n.$$

Remarks. It can be shown that in the group $GL_3(\mathbf{Z}_2)$, which has order 168, every maximal subgroup has index either 7 or 8. We know (see **385**) that $GL_3(\mathbf{Z}_2)$ is simple. It is in fact the only known non-abelian simple group in which every maximal subgroup has prime power index.

642 Suppose that G has a non-trivial abelian normal subgroup. Then the following two statements are equivalent:
 (i) G has a maximal subgroup M such that $M_G = 1$.
 (ii) $\Phi(G) = 1$ and G has a unique minimal normal subgroup.

643 Suppose that G has an abelian minimal normal subgroup L such that $O_q(G/L)$ is non-trivial for some prime q. Then the following two statements are equivalent:
 (i) G has a maximal subgroup M such that $M_G = 1$.
 (ii) $C_G(L) = L$.
(Hint. To prove that (ii) $\Rightarrow$ (i), apply 7.65, **621** and **642**.)

644 Suppose that $O_p(G) \neq 1$, and that G has a maximal subgroup M such that $M_G = 1$. Then
 (i) $O_p(G) = F(G)$,
 (ii) $O_p(G)$ is the unique maximal abelian normal subgroup of G, and
 (iii) $O_p(G)$ is a maximal abelian subgroup of G.

(See **235, 251, 400**. See also **645**.)

645 Let $G = GL_2(\mathbf{Z}_3)$ and $K = Z(G)$. Show that
 (i) K is the unique maximal abelian normal subgroup of G,
 (ii) $K < F(G)$, and
 (iii) K is not a maximal abelian subgroup of G.
(cf. **644**. Hint. Apply **193, 289** and **634**.)

*646 Suppose that G is non-trivial and supersoluble (see **389**). Then every maximal subgroup of G has prime index in G. (Remark. B. Huppert [a61] proved conversely that a non-trivial finite group in which every maximal subgroup has prime index is necessarily supersoluble. See 11.16 and the following remarks.)

647 (Ore [a76]). Suppose that G is a non-trivial soluble group, and let M and M^* be maximal subgroups of G. Then M and M^* are conjugate subgroups of G if and only if $M_G = M_G^*$.

648 Suppose that G is a non-trivial soluble group. Then any two faithful primitive actions of G on sets are equivalent. (See **611**, 4.19. Hint. Apply **611, 647**, 4.20, 4.21.)

*649 Suppose that the group H acts on the abelian group A, say with action φ. Let $\bar{H} = \operatorname{Im} \varphi \leqslant \operatorname{Aut} A$ (see 9.3). We say that φ is *irreducible* if the only $\bar{H}$-invariant subgroups of A are 1 and A.

Suppose that $A \neq 1$ and let $J = H_\varphi \times A$. Then φ is irreducible if and only if A is a minimal normal subgroup of J. In particular, if φ is irreducible and A is finite then A is an elementary abelian p-group for some p.

650 (i) Let F be any field. Then the action of $F^\times$ on F^+ by multiplication (as in 9.2(i)) is irreducible (**649**).

(ii) For any prime p and any positive integer n, there is a finite soluble group G with a maximal subgroup of index p^n in G. (cf. 11.15. Hints. Apply **601** and **649**. The existence of a field with p^n elements may be assumed. For the proof of existence of such a field, see, for instance, Herstein [b19] p. 316, lemma 7.4, or Lang [b28] p. 182, §5, or Rotman [b34] p. 155, theorem 8.6.)

651 The following three statements are equivalent:

(i) $O_p(G) \neq 1$ and G has a maximal subgroup M such that $M_G = 1$.

(ii) $O_p(G) \neq 1$ and there is a faithful primitive action of G on some set (see **611**).

(iii) There is a non-trivial elementary abelian p-group A and a subgroup H of Aut A such that the natural action of H on A is irreducible and G is isomorphic to the relative holomorph HA of A (see **649**).

(Hints. To prove that (i)$\Leftrightarrow$(ii), apply **611**. To prove that (i) $\Rightarrow$ (iii), apply 9.13, 9.14, 11.14 and **649**. To prove that (iii) $\Rightarrow$ (i), apply **486, 499, 601, 649**.)

652 The following two statements are equivalent:

(i) G has a non-trivial abelian normal subgroup L, and G has a subgroup M such that $|G : M| = p$ and $M_G = 1$.

(ii) G is isomorphic to a relative holomorph of a group of order p.

(Hints. To prove that (i) $\Rightarrow$ (ii), apply 9.13, 9.14 and 11.14. To prove that (ii) $\Rightarrow$ (i), apply **486** and **499**. Remark. If G satisfies (i) and (ii) then, in particular, G is metacyclic: see **152** and **512**.)

653 Let the group H act on the group K, say with action φ. Let $\bar{H} = \operatorname{Im} \varphi \leqslant \operatorname{Aut} K$.

(i) If $J \trianglelefteq H$, then $C_K(J)$ is an $\bar{H}$-invariant subgroup of K (see **478**).

(ii) Suppose that H is finite, K is finite and elementary abelian, say of order p^n, and φ is irreducible (**649**). Then, for each prime q, either $O_q(H) \leqslant \operatorname{Ker} \varphi$ or $p^n \equiv 1 \bmod q$.

654 (i) Let A be a non-trivial abelian normal subgroup of G. Then the action of G on A by conjugation is irreducible if and only if A is a minimal normal subgroup of G (see 9.2(iii), **649**).

(ii) Suppose that A is an abelian minimal normal subgroup of G. Then $|A| = p^n$ for some prime p and positive integer n and, for each prime q, either $O_q(G/C_G(A))$ is trivial or $p^n \equiv 1 \bmod q$. In particular, $O_p(G/C_G(A))$ is trivial. (Hints. Note that the action of G on A by conjugation determines a faithful action of $G/C_G(A)$ on A: see **476**. Apply **653**.)

655 Let V be a vector space of finite dimension n over a field F, with $n > 0$.

(i) The natural action of GL(V) on V^+ is irreducible. Moreover, $\mathscr{A}(V)$, the affine group of V, has a unique minimal normal subgroup A, and $A \cong V^+$ (see **484, 486, 649**).

(ii) Suppose that $F = \mathbf{Z}_p$ for some prime p: then V^+ is elementary abelian of order p^n and GL(V) is finite. Let $1 < H \leqslant \operatorname{GL}(V)$ and suppose that the natural action of H on V^+ is irreducible. If H is soluble then $p^n \equiv 1 \bmod q$ for some prime divisor q of $|H|$, while if H is nilpotent then $p^n \equiv 1 \bmod q$ for every prime divisor q of $|H|$.

(iii) If $F = \mathbf{Z}_p$ then there is a cyclic subgroup H of GL(V) such that $|H| = p^n - 1$

and the natural action of H on V^+ is irreducible.
(Hints. For (ii), apply 653. For (iii), note that V^+ is isomorphic to the additive group of a field with p^n elements and apply 475, 9.14, 9.15(ii) and 650(i).)

656 Suppose that G has a maximal subgroup M such that $|G:M| = 6$. Then G has a chief factor isomorphic to either A_5 or A_6. This can be proved by the following argument:
   (i) Note first that in order to establish the result, it will suffice to assume that $M_G = 1$ and deduce that G has a minimal normal subgroup isomorphic to either A_5 or A_6. Suppose then that $M_G = 1$, and let K be a minimal normal subgroup of G.
   (ii) Use 11.14 to show that K is non-abelian.
   (iii) Note that, by 4.14, $|G|$ divides $2^4 \times 3^2 \times 5$. Then use 8.10, together with 5.17 and 5.19, to show that K is simple.
   (iv) By 184, K can be embedded in A_6. Hence, if $K \neq A_6$, $|K| \leqslant 180$.
   (v) Finally, apply 5.30 and 296.
(Remark. Both A_5 and A_6 have maximal subgroups of index 6: see 5.25, 615.)

657 (a) A group of odd order cannot have a maximal subgroup of index 9. This can be proved by the following argument:
   (i) Suppose the result false and let G be a group of least possible odd order with a maximal subgroup M such that $|G:M| = 9$. Then $M_G = 1$.
   (ii) Let K be a minimal normal subgroup of G. Apply 4.36, 11.2, 11.14 and 636 to show that K is non-abelian.
   (iii) Note that, by 4.14, $|G|$ divides $3^4 \times 5 \times 7$. Then use 8.10 to show that K is simple.
   (iv) Note that $|K|$ is divisible by 9. Use Sylow's theorem to show that the number of distinct Sylow 3-subgroups of K is 7, hence that K has a subgroup of index 7.
   (v) Deduce, by means of 4.14, that $|K|$ divides $3^2 \times 5 \times 7$.
   (vi) Derive a final contradiction by applying 5.19 and 570.
(Remark. By the Feit–Thompson theorem, every group of odd order is soluble: see 383. Then 11.14 is immediately applicable to give the result. However, it is unnecessary to invoke the Feit–Thompson theorem, as the argument outlined above shows.)
   (b) For any prime p such that $p \equiv -1 \pmod 3$, there is a group of order $3p^2$ with a maximal subgroup of order 3 and index p^2. (cf. 604. Hints. Let A be an elementary abelian group of order p^2. Note that, by 601, it is enough to show that there is a relative holomorph G of A of order $3p^2$ such that A is a minimal normal subgroup of G. See also 228. If G is a relative holomorph of A of order $3p^2$, use 4.36 and 4.38 to show that if A is not minimal normal in G then there is a subgroup B of A such that $|B| = p$ and $B \leqslant Z(G)$.)

We shall now prove an interesting partial converse to 11.15. We follow the proof given in Huppert [b21] p. 718, theorem 6.9.4.

11.16 Theorem (P. Hall). *Suppose that for every maximal subgroup M of G, $|G:M|$ is either a prime or the square of a prime. Then G is soluble.*
Proof. We argue by induction on $|G|$. We may suppose $G \neq 1$. Let K be a minimal normal subgroup of G. Any maximal subgroup of G/K is of the form M/K, where M is a maximal subgroup of G. Hence every maximal subgroup of G/K has index either a prime or the square of a prime. Therefore, by the induction hypothesis, G/K is soluble.
   Let p be the largest prime divisor of $|K|$ and let P be a Sylow p-subgroup

of K. If $N_G(P) = G$ then, since K is minimal normal in $G, K = P$. In this case K is soluble (7.44).

Now suppose that $N_G(P) < G$. Then there is a maximal subgroup M of G such that $N_G(P) \leqslant M$. By Frattini's lemma (5.13),

$$G = N_G(P)K.$$

Also $M \cap K \trianglelefteq M$ and, since $P \leqslant M \cap K \leqslant K, P$ is a Sylow p-subgroup of $M \cap K$ and (again by 5.13)

$$M = N_M(P)(M \cap K).$$

Now it follows, by 3.40 and Sylow's theorem, that

$$|G : N_G(P)| = |K : N_K(P)| \equiv 1 \bmod p \qquad \text{(i)}$$

and

$$|M : N_M(P)| = |M \cap K : N_{M \cap K}(P)| \equiv 1 \bmod p. \qquad \text{(ii)}$$

Since $N_G(P) \leqslant M \leqslant G = N_G(P)K$,

$$G = MK,$$

and therefore

$$|G : M| = |K : M \cap K|. \qquad \text{(iii)}$$

Moreover,

$$N_M(P) = M \cap N_G(P) = N_G(P)$$

and $$|G : N_G(P)| = |G : M||M : N_G(P)|.$$

Therefore we deduce from (i) and (ii) that

$$|G : M| \equiv 1 \bmod p. \qquad \text{(iv)}$$

By hypothesis, there is a prime q such that

$$|G : M| \text{ is either } q \text{ or } q^2. \qquad \text{(v)}$$

Certainly $q \neq p$ and, by (iii), q divides $|K|$. Therefore, by choice of p, it follows that

$$q < p.$$

Together with (iv) and (v), this implies that

$$|G : M| = q^2 \equiv 1 \bmod p.$$

This last congruence is possible with $q < p$ only if $p = 3$ and $q = 2$.

Thus $$|G : M| = 4.$$

Now (iii) shows that K has a subgroup of index 4. Therefore, by 4.14, K has a proper normal subgroup L such that K/L can be embedded in Σ_4, which is soluble (**364**). Then K/L is soluble (**7.46**), and so K/L has a non-trivial abelian quotient group (**374**). Hence, by **3.30**, K has a non-trivial abelian quotient group, so that

$$K' < K.$$

Since K' is characteristic in K (**3.51**), **3.15** shows that

$$K' \trianglelefteq G.$$

Because K is minimal normal in G, it follows that

$$K' = 1.$$

Thus K is abelian.

Now in any case both K and G/K are soluble. Therefore, by **7.47**, G is soluble. This completes the induction argument.

Remarks. This result shows in particular that if every maximal subgroup of G has prime index in G then G is soluble. B. Huppert [a61] proved even more, that G is supersoluble (**389**); this is the converse of the result of **646**. The proof requires rather more information about irreducible group actions (**649**) than is included in this book: see Huppert [b21] p. 718, theorem 6.9.5, or Schenkman [b35] p. 236, theorem 7.7.c, or Scott [b36] p. 226, 9.3.8.

We shall now prove some fundamental theorems of P. Hall on the arithmetical structure of finite soluble groups.

11.17 Definition. For any set ϖ of prime numbers, we denote by ϖ' the set of all primes which do not belong to ϖ.

Let $H \leqslant G$. Then H is said to be a *Hall ϖ-subgroup* of G if $|H|$ is a ϖ-number and $|G : H|$ is a ϖ'-number (see **3.41**).

Note that a Hall ϖ-subgroup of G is a Hall subgroup of G in the sense of **10.17**. A Hall p-subgroup of G is exactly the same thing as a Sylow p-subgroup of G; and a Hall p'-subgroup of G is exactly the same thing as a p-complement of G (see **10.19**).

Note also that if G has a Hall ϖ-subgroup H then $|H|$ is determined by $|G|$: namely, $|H|$ must be the largest ϖ-number which divides $|G|$.

*658 Suppose that H is a Hall ϖ-subgroup of G.
   (i) If $J \leqslant G$ with $|J| = |H|$, then J is a Hall ϖ-subgroup of G.
   (ii) For every element $g \in G$, H^g is a Hall ϖ-subgroup of G.
   (iii) If $H \leqslant L \leqslant G$, then H is a Hall ϖ-subgroup of L.

*659 Let $H \leqslant G$ and $K \trianglelefteq G$.
   (i) If H is a ϖ-subgroup of G then HK/K is a ϖ-subgroup of G/K.
   (ii) If H is a Hall ϖ-subgroup of G then $H \cap K$ is a Hall ϖ-subgroup of K and HK/K is a Hall ϖ-subgroup of G/K.

660 Suppose that H is a Hall ϖ-subgroup of G. Then $O_\varpi(G) \leqslant H \leqslant O^\varpi(G)$ and $HO^\varpi(G) = G$.

661 G is nilpotent if and only if every Hall subgroup of G is normal in G (cf. 11.3).

Sylow's theorem guarantees the existence of Sylow p-subgroups for every prime p in every finite group G. The following theorem of Hall, established in 1928, guarantees the existence of Hall ϖ-subgroups for every set ϖ of primes in every finite *soluble* group G. It also establishes an analogue of the statement in 5.9(b).

11.18 Theorem (P. Hall [a47]). *Suppose that G is soluble and let ϖ be any set of primes. Then*
  (i) *G has a Hall ϖ-subgroup.*
  (ii) *If H is a Hall ϖ-subgroup of G and V is any ϖ-subgroup of G then $V \leqslant H^g$ for some $g \in G$. In particular, the Hall ϖ-subgroups of G form a single conjugacy class of subgroups of G.*
Proof. We prove (i) and the first statement of (ii) together, by induction on $|G|$. Then the second statement of (ii) obviously follows, since all Hall ϖ-subgroups of G have the same order and any subgroup of this order is a Hall ϖ-subgroup of G (see **658**).

The theorem is clear if $|G| = 1$. Suppose that $|G| > 1$. Since all subgroups and all quotient groups of G are soluble (7.46), the induction hypothesis implies that the theorem is true for every proper subgroup of G and for every quotient of G by a non-trivial normal subgroup.

Let $R = O_\varpi(G)$ and suppose first that $R \neq 1$. Then, by the induction hypothesis, G/R has a Hall ϖ-subgroup H/R, where $H \leqslant G$. Then $|H| = |H/R| . |R|$, which is a ϖ-number, and $|G : H| = |G/R : H/R|$, which is a ϖ'-number. Thus H is a Hall ϖ-subgroup of G. Now let V be any ϖ-subgroup of G. Then VR/R is a ϖ-subgroup of G/R (**659**) and so, by the induction hypothesis,

$$VR/R \leqslant (H/R)^{Rg} = H^g/R$$

for some $g \in G$ (**230**). Then $V \leqslant VR \leqslant H^g$. This completes the argument in this case.

Now suppose that $R = 1$. Let $K = O_{\varpi'}(G)$ and let L be a minimal normal subgroup of G. By 7.56, L is an elementary abelian p-group for some prime p, and since $R = 1$, $p \notin \varpi$. Hence

$$L \leqslant K.$$

If $K = G$ then G is a ϖ'-group and the theorem is clearly true : in this case 1 is the unique ϖ-subgroup of G. Therefore we may suppose also that

$$K < G.$$

Then there is a chief factor of G of the form J/K, and, by 7.56, J/K is an elementary abelian q-group for some prime q. If $q \in \varpi'$ then, since $|J| = |J/K| . |K|$, J would be a normal ϖ'-subgroup of G with $K < J$; contrary to the definition of K. Hence $q \in \varpi$. Let Q be a Sylow q-subgroup of J. Then, since J/K is a q-group, $J = QK$ (**252**). Since $Q \neq 1$ and $R = 1, N_G(Q) < G$. Hence, by the induction hypothesis, $N_G(Q)$ has a Hall ϖ-subgroup H. Now

$$|G : H| = |G : N_G(Q)| |N_G(Q) : H|.$$

By Frattini's lemma (5.13),

$$G = N_G(Q)J = N_G(Q)QK = N_G(Q)K.$$

Therefore, by 3.40,

$$|G : N_G(Q)| = |K : N_G(Q) \cap K|,$$

which is a ϖ'-number. Since $|N_G(Q) : H|$ is also a ϖ'-number, by definition of H, $|G : H|$ is a ϖ'-number. Thus H is a Hall ϖ-subgroup of G.

Let V be any ϖ-subgroup of G. Then VL/L is a ϖ-subgroup of G/L and HL/L is a Hall ϖ-subgroup of G/L (**659**). Hence, by the induction hypothesis,

$$VL/L \leqslant (HL/L)^{Lx} = H^x L/L$$

for some $x \in G$ (**230**). Then $V \leqslant VL \leqslant H^x L \leqslant G$. Now H^x is a Hall ϖ-subgroup of $H^x L$(**658**). Therefore if $H^x L < G$ it follows, by the induction hypothesis, that

$$V \leqslant (H^x)^y = H^{xy}$$

for some $y \in H^x L$. This completes the argument in this case.

Now suppose that $H^x L = G$. Then also

$$HL = (H^x L)^{x^{-1}} = G.$$

Since H is a ϖ-subgroup of G and $p \notin \varpi$, $H \cap L = 1$. Thus H is a complement to L in G. Hence, by **601**, H is a maximal subgroup of G. Moreover, since H_G is a normal ϖ-subgroup of G, $H_G = 1$. Clearly $L = O_p(G)$. Let

$$W = VL.$$

Then

$$L \leqslant W \leqslant G = HL,$$

so that, by Dedekind's rule (7.3),

$$W = (W \cap H)L.$$

Now $W \cap H$ is a ϖ-subgroup of W. Also $(W \cap H) \cap L = 1$ and so, by 3.40,

$$|W : W \cap H| = |L|,$$

a ϖ'-number. Thus $W \cap H$ is a Hall ϖ-subgroup of W. Since also V is clearly a Hall ϖ-subgroup of W, it follows, by the induction hypothesis, that if $W < G$ then

$$V = (W \cap H)^w$$

for some $w \in W$. Then $\qquad\qquad V \leqslant H^w$.

Finally, if $W = G$ then V is a complement to L in G and therefore, since H is soluble and $H \neq 1$, 11.14(vi) shows that

$$V = H^g \quad \text{for some } g \in G.$$

This completes the induction argument.

Remark. The last part of the proof can be shortened slightly by applying 10.29.

11.19 Definition. Let G be a non-trivial group: suppose that $|G| = p_1^{m_1} p_2^{m_2} \dots p_s^{m_s}$, where $s, m_1, \dots, m_s$ are positive integers and $p_1, \dots, p_s$ distinct primes. Let $S = \{1, 2, \dots, s\}$. Then the possible orders of non-trivial Hall subgroups of G are the $2^s - 1$ distinct numbers $\prod_{j \in T} p_j^{m_j}$, where T ranges over the $2^s - 1$ distinct non-empty subsets of S.

Now suppose that G is soluble. Then Hall's theorem 11.18 guarantees the existence in G of Hall subgroups of all possible orders. In particular, for each $i \in S$, G has a subgroup H_i of order $\prod_{j \in S \setminus \{i\}} p_j^{m_j}$ and index $p_i^{m_i}$; that is, a p_i-complement (10.19).

Any set $\mathscr{S} = \{H_1, H_2, \dots, H_s\}$ of s subgroups of G, with H_i a p_i-complement of G for each $i = 1, \dots, s$, will be called a *complement system* of G.

662 Let G be a non-trivial soluble group. Then, for every prime divisor p of $|G|$, G has a maximal subgroup M such that $|G : M|$ is a power of p (cf. 11.15; see also **677**).

663 Suppose that G is a non-trivial group such that for every maximal subgroup M of G, $|G : M| \leqslant 5$. Prove that G is a soluble ϖ-group, where $\varpi = \{2, 3, 5\}$ and that G has a normal Sylow 5-subgroup (possibly trivial). (Hints. Apply 11.16. To prove that G has a normal Sylow 5-subgroup, see (i), (ii), and (iv) in the proof of 11.16.)

664 Suppose that G is soluble and that $|G|$ has s distinct prime divisors, where $s \geqslant 2$. Show that for some prime divisor p of $|G|$,

$$|G| < |H|^{s/s-1},$$

where H is a p-complement of G (cf. **503**).

665 Suppose that G is soluble and that $|G|$ has s distinct prime divisors. Suppose also that for every prime divisor p of $|G|$, the p-complements of G are nilpotent.
(i) Prove that if $s \geqslant 3$ then G is nilpotent. (See also **672**. Hint. Prove that every Sylow subgroup of G is normal in G.)
(ii) Show by an example that if $s = 2$ then G need not be nilpotent.

Hall proved a remarkable converse to 11.18, namely that if G has a complement system then G is necessarily soluble. We shall prove this result in 11.26. First we establish a few properties of complement systems.

We show that any complement system of G determines in a very simple way proper Hall subgroups of all possible orders.

11.20 Theorem. *Let G be a non-trivial soluble group, with*
$|G| = p_1^{m_1} p_2^{m_2} \dots p_s^{m_s}$, *where* $s, m_1, \dots, m_s$ *are positive integers and* $p_1, \dots, p_s$
distinct primes. Let $S = \{1, 2, \dots, s\}$, *and for each* $i \in S$, *let* H_i *be a* p_i-*comple-*
ment of G. *For each non-empty subset* T *of* S, *let*

$$H_T = \bigcap_{j \in T} H_j.$$

Then, for each non-empty proper subset T *of* S, $H_{S \backslash T}$ *is a Hall subgroup of*
G *of order* $\prod\limits_{j \in T} p_j^{m_j}$ *and* $H_S = 1$.

Proof. An equivalent statement is that for each non-empty subset U of
S, H_U is a Hall subgroup of G of *index* $\prod\limits_{j \in U} p_j^{m_j}$. This is established by a
straightforward argument, using induction on $|U|$ and the result of **100**.

11.21 Definition. Let G be a non-trivial soluble group and let $p_1, \dots, p_s$
be the distinct prime divisors of $|G|$. For each $i = 1, \dots, s$, let H_i be a
p_i-complement of G, and let $\mathscr{S} = \{H_1, \dots, H_s\}$, a complement system of G.
   For each $g \in G$ let

$$\mathscr{S}^g = \{H_1^g, \dots, H_s^g\}.$$

Since $|H_i^g| = |H_i|$ for each $i = 1, \dots, s, \mathscr{S}^g$ is also a complement system of
G. Now it is clear that G acts by conjugation on the set of all complement
systems of G. The stabilizer in G of $\mathscr{S}$ for this action is the subgroup

$$\{g \in G : \mathscr{S}^g = \mathscr{S}\} = \{g \in G : H_i^g = H_i \text{ for each } i = 1, \dots, s\}$$

$$= \bigcap_{i=1}^{s} N_G(H_i).$$

We denote this subgroup by $N_G(\mathscr{S})$; it is called a *system normalizer* of
G.
   Now let $\mathscr{S}^* = \{H_1^*, \dots, H_s^*\}$ be another complement system of G,
with H_i^* a p_i-complement of G for each $i = 1, \dots, s$. By 11.18 (ii), there are
elements $g_1, \dots, g_s \in G$ such that $H_i^* = H_i^{g_i}$ for each $i = 1, \dots, s$. We shall
now show that we can choose $g_1 = g_2 = \dots = g_s$.

11.22 Theorem (P. Hall [a50]). *Let* $\mathscr{S}$ *and* $\mathscr{S}^*$ *be complement systems*
of the non-trivial soluble group G. *Then* $\mathscr{S}^* = \mathscr{S}^g$ *for some* $g \in G$.

Proof. Let $p_1, \dots, p_s$ be the distinct prime divisors of $|G|$ and let $\mathscr{S} = \{H_1, \dots, H_s\}$, where H_i is a p_i-complement of G. Since, by 11.18 (ii),
the p_i-complements of G form a single conjugacy class of subgroups of
G, 4.33 shows that for each $i = 1, \dots, s$, the number of distinct p_i-comple-
ments of G is equal to $|G : N_G(H_i)|$. Hence the number of distinct comple-
ment systems of G is equal to

$$\prod_{i=1}^{s} |G : N_G(H_i)| = n, \text{ say}.$$

For the action of G by conjugation on the set of all complement systems of G, it follows from 4.11 that the length of the orbit of $\mathscr{S}$ is

$$|G : N_G(\mathscr{S})| = \left|G : \bigcap_{i=1}^{s} N_G(H_i)\right|.$$

Since $H_i \leqslant N_G(H_i) \leqslant G, |G : N_G(H_i)|$ is a power of p_i for each $i = 1, \ldots, s$. Hence

$$(|G : N_G(H_i)|, |G : N_G(H_j)|) = 1$$

whenever $i \neq j$. Hence, by repeated application of **100**,

$$|G : N_G(\mathscr{S})| = \prod_{i=1}^{s} |G : N_G(H_i)| = n.$$

Thus the orbit of $\mathscr{S}$ must include every complement system of G (that is, the action is transitive). Hence $\mathscr{S}^* = \mathscr{S}^g$ for some $g \in G$.
Remark. Let G be a non-trivial soluble group, let $p_1, \ldots, p_s$ be the distinct prime divisors of G, and, for each $i = 1, \ldots, s$, let P_i and P_i^* be Sylow p_i-subgroups of G. It is *not* in general true that there is an element $g \in G$ such that $P_i^* = P_i^g$ for each $i = 1, \ldots, s$: see **671**.

11.23 Corollary. *Let G be a non-trivial soluble group. Then the system normalizers of G form a single conjugacy class of nilpotent subgroups of G.*
Proof. Let $\mathscr{S} = \{H_1, \ldots, H_s\}$ be a complement system of G. The corresponding system normalizer of G is the subgroup

$$L = N_G(\mathscr{S}) = \bigcap_{i=1}^{s} N_G(H_i).$$

For any element $g \in G$,

$$L^g = \bigcap_{i=1}^{s} N_G(H_i)^g = \bigcap_{i=1}^{s} N_G(H_i^g) = N_G(\mathscr{S}^g),$$

the system normalizer corresponding to the complement system $\mathscr{S}^g$. This remark, together with 11.22, shows that the system normalizers of G form a single conjugacy class of subgroups of G.

Suppose that for each $i = 1, \ldots, s$, H_i is a p_i-complement of G. If $s = 1$ then G is a p_1-group, $H_1 = 1$ and $L = N_G(H_1) = G$, which in this case is nilpotent. Suppose now that $s > 1$ and, for each $i = 1, \ldots, s$, let

$$P_i = \bigcap_{j \neq i} H_j.$$

By 11.20, P_i is a Sylow p_i-subgroup of G. Moreover, $L = N_G(\mathscr{S}) \leqslant N_G(P_i)$ for each i. Since $P_i \trianglelefteq N_G(P_i), L \cap P_i \trianglelefteq L$; and it is easy to check that $L \cap P_i$ is a Sylow p_i-subgroup of L. This shows that every Sylow subgroup of L is normal in L, because $p_1, \ldots, p_s$ are the only prime divisors of $|G|$. Hence, by 11.3, L is nilpotent.

11.24. In his paper [a51], P. Hall proved many other interesting results about system normalizers. For instance, he showed that any system normalizer of a non-trivial soluble group G *covers* every central chief factor of G and *avoids* every chief factor of G which is not central (see **324**). It follows that the order of a system normalizer of G is equal to the product of the orders of the central chief factors in any particular chief series of G. For further information, see Huppert [b21] §6.11.

666 Let $\mathscr{S}$ be a complement system of the non-trivial soluble group G. Then the number of distinct complement systems of G is equal to $|G : N_G(\mathscr{S})|$.

667 Find the order of the system normalizers in each of the soluble groups Σ_3, A_4, Σ_4.

668 Let G be a non-trivial nilpotent group. Then the unique system normalizer of G is G itself.

669 Let $K \lhd G$, a non-trivial soluble group, and let $\bar{G} = G/K$. Let the distinct prime divisors of $|G|$ be $p_1, \ldots, p_s$ and suppose that $p_1, \ldots, p_r$ are the prime divisors of $|\bar{G}|$, where $r \leqslant s$. For each $i = 1, \ldots, s$, let H_i be a p_i-complement of G, let $\mathscr{S}$ be the complement system $\{H_1, \ldots, H_s\}$ of G and let $\bar{\mathscr{S}} = \{H_1 K/K, \ldots, H_r K/K\}$. Then $\bar{\mathscr{S}}$ is a complement system of $\bar{G}$ and

$$N_{\bar{G}}(\bar{\mathscr{S}}) = N_G(\mathscr{S})K/K.$$

(Hints. It is enough to assume that K is a minimal normal subgroup of G. Then K is a p_j-group for some $j \in \{1, \ldots, s\}$ and $K \leqslant H_i$ whenever $i \neq j$. Note that if $r < s$ then $r = s - 1$ and $j = s$. To show that $N_{\bar{G}}(\bar{\mathscr{S}}) \leqslant N_G(\mathscr{S})K/K$, note that H_j is a p_j-complement of $H_j K$ and use **11.18**.)

670 Let G be a non-trivial soluble group. Prove that if L is a system normalizer of G then $L^G = G$ (where L^G denotes the normal closure of L in G : see **180**). In particular, the system normalizers of G are non-trivial. (Hint. Suppose the result false and derive a contradiction by applying **668** and **669**.)

671 Let $p \geqslant 5$ and let G be the natural wreath product $C_p \wr \Sigma_3$ (see 9.19, 9.20) : then $|G| = 6p^3$. Let A be the base group of G : thus

$$A = \langle a_1 \rangle \times \langle a_2 \rangle \times \langle a_3 \rangle,$$

where each a_i has order p; and A is the unique Sylow p-subgroup of G. Let $\sigma = (123)$ and $\tau = (12)$. Then $\langle \sigma \rangle$ and $\langle \sigma^{a_1} \rangle$ are Sylow 3-subgroups of G and $\langle \tau \rangle$ is a Sylow 2-subgroup of G.

Show that there is no element $g \in G$ such that

$$\{\langle \sigma \rangle, \langle \tau \rangle, A\}^g = \{\langle \sigma^{a_1} \rangle, \langle \tau \rangle, A\}$$

(cf. 11.22).

In proving Hall's converse to 11.18, we shall use the following lemma.

11.25 Lemma (H. Wielandt [a103], 1960). *Suppose that G has three soluble subgroups H_1, H_2, H_3, and suppose that $(|G:H_i|, |G:H_j|) = 1$ whenever $i, j \in \{1, 2, 3\}$ with $i \neq j$. Then G is soluble.*

Proof. We argue by induction on $|G|$. Since $(|G:H_1|, |G:H_2|) = 1$, **100** shows that

$$G = H_1 H_2.$$

Since H_2 is soluble, we may suppose that $H_1 \neq 1$. Let L_1 be a minimal normal subgroup of H_1. Then, since H_1 is soluble, L_1 is an elementary abelian p-group for some prime p(7.56). By hypothesis, p does not divide both $|G : H_2|$ and $|G : H_3|$. We suppose, without loss of generality, that p does not divide $|G : H_2|$.

Let $$J = H_1 \cap H_2.$$
Then $$|G : H_2| = |H_1 H_2|/|H_2| = |H_1 : J|,$$

by **98**. Thus p does not divide $|H_1 : J|$. Since $J \leqslant H_1$ and $L_1 \trianglelefteq H_1$, $J \leqslant J L_1 \leqslant H_1$. Hence p does not divide $|JL_1 : J|$. However, by **98** (or 3.40),

$$|JL_1 : J| = |L_1 : J \cap L_1|,$$

and, since L_1 is a p-group, $|L_1 : J \cap L_1|$ is a power of p. Therefore

$$|JL_1 : J| = 1,$$

that is, $$L_1 \leqslant J.$$

Now let $L = L_1^G$, the normal closure of L_1 in G (**180**). Then (cf. **264**)

$$L = \langle x^g : x \in L_1, g \in G \rangle$$
$$= \langle x^{h_1 h_2} : x \in L_1, h_1 \in H_1, h_2 \in H_2 \rangle \text{ (since } G = H_1 H_2)$$
$$\leqslant H_2,$$

since $L_1 \trianglelefteq H_1$ and $L_1 \leqslant J \leqslant H_2$. Since H_2 is soluble, L is soluble (7.46). Moreover, $1 < L_1 \leqslant L \trianglelefteq G$.

We may now apply the induction hypothesis to G/L. For each $i = 1, 2, 3$, $H_i L/L \leqslant G/L$ and $H_i L/L \cong H_i/(H_i \cap L)$(3.40); this is soluble, since H_i is soluble (7.46). Whenever $i \neq j$,

$$(|G/L : H_i L/L|, |G/L : H_j L/L|) = (|G : H_i L|, |G : H_j L|);$$

this number divides $(|G : H_i|, |G : H_j|)$ and is therefore equal to 1. Hence by the induction hypothesis, G/L is soluble.

Finally, since L and G/L are both soluble, it follows by 7.47 that G is soluble. This completes the induction argument.

The main theorem is the following.

11.26 Theorem (P. Hall [a49]). *Suppose that G has a p-complement for each prime divisor p of $|G|$. Then G is soluble.*

Suppose that $|G| = p_1^{m_1} p_2^{m_2}$, where p_1 and p_2 are distinct primes and m_1 and m_2 non-negative integers. Then a Sylow p_1-subgroup of G is a p_2-

complement of G and a Sylow p_2-subgroup of G is a p_1-complement of G. Hence it follows from Sylow's theorem that G has a p-complement for each prime divisor p of $|G|$ and therefore, according to 11.26, G must be soluble. Thus 11.26 includes as a special case a theorem of Burnside to which we have referred before (see 4.29, **382**, **572**).

11.27 Theorem (Burnside [a11], 1904). *Suppose that* $|G| = p^m q^n$, *where p and q are distinct primes and m and n are non-negative integers. Then G is soluble.*

The proof of 11.26 depends on this theorem of Burnside. Burnside's proof relies on the theory of group characters and we do not include it here: see for instance Curtis and Reiner [b7] p. 239, theorem 34.1, or Gorenstein [b13] p. 131, theorem 4.3.3, or Huppert [b21] p. 492, theorem 5.7.3, or Schenkman [b35] p. 263, theorem 8.5.f, or Scott [b36] p. 334, 12.3.3. There is also a more recent proof of 11.27 which is independent of character theory, but this too involves techniques not included in this book: see the papers of H. Bender [a7], D. M. Goldschmidt [a40] and H. Matsuyama [a73]; see also Gagen [b12].

Proof of 11.26. We may suppose $G \neq 1$. Let $|G| = p_1^{m_1} p_2^{m_2} \dots p_s^{m_s}$, where $s, m_1, \dots, m_s$ are positive integers and $p_1, \dots, p_s$ distinct primes. We argue by induction on s. If $s \leqslant 2$, the result is true, by 11.27. Now assume that $s > 2$. For each $i = 1, \dots, s$, let H_i be a p_i-complement of G. Let $i, j \in \{1, 2, \dots, s\}$ with $i \neq j$. Then

$$(|G:H_i|, |G:H_j|) = 1,$$

and so, by **100**,

$$|H_i : H_i \cap H_j| = |G : H_i \cap H_j| / |G : H_i| = |G : H_j| = p_j^{m_j}.$$

Hence $H_i \cap H_j$ is a p_j-complement of H_i. Thus H_i has a p-complement for each prime divisor p of $|H_i|$. Since $|H_i|$ has just $s - 1$ distinct prime divisors, it follows by the induction hypothesis that H_i is soluble. This is true for each $i = 1, \dots, s$. Since $s \geqslant 3$, 11.25 now applies to show that G is soluble. This completes the induction argument.

Remark. 11.18 and 11.26 together give an arithmetical characterization of finite soluble groups, namely the equivalence of the following three statements:
  (i) G is soluble.
  (ii) G has a Hall ϖ-subgroup for every set ϖ of primes.
  (iii) G has a p-complement for every prime divisor p of $|G|$.

672 Suppose that G has three nilpotent subgroups H_1, H_2, H_3 such that $(|G:H_i|, |G:H_j|) = 1$ whenever $i, j \in \{1, 2, 3\}$ with $i \neq j$. Then G is nilpotent. Moreover, if H_1, H_2 and H_3 are abelian then G is abelian. (cf. 11.25. This improves the result of **665**(i). Hint. Prove that every Sylow subgroup of G is normal in G.)

673 Show by an example that an insoluble group G can have soluble subgroups H_1 and H_2 such that $(|G:H_1|,|G:H_2|) = 1$. (cf. 11.25. Hint. See 5.25.)

674 (i) Suppose that G satisfies the converse of Lagrange's theorem; that is, that G has a subgroup of order m for every divisor m of $|G|$. Then G is soluble (see also **675, 676**).

(ii) (D. H. McLain [a74]). Let G be a soluble group of order $p_1^{m_1} p_2^{m_2} \ldots p_s^{m_s}$, where $s, m_1, \ldots, m_s$ are positive integers and $p_1, \ldots, p_s$ are distinct primes. Let A be any abelian group of order $p_1^{m_1-1} p_2^{m_2-1} \ldots p_s^{m_s-1}$. Then the group $G \times A$ satisfies the converse of Lagrange's theorem. (Hint. Apply 11.18 and **135**.)

675 Let G be a non-trivial group such that every subgroup of G satisfies the converse of Lagrange's theorem (see **674**). Then G has a normal Sylow p-subgroup, where p is the largest prime divisor of $|G|$. (See also **676**. Hint. Argue by induction on $|G|$. If G is not a p-group, use 4.18.)

676 (Ore [a76], G. Zappa [a104], D. H. McLain [a74]). The following three statements are equivalent:

(i) G is supersoluble (**389**).

(ii) Whenever $J < H \leqslant G$, with J a maximal subgroup of H, $|H:J|$ is a prime number.

(iii) Each subgroup of G satisfies the converse of Lagrange's theorem (see **674, 675**).

(Hints. For (i) $\Rightarrow$ (ii), see **646**. For (ii) $\Rightarrow$ (iii), argue by induction on $|G|$ and note that every subgroup of G satisfies the same hypothesis as G. To show that G itself satisfies the converse of Lagrange's theorem, use 11.16 and **662** to show that for any prime divisor p of $|G|$, G has a subgroup of index p. For (iii) $\Rightarrow$ (i), argue by induction on $|G|$. If $|G| > 1$, note that, by **675**, G has a normal Sylow p-subgroup P for some prime divisor p of $|G|$. Use **674**, 11.18 and the induction hypothesis to show that G/P is supersoluble. By hypothesis, G has a subgroup M of index p. Use 5.6 and **389**(v) to show that $(M \cap P) \lhd G$ and $G/(M \cap P)$ is supersoluble. Choose $K \lhd G$ such that $K \leqslant M \cap P$, G/K is supersoluble and $|K|$ is as small as possible. If $K \neq 1$, note that, by **391**, $[K, P] < K$. Then consider a chief factor K/L of M with $[K, P] \leqslant L < K$ and use **389**(v) to derive a contradiction to the choice of K.)

677 Suppose that for every non-trivial subgroup H of G, H has a proper subgroup of index a power of p for every prime divisor p of $|H|$. Then G is soluble (cf. **662, 678**).

678 Show by an example that an insoluble group G can have a subgroup of index p for every prime divisor p of $|G|$. (cf. **662**, 11.16, 11.26. Hint. Let H be a non-abelian simple group and let $G = H \times K$, where K is a suitable cyclic group.)

679 Suppose that G has abelian subgroups A and B such that $(|G:A|,|G:B|) = 1$. Then G is soluble. This can be proved by the following argument.

Suppose the result false, and let G be an insoluble group of least possible order with abelian subgroups A, B such that $(|G:A|,|G:B|) = 1$. Then

(i) A and B are proper subgroups of G, and $G = AB$.

(ii) Let $1 < K \trianglelefteq G$. Then G/K is soluble. (Hint. Use the minimality of G.)

(iii) G is simple. (Hint. If $K < G$, use the minimality of G.)

(iv) $A \cap B = 1$. (Hint. Apply **264**(ii).)

(v) A is a maximal subgroup of G. (Hint. Apply **264**(ii) again.)

(vi) Derive a contradiction by applying **632**.

(cf. **672, 673**. Remarks. Other methods yield stronger results than this. For instance, by brief but ingenious commutator calculations, N. Itô [a63] has shown that whenever G has abelian subgroups A and B such that $G = AB$ then G is soluble, and in fact $G'' = 1$. This remains true when G is infinite. See also Huppert [b21] p. 674, theorem 6.4.4. For finite G, a deeper result of H. Wielandt [a101] and O. H.

Kegel [a67] shows that whenever G has nilpotent subgroups G_1 and G_2 such that $G = G_1 G_2$ then G is soluble. In particular, if G has nilpotent subgroups G_1 and G_2 such that $(|G : G_1|, |G : G_2|) = 1$ then G is soluble. See Huppert [b21] §6.4, or Schenkman [b35] p. 269, theorem 9.2.e, or Scott [b36] §13.2.)

11.28. We end with some remarks on more recent results in the theory of finite soluble groups. We can only touch on some of the more important developments and refer for further details to Huppert [b21] chapter 6 and the references given there.

In 1961, R. W. Carter [a13] showed that every finite soluble group possesses nilpotent self-normalizing subgroups, and that these so-called *Carter subgroups* form a single conjugacy class. Every Carter subgroup contains a system normalizer (11.21) and every system normalizer is contained in a Carter subgroup, but the classes of Carter subgroups and system normalizers coincide only in special cases.

In 1963, W. Gaschutz [a36] established a wide-ranging generalization, containing the main parts of both Carter's result and Hall's theorem 11.18. A class $\mathfrak{F}$ of finite groups is called a *formation* if it has the following two properties: (i) every quotient group of every $\mathfrak{F}$-group is an $\mathfrak{F}$-group, (ii) every finite group has an $\mathfrak{F}$-residual (see 3.45). Examples of formations are the class of finite ϖ-groups for any set ϖ of primes (see 3.44), the class of finite abelian groups (see 3.52), the class of finite nilpotent groups and the class of finite soluble groups (see 7.50). A formation $\mathfrak{F}$ is said to be *saturated* if, whenever G is a finite group such that $G/\Phi(G)$ is an $\mathfrak{F}$-group, G is itself an $\mathfrak{F}$-group (Gaschutz and U. Lubeseder [a38]). For example, the formations of finite ϖ-groups, finite nilpotent groups and finite soluble groups are all saturated (see **620, 621**, 11.5 and 7.47), but the formation of finite abelian groups is not saturated (see 11.10).

Gaschütz proved that for any saturated formation $\mathfrak{F}$, any finite soluble group G possesses $\mathfrak{F}$-subgroups with certain special properties and these subgroups form a single conjugacy class. The subgroups in question are now called the $\mathfrak{F}$-*projectors* of G, and (by a result of T. O. Hawkes [a53]) are characterized by the following property: a subgroup H of G is an $\mathfrak{F}$-projector of G if and only if, whenever $K \trianglelefteq G, HK/K$ is a maximal $\mathfrak{F}$-subgroup of G/K. When $\mathfrak{F}$ is the saturated formation of finite ϖ-groups, the $\mathfrak{F}$-projectors of G are the Hall ϖ-subgroups of G; and when $\mathfrak{F}$ is the saturated formation of finite nilpotent groups, the $\mathfrak{F}$-projectors of G are the Carter subgroups of G.

In 1967, R. W. Carter and T. O. Hawkes [a15] showed that, for any saturated formation $\mathfrak{F}$ which contains the formation of finite nilpotent groups, it is possible to define in any finite soluble group G a class of $\mathfrak{F}$-subgroups which are analogous to Hall's system normalizers and which are called the $\mathfrak{F}$-*normalizers* of G. These form a single conjugacy class of subgroups of G. Every $\mathfrak{F}$-projector of G contains an $\mathfrak{F}$-normalizer of G and every $\mathfrak{F}$-normalizer is contained in an $\mathfrak{F}$-projector of G. When $\mathfrak{F}$ is the formation of finite nilpotent groups, the $\mathfrak{F}$-normalizers of G are simply its system normalizers.

Another significant development is the discovery in 1967 by B. Fischer, W. Gaschütz and B. Hartley [a24] of subgroups which are in a sense dual to projectors. A class $\mathfrak{K}$ of finite groups is called a *Fitting class* if it has the following two properties: (i) every normal subgroup of every $\mathfrak{K}$-group is a $\mathfrak{K}$-group, (ii) every finite group has a $\mathfrak{K}$-radical (3.45). These two properties are dual to the two properties defining a formation. The class of finite ϖ-groups is a Fitting class (see 3.43), and so is the class of finite nilpotent groups (see 7.63).

A subgroup V of G is said to be a $\mathfrak{K}$-*injector* of G if, whenever K is a subnormal subgroup of G, $V \cap K$ is a maximal $\mathfrak{K}$-subgroup of K. Fischer, Gaschütz and Hartley proved that for any Fitting class $\mathfrak{K}$, any finite soluble group G possesses $\mathfrak{K}$-injectors and these form a single conjugacy class of subgroups of G. (Here no extra condition on the Fitting class corresponding to saturation for formations is needed.) When $\mathfrak{K}$ is the Fitting class of finite ϖ-groups, the $\mathfrak{K}$-injectors of G are again the Hall ϖ-subgroups of G; but when $\mathfrak{K}$ is the Fitting class of finite nilpotent groups, the $\mathfrak{K}$-injectors of G are usually distinct from the Carter subgroups of G.

The precise significance of these various distinguished conjugacy classes of subgroups in the theory of finite soluble groups is still an active subject of investigation.

REFERENCES

Articles

[a1] J. L. Alperin, Sylow intersections and fusion, *J. Algebra* **6** (1967), 222–41.

[a2] J. L. Alperin and D. Gorenstein, Transfer and fusion in finite groups, *J. Algebra* **6** (1967), 242–55.

[a3] R. Baer, Erweiterung von Gruppen und ihren Isomorphismen, *Math. Z.* **38** (1934), 375–416.

[a4] R. Baer, The subgroup of the elements of finite order of an abelian group, *Ann. of Math.* (2) **37** (1936), 766–81.

[a5] R. Baer, Das Hyperzentrum einer Gruppe III, *Math. Z.* **59** (1953), 299–338.

[a6] R. Baer, Nilpotent characteristic subgroups of finite groups, *Amer. J. Math.* **75** (1953), 633–64.

[a7] H. Bender, A group theoretic proof of Burnside's $p^a q^b$-theorem, *Math. Z.* **126** (1972), 327–38.

[a8] R. Brauer and K. A. Fowler, On groups of even order, *Ann. of Math.* (2) **62** (1955), 565–83.

[a9] W. Burnside, On some properties of groups of odd order II, *Proc. London Math. Soc.* **33** (1901), 257–68.

[a10] W. Burnside, On an unsettled question in the theory of discontinuous groups, *Quart. J. Math.* **33** (1902), 230–8.

[a11] W. Burnside, On groups of order $p^a q^b$, *Proc. London Math. Soc.* (2) **1** (1904), 388–92.

[a12] W. Burnside, On the theory of groups of finite order, *Proc. London Math. Soc.* (2) **7** (1909), 1–7.

[a13] R. W. Carter, Nilpotent self-normalizing subgroups of soluble groups, *Math. Z.* **75** (1961), 136–9.

[a14] R. W. Carter, Simple groups and simple Lie algebras, *J. London Math. Soc.* **40** (1965), 193–240.

[a15] R. W. Carter and T. O. Hawkes, The $\mathfrak{F}$-normalizers of a finite soluble group, *J. Algebra* **5** (1967), 175–202.

[a16] A. Cayley, On the theory of groups as depending on the symbolical equation $\theta^n = 1$, *Philos. Mag.* (4) **7** (1854), 40–7 (*Collected Mathematical Papers*, Cambridge 1889, vol. 2, pp. 123–30).

[a17] C. Chevalley, Sur certains groupes simples, *Tôhoku Math. J.* (2) **7** (1955), 14–66.

[a18] A. L. S. Corner, On a conjecture of Pierce concerning direct decompositions of abelian groups, *Proc. Colloq. Abelian Groups*, Akadémiai Kiadó, Budapest (1964), 43–8.

[a19] L. E. Dickson, A new system of simple groups, *Math. Ann.* **60** (1905), 137–50.

[a20] A. P. Ditsman, On p-groups (in Russian), *Dokl. Akad. Nauk SSSR* **15** (1937), 71–6.

[a21] W. Feit, The current situation in the theory of finite simple groups, *Actes Congrès Intern. Math.* (Nice, 1970), vol. 1, 55–93.

[a22] W. Feit and J. G. Thompson, A solvability criterion for finite groups and some consequences, *Proc. Nat. Acad. Sci. U.S.A.* **48** (1962), 968–70.

[a23] W. Feit and J. G. Thompson, Solvability of groups of odd order, *Pacific J. Math.* **13** (1963), 775–1029.

[a24] B. Fischer, W. Gaschütz and B. Hartley, Injektoren endlicher auflösbarer Gruppen, *Math. Z.* **102** (1967), 337–9.

[a25] H. Fitting, Die Theorie der Automorphismenringe Abelscher Gruppen und ihr Analogon bei nicht kommutativen Gruppen, *Math. Ann.* **107** (1933), 514–42.

[a26] H. Fitting, Über die direkten Produktzerlegungen einer Gruppe in direkt unzerlegbare Faktoren, *Math. Z.* **39** (1934), 16–30.

[a27] H. Fitting, Beiträge zur Theorie der Gruppen endlicher Ordnung, *Jber. Deutsch. Math. Verein.* **48** (1938), 77–141.

[a28] S. B. Fomin, Über periodische Untergruppen der unendlichen abelschen Gruppen, *Mat. Sb.* **2** (1937), 1007–9.

[a29] G. Frattini, Intorno alla generazione dei gruppi di operazioni, *Rend. Atti Accad. Lincei* (4) **1** (1885), 281–5, 455–7.

[a30] F. G. Frobenius, Über auflösbare Gruppen III, S.-B. Preuss. Akad. Berlin (1901), 849–57 (*Gesamm. Abh.*, Springer 1968, vol. 3, pp. 180–8).

[a31] F. G. Frobenius, Über auflösbare Gruppen IV, S.-B. Preuss. Akad. Berlin (1901), 1216–30 (*Gesamm. Abh.*, Springer 1968, vol. 3, pp. 189–203).

[a32] F. G. Frobenius, Über auflösbare Gruppen V, S.-B. Preuss. Akad. Berlin (1901), 1324–9 (*Gesamm. Abh.*, Springer 1968, vol. 3, pp. 204–9).

[a33] F. G. Frobenius and L. Stickelberger, Über Gruppen von vertauschbaren Elementen, *J. reine angew. Math.* **86** (1879), 217–62 (*Gesamm. Abh.*, Springer 1968, vol. 1, pp. 545–90).

[a34] W. Gaschütz, Zur Erweiterungstheorie der endlichen Gruppen, *J. reine angew. Math.* **190** (1952), 93–107.

[a35] W. Gaschütz, Über die Φ-Untergruppe endlicher Gruppen, *Math. Z.* **58** (1953), 160–70.

[a36] W. Gaschütz, Zur Theorie der endlichen auflösbaren Gruppen, *Math. Z.* **80** (1963), 300–5.

[a37] W. Gaschütz, Nichtabelsche p-Gruppen besitzen äussere p-Automorphismen, *J. Algebra* **4** (1966), 1–2.

[a38] W. Gaschütz and U. Lubeseder, Kennzeichnung gesättigter Formationen, *Math. Z.* **82** (1963), 198–9.

[a39] G. Glauberman, A characteristic subgroup of a p-stable group, *Canad. J. Math.* **20** (1968), 1101–35.

[a40] D. M. Goldschmidt, A group theoretic proof of the $p^a q^b$ theorem for odd primes, *Math. Z.* **113** (1970), 373–5.

[a41] E. S. Gold, On nil-algebras and finitely approximable p-groups (in Russian), *Izv. Akad. Nauk SSSR Ser. Mat.* **28** (1964), 273–6 (*Amer. Math. Soc. Translations* (2) **48** (1965), 103–6).

[a42] E. S. Gold and I. R. Shafarevich, On class field towers (in Russian), *Izv. Akad. Nauk SSSR Ser. Mat.* **28** (1964), 261–72 (*Amer. Math. Soc. Translations* (2) **48** (1965), 91–102).

[a43] D. Gorenstein, Finite groups in which Sylow 2-subgroups are abelian and centralizers of involutions are solvable, *Canad. J. Math.* **17** (1965), 860–906.

[a44] D. Gorenstein, Finite simple groups and their classification, *Israel J. Math.* **19** (1974), 5–66.

[a45] O. Grün, Beiträge zur Gruppentheorie I, *J. reine angew. Math.* **174** (1935), 1–14.

[a46] M. Hall Jr., Solution of the Burnside problem for exponent six, *Illinois J. Math.* **2** (1958), 764–86.

[a47] P. Hall, A note on soluble groups, *J. London Math. Soc.* **3** (1928), 98–105.

[a48] P. Hall, A contribution to the theory of groups of prime-power order, *Proc. London Math. Soc.* (2) **36** (1933), 29–95.

[a49] P. Hall, A characteristic property of soluble groups, *J. London Math. Soc.* **12** (1937), 198–200.

[a50] P. Hall, On the Sylow systems of a soluble group, *Proc. London Math. Soc.* (2) **43** (1937), 316–23.

[a51] P. Hall, On the system normalizers of a soluble group, *Proc. London Math. Soc.* (2) **43** (1937), 507–28.

[a52] P. Hall and G. Higman, On the *p*-length of *p*-soluble groups and reduction theorems for Burnside's problem, *Proc. London Math. Soc.* (3) **6** (1956), 1–42.

[a53] T. O. Hawkes, On formation subgroups of a finite soluble group, *J. London Math. Soc.* **44** (1969), 243–50.

[a54] H. Heineken and I. J. Mohamed, A group with trivial centre satisfying the normalizer condition, *J. Algebra* **10** (1968), 368–76.

[a55] I. N. Herstein, A remark on finite groups, *Proc. Amer. Math. Soc.* **9** (1958), 255–7.

[a56] D. G. Higman, Focal series in finite groups, *Canad. J. Math.* **5** (1953), 477–97.

[a57] G. Higman, A finitely generated infinite simple group, *J. London Math. Soc.* **26** (1951), 61–4.

[a58] G. Higman, B. H. Neumann and H. Neumann, Embedding theorems for groups, *J. London Math. Soc.* **24** (1949), 247–54.

[a59] O. Hölder, Zurückführung einer beliebigen algebraischen Gleichung auf eine Kette von Gleichungen, *Math. Ann.* **34** (1889), 26–56.

[a60] O. Hölder, Bildung zusammengesetzter Gruppen, *Math. Ann.* **46** (1895), 321–422.

[a61] B. Huppert, Normalteiler und maximale Untergruppen endlicher Gruppen, *Math. Z.* **60** (1954), 409–34.

[a62] N. Itô, Note on (LM)-groups of finite orders, *Kōdai Math. Sem. Rep.* (1951), 1–6.

[a63] N. Itô, Über das Produkt von zwei abelschen Gruppen, *Math. Z.* **62** (1955), 400–1.

[a64] K. Iwasawa, Ueber die Struktur der endlichen Gruppen, deren echte Untergruppen sämtlich nilpotent sind, *Proc. Phys.-Math. Soc. Japan* (3) **23** (1941), 1–4.

[a65] Z. Janko, A new finite simple group with abelian Sylow 2-subgroups and its characterization, *J. Algebra* **3** (1966), 147–86.

[a66] C. Jordan, Commentaire sur Galois, *Math. Ann.* **1** (1869), 141–60 (*Oeuvres*, Gauthier-Villars 1961, vol. 1, pp. 211–30).

[a67] O. H. Kegel, Produkte nilpotenter Gruppen, *Arch. Math.* **12** (1961), 90–3.

[a68] W. Krull, Über verallgemeinerte endliche Abelsche Gruppen, *Math. Z.* **23** (1925), 161–96.

[a69] E. Landau, Über die Klassenzahl der binären quadratischen Formen von negativer Discriminante, *Math. Ann.* **56** (1903), 671–6.

[a70] F. Levi and B. L. van der Waerden, Über eine besondere Klasse von Gruppen, *Abh. Math. Sem. Univ. Hamburg* **9** (1933), 154–8.

[a71] E. Mathieu, Mémoire sur l'étude des fonctions de plusieurs quantités, *J. Math. Pures Appl.* (2) **6** (1861), 241–323.

[a72] E. Mathieu, Sur la fonction cinq fois transitive de 24 quantités, *J. Math. Pures Appl.* (2) **18** (1873), 25–46.

[a73]   H. Matsuyama, Solvability of groups of order $2^a p^b$, *Osaka J. Math.* **10** (1973), 375–8.

[a74]   D. H. McLain, The existence of subgroups of given order in finite groups, *Proc. Cambridge Philos. Soc.* **53** (1957), 278–85.

[a75]   P. S. Novikov and S. I. Adyan, Infinite periodic groups (in Russian), *Izv. Akad. Nauk SSSR Ser. Math.* **32** (1968), 212–44, 251–524, 709–31 (translation in *Math. USSR Izv.* **2** (1968), 209–36, 241–479, 665–85).

[a76]   O. Ore, Contributions to the theory of groups of finite order, *Duke Math. J.* **5** (1939), 431–60.

[a77]   R. Rado, A proof of the basis theorem for finitely generated Abelian groups, *J. London Math. Soc.* **26** (1951), 74–5, 160.

[a78]   R. Remak, Über die Zerlegung der endlichen Gruppen in direkte unzerlegbare Faktoren, *J. reine angew. Math.* **139** (1911), 293–308.

[a79]   R. Remak, Über minimale invariante Untergruppen in der Theorie der endlichen Gruppen, *J. reine angew. Math.* **162** (1930), 1–16.

[a80]   R. Remak, Über die Darstellung der endlichen Gruppen als Untergruppen direkter Produkte, *J. reine angew. Math.* **163** (1930), 1–44.

[a81]   J. E. Roseblade, On groups in which every subgroup is subnormal, *J. Algebra* **2** (1965), 402–12.

[a82]   I. N. Sanov, Solution of Burnside's problem for exponent 4 (in Russian), *Leningrad State Univ. Annals (Uchenye Zapiski) Math. Ser.* **10** (1940), 166–70.

[a83]   O. J. Schmidt, Sur les produits directs, *Bull. Soc. Math. France* **41** (1913), 161–4.

[a84]   O. J. Schmidt, Über Gruppen, deren sämtliche Teiler spezielle Gruppen sind, *Rec. Math. Moscou* **31** (1924), 366–72.

[a85]   O. J. Schmidt, Über unendliche Gruppen mit endlicher Kette, *Math. Z.* **29** (1928), 34–41.

[a86]   O. J. Schmidt, Infinite soluble groups (in Russian), *Mat. Sb.* **17** (1945), 145–62.

[a87]   O. Schreier, Die Untergruppen der freien Gruppen, *Abh. Math. Sem. Univ. Hamburg* **5** (1927), 161–83.

[a88]   O. Schreier, Über den Jordan-Hölderschen Satz, *Abh. Math. Sem. Univ. Hamburg* **6** (1928), 300–2.

[a89]   O. Schreier and B. L. van der Waerden, Die Automorphismen der projektiven Gruppen, *Abh. Math. Sem. Univ. Hamburg* **6** (1928), 303–22.

[a90]   I. Schur, Neuer Beweis eines Satzes über endliche Gruppen, *S.-B. Preuss. Akad. Berlin* (1902), 1013–19 (*Gesamm. Abh.*, Springer 1973, vol. 1, pp.79–85).

[a91]   I. Schur, Über die Darstellung der endlichen Gruppen durch gebrochene lineare Substitutionen, *J. reine angew. Math.* **127** (1904), 20–50 (*Gesamm. Abh.*, Springer 1973, vol. 1, pp. 86–116).

[a92]   M. Suzuki, A new type of simple groups of finite order, *Proc. Nat. Acad. Sci. U.S.A.* **46** (1960), 868–70.

[a93]   M. Suzuki, On a class of doubly transitive groups, *Ann. of Math.* (2) **75** (1962), 105–45.

[a94]   M. Suzuki, On the existence of a Hall normal subgroup, *J. Math. Soc. Japan* **15** (1963), 387–91.

[a95]   L. Sylow, Théorèmes sur les groupes de substitutions, *Math. Ann.* **5** (1872), 584–94.

[a96]   J. G. Thompson, Finite groups with fixed-point-free automorphisms of prime order, *Proc. Nat. Acad. Sci. U.S.A.* **45** (1959), 578–81.

[a97]   J. G. Thompson, Normal p-complements for finite groups, *J. Algebra* **1** (1964), 43–6.

[a98] J. G. Thompson, Nonsolvable finite groups all of whose local subgroups are solvable, *Bull. Amer. Math. Soc.* **74** (1968), 383–437, *Pacific J. Math.* **33** (1970), 451–536, *ibid.* **39** (1971), 483–534, *ibid.* **48** (1973), 511–92, *ibid.* **50** (1974), 215–97, *ibid.* **51** (1974) 573–630.

[a99] J. Wiegold, Multiplicators and groups with finite central factor-groups, *Math. Z.* **89** (1965), 345–7.

[a100] H. Wielandt, Eine Verallgemeinerung der invarianten Untergruppen, *Math. Z.* **45** (1939), 209–44.

[a101] H. Wielandt, Über Produkte von nilpotenten Gruppen, *Illinois J. Math.* **2** (1958), 611–8.

[a102] H. Wielandt, Ein Beweis für die Existenz der Sylowgruppen, *Arch. Math.* **10** (1959), 401–2.

[a103] H. Wielandt, Über die Normalstruktur von mehrfach faktorisierten Gruppen, *J. Austral. Math. Soc.* **1** (1960), 143–6.

[a104] G. Zappa, Remark on a recent paper of O. Ore, *Duke Math. J.* **6** (1940), 511–2.

[a105] G. Zappa, Generalizzazione di un teorema di Kochendörffer, *Matematiche (Catania)* **13** (1958), 61–4.

[a106] G. Zappa, Sur les systèmes distingués de représentants et sur les compléments normaux des sous-groupes de Hall, *Acta Math. Acad. Sci. Hungar.* **13** (1962), 227–30.

[a107] H. Zassenhaus, Zum Satz von Jordan-Hölder-Schreier, *Abh. Math. Sem. Univ. Hamburg* **10** (1934), 106–8.

[a108] H. Zassenhaus, Über endliche Fastkörper, *Abh. Math. Sem. Univ. Hamburg* **11** (1936), 187–220.

Books

[b1] E. Artin, *Galois theory* (edited and supplemented by A. N. Milgram), 2nd ed., Notre Dame 1944.

[b2] E. Artin, *Geometric algebra*, Interscience 1957.

[b3] W. Burnside, *Theory of groups of finite order*, 2nd ed., Cambridge 1911 (Dover reprint 1955).

[b4] R. W. Carter, *Simple groups of Lie type*, Wiley-Interscience 1972.

[b5] H. S. M. Coxeter, *Introduction to geometry*, Wiley 1961.

[b6] H. S. M. Coxeter and W. O. J. Moser, *Generators and relations for discrete groups*, 3rd ed., Springer 1972.

[b7] C. W. Curtis and I. Reiner, *Representation theory of finite groups and associative algebras*, Interscience 1962.

[b8] L. E. Dickson, *Linear groups with an exposition of the Galois field theory*, Teubner 1901 (Dover reprint 1958, with an introduction by W. Magnus).

[b9] J. Dieudonné, *Sur les groupes classiques*, 3rd ed., Hermann 1967.

[b10] J. Dieudonné, *La géometrie des groupes classiques*, 3rd ed., Springer 1971.

[b11] L. Fuchs, *Infinite abelian groups* (2 vols.), Academic 1970, 1973.

[b12] T. M. Gagen, *Topics in finite groups*, London Math. Soc. Lecture Note Series 16, Cambridge 1975.

[b13] D. Gorenstein, *Finite groups*, Harper and Row, 1968.

[b14] J. A. Green, *Sets and groups*, Routledge and Kegan Paul, 1965.

[b15] P. A. Griffith, *Infinite abelian group theory*, Chicago 1970.

[b16] K. W. Gruenberg, *Cohomological topics in group theory*, Springer 1970.

[b17] M. Hall Jr., *The theory of groups*, Macmillan 1959.

[b18] B. Hartley and T. O. Hawkes, *Rings, modules and linear algebra*, Chapman and Hall 1970.

[b19] I. N. Herstein, *Topics in algebra*, Blaisdell 1964.

[b20] I. N. Herstein, *Noncommutative rings*, Carus Mathematical Monographs No. 15, Math. Assoc. of Amer. 1968.

[b21] B. Huppert, *Endliche Gruppen I*, Springer 1967.

[b22] D. L. Johnson, *Presentations of groups*, London Math. Soc. Lecture Note Series 22, Cambridge 1976.

[b23] C. Jordan, *Traité des substitutions et des équations algébriques*, Gauthier–Villars 1870 (Blanchard reprint 1957).

[b24] I. Kaplansky, *Infinite abelian groups*, 2nd ed., Michigan 1969.

[b25] I. Kaplansky, *Fields and rings*, Chicago 1969.

[b26] F. Klein and R. Fricke, *Vorlesungen über die Theorie der elliptischen Modulfunctionen*, vol. 1, Teubner 1890.

[b27] A. G. Kurosh, *The theory of groups* (2 vols., translated from Russian and edited by K. A. Hirsch), 2nd English ed., Chelsea 1960.

[b28] S. Lang, *Algebra*, Addison-Wesley 1965.

[b29] W. Ledermann, *Introduction to group theory*, Oliver and Boyd 1973.

[b30] I. D. Macdonald, *The theory of groups*, Oxford 1968.

[b31] S. MacLane, *Homology*, Springer 1963.

[b32] D. S. Passman, *Permutation groups*, Benjamin 1968.

[b33] M. B. Powell and G. Higman, Editors, *Finite simple groups*, Academic 1971.

[b34] J. J. Rotman, *The theory of groups, an introduction*, 2nd ed., Allyn and Bacon 1973.

[b35] E. Schenkman, *Group theory*, Van Nostrand 1965.

[b36] W. R. Scott, *Group theory*, Prentice-Hall 1964.

[b37] J.-P. Serre, *Representations linéaires des groupes finis*, Hermann 1967.

[b38] H. Wielandt, *Finite permutation groups* (translated from German by R. Bercov), Academic 1964.

[b39] R. J. Wilson, *Introduction to graph theory*, Oliver and Boyd 1972.

[b40] H. Wussing, *Die Genesis des abstrakten Gruppenbegriffes*, VEB Deutscher Verlag der Wissenschaften 1969.

[b41] H. Zassenhaus, *The theory of groups*, 2nd English ed., Chelsea 1958.

INDEX OF NOTATION

$\text{Dr} \prod\limits_{i=1}^{n} G_i$ $G_1 \times G_2 \times \ldots \times G_n$ 164

$\sum\limits_{i=1}^{n} \varphi_i$ sum of homomorphisms $\varphi_1, \ldots, \varphi_n$ of a group into a group   177

G^X set of maps of set X into G 186

$\text{Dr } G^X$ direct power of G with index set X 186, 188

$\text{Cr } G^X$ cartesian power of G with index set X 188

$T(G)$ torsion subgroup of G (where G is abelian)   196

G^n subgroup of n-th powers of elements in G (where G is abelian)   197

$H_\varphi \times K$ semidirect product of K by H with action φ 208

$\text{Hol } K$ holomorph of K 210

$\text{Dih } A$ generalized dihedral group corresponding to abelian group A 210

$\mathscr{A}(V)$ affine group of vector space V 211

$G \wr H$ a wreath product of G by H 220

ϖ' set of prime numbers which do not belong to ϖ 231, 283

T/U special element of abelian section H/J of G corresponding to right transversals T, U to H in G 233

$\text{Foc}_G(H)$ focal subgroup of H in G (where $H \leqslant G$)   254

$x \underset{G}{=} y$ x and y are conjugate in G (where $x, y \in G$)   258

$\Phi(G)$ Frattini subgroup of G 266

$N_G(\mathscr{S})$ system normalizer of G corresponding to complement system $\mathscr{S}$ of G (where G is soluble)   287

INDEX OF SUBJECTS